Photocatalysis for Organic Synthesis
Principles and Applications

Photocatalysis for Organic Synthesis
Principles and Applications

Dr. Ajay Kumar

ARCLER
PRESS

www.arclerpress.com

Photocatalysis for Organic Synthesis: Principles and Applications

Dr. Ajay Kumar

Arcler Press
4164 Lakeshore Road
Burlington, ON L7L 1A4
Canada
www.arclerpress.com
Email: orders@arclereducation.com

ISBN: 978-1-77956-948-6 (Hardcover)

ABOUT THE AUTHOR

Dr. Ajay Kumar, Associate Professor at the School of Pharmaceutical Sciences, Chhatrapati Shahu Ji Maharaj University (accredited with NAAC A++, categorized in UGC grade 1 and has QS ranking), Kanpur, is a distinguished academician in School of Pharmaceutical Sciences (ranked NIRF in 2024). Dr. Ajay Kumar is M.Pharm. in Pharmaceutical Chemistry, and completed Ph.D.(Pharmaceutical Sciences) from Sam Higginbottom Institute of Agriculture, Technology and Sciences in 2016. His area of research spans Pharmaceutical and Medicinal chemistry, focusing on the design, synthesis, and evaluation of novel antimicrobial, antitubercular, anticancer, and anti-inflammatory compounds with potential therapeutic applications. He has guided more than 15 M.Pharm. and currently supervising 05 Ph.D. research scholars and published approximately 27 papers in Scopus indexed impactful journals. He is a life member of APTI. Apart from sound academician, he has quite experience in university administrative work.

TABLE OF CONTENTS

LIST OF FIGURES

PREFACE

Photocatalysis is a rapidly evolving area of modern chemistry that utilizes light energy to promote chemical conversions with the aid of catalysts. The process offers a unique avenue for speeding up reactions under mild conditions, sometimes replacing harsh reagents and high temperatures with cleaner and more energy-conserving alternatives. It bridges the domain of photochemistry and catalytic science, enabling precise control of reaction pathways and selectivity. Since photocatalysis utilizes photons as a renewable energy source, it aligns with the sustainable chemistry mission to minimize waste generation and maximize resource utilization. Its applicability for reaction types has also made it a flexible tool in the construction of new synthetic methodologies in industrial and academic laboratories.

The philosophy of photocatalysis extends beyond the generation of reactive intermediates; it encompasses materials design, understanding of reaction mechanisms, and the incorporation of advanced reactor systems. Such a multi-faceted approach enables chemists to construct complex molecules better, whether in pharmaceutical lead discovery, agrochemical synthesis, or fine chemical production. The topic is not so much the manipulation of reaction conditions as it is about revolutionizing how chemical processes are conceived. Photocatalysis offers a platform for combining renewable energy inputs with transformative chemical applications, providing a route for reactions that were previously challenging or inefficient under regular conditions.

Photocatalysis in Organic Synthesis: Principles and Applications aims to provide readers with an equal portion of theoretical background, mechanistic insight, and practicality concerning the application of photocatalysis in modern organic synthesis. The book begins with the fundamental principles of photocatalytic processes and proceeds to advanced applications, expert methods, and recent developments. Each chapter builds upon the previous material to present a logical continuity from simple ideas to higher-level concepts, making it suitable for both learners seeking a foundation and experienced researchers looking to expand their knowledge.

Chapter 1 presents an overview of photocatalysis, including its fundamental ideas, terminology, and scope of research. It emphasizes the historical context of the field, the grounds for choosing light-fueled catalysis, and the broad range of potential applications in organic synthesis. The chapter is also committed to environmental and economic motives for the increased interest in this approach.

Chapter 2 establishes the background in photochemistry by explaining how light behaves around matter, electronic excitation concepts, and the roles of singlet and triplet states. Chapter 2 summarizes the major photophysical processes that provide the background for chemical changes to occur, including energy transfer, electron transfer, and quenching processes, which form a sound scientific basis for describing photocatalytic systems.

Chapter 3 covers some of the photocatalysts that have been used in organic synthesis, from the metal complexes and semiconductor materials to organic dyes and hybrid systems. It explains their structural features, photophysical properties, and use in various types of reactions, accompanied by advice on photocatalyst choice and optimization for specific uses.

Chapter 4 addresses the mechanisms of photocatalytic reactions, describing how light energy is harnessed to drive chemical reactivity. The mechanisms addressed are oxidative and reductive quenching cycles, radical formation, and twin catalytic cycles. A mechanistic understanding is critical to the development of efficient and selective photocatalytic conversions.

Chapter 5 addresses light sources and photocatalytic reactor design, which provides insight into the influence of reaction vessels, light wavelengths, and irradiation setup on efficiency and selectivity. It includes real-world information on scaling up photocatalytic processes and integrating them into continuous flow equipment.

Chapter 6 highlights photocatalysis in C–C bond assembly, a building block for organic synthesis. It illustrates various methodologies of carbon–carbon bond formation under light-provoked catalytic conditions, including cross-coupling, cycloaddition, and radical-based coupling reactions, illustrated with representative examples and mechanistic rationales.

Chapter 7 describes energy transfer and dual catalysis, emphasizing how photocatalysts can be hybridized with other catalytic systems to achieve synergistic objectives. This includes the combination of photocatalysis with transition-metal catalysis or organocatalysis to access transformations that are challenging under single catalytic regimes.

Chapter 8 provides an overview of photocatalysis in flow chemistry, wherein reactions are conducted in continuously running systems as opposed to batch setups. It is advantageous in terms of scalability, reproducibility, and safety, making it highly appealing for the industrial adoption of photocatalytic methods.

Chapter 9 is dedicated to gold's role in photocatalysis and the ability of gold-based catalysts to participate in light-induced processes. It covers both homogeneous and heterogeneous systems, with a focus on unique patterns of reactivity and new applications in organic synthesis.

Chapter 10 concludes with examples in green and sustainable chemistry, demonstrating how photocatalysis is engaged with waste minimization, energy efficiency, and the use of renewable resources. Examples include biomass valorization, the use of oxygen from air as an oxidant, and photoredox conversions that utilize fewer toxic reagents.

This book will be applicable to graduate students, scientists in organic chemistry, catalysis, and chemical engineering who are interested in learning and applying photocatalytic methods in synthesis. It will also be beneficial to researchers in the drug, agrochemical, and fine chemical industries interested in integrating energy-efficient and sustainable practices into process development.

CHAPTER 1

INTRODUCTION TO PHOTOCATALYSIS

Contents

INTRODUCTION

The photocatalysis process is the catalysis of a chemical reaction through the employment of a light-excited catalyst, typically ultraviolet or visible radiation. The method utilizes the energy released by photons to cause electron excitation in a photocatalyst, enabling numerous chemical reactions. Photocatalysts are typically semiconductor materials, such as titanium dioxide, which function by absorbing light energy to produce electron-hole pairs. When these couples are brought in contact with target substrates, they are capable of catalytically degrading organic contaminants or facilitating energy conversion processes, such as water splitting for hydrogen production. What makes the technology relevant is that it is capable of harnessing renewable energy sources for both environmental cleanup and energy generation, thus making it a crucial element in the advancement of sustainable operations within the municipal and industrial sectors. Practically, photocatalysis is extremely promising, particularly in the treatment of wastewater pollutants and the degradation of emerging contaminants. Photocatalysis has been found to break down organic pollutants in their complex state into simpler ones, which are less toxic under light exposure, thus acting as an advanced oxidation process (AOP). However, caution must be exercised in applying photocatalysis, as the degradation of such pollutants can sometimes lead to the creation of by-products that are equal to or even more toxic than the original pollutants. Research emphasizes the importance of knowing such by-products and points out that while promising, photocatalysis is not always universally effective. The efficacy and feasibility of photocatalytic treatment depend on a range of factors, including the type of wastewater and the specific photocatalyst used (Jos et al., 2020). Photocatalysis is more than just the elimination of pollutants; it encompasses a broader range of potential applications, from energy conversion to sustainable development policy. While scientific studies push on, more emphasis lies on the development of photocatalysts that increase efficiency while minimizing costs and environmental impacts. Hydrogen production via photocatalytic water splitting is an excellent example, bringing hydrogen's clean energy potential and photocatalysis' solar energy capability together. But to achieve a cost-effective and feasible hydrogen economy, it is essential to continuously boost the efficiency of photocatalytic systems, address the reusability challenges of catalysts, and ensure the economy's feasibility. Ultimately, this synergy of advancements in photocatalysis drives a shift towards cleaner sources of energy, thereby supporting global efforts to combat climate change and promote a circular economy (Yue et al., 2021).

Explanation of Photocatalysis

Photocatalysis operates by engaging light, most often ultraviolet or visible light, with a semiconductor material, thereby catalyzing chemical reactions that would otherwise require a significant energy input or might not occur at all. Essentially, photocatalysis involves stimulating electrons in the semiconductor's conduction band, resulting in the formation of electron-hole pairs. This can be done by breaking down environmental pollutants, such as organic dyes and pesticides, into less harmful products and, therefore, is a significant technology in addressing environmental issues. Its efficiency also relies a lot on factors such as the band gap of the semiconductor, surface area, and the application of co-catalysts, which enhance the overall reactivity and efficiency of the process. Therefore, the understanding of these variables is essential to optimize photocatalytic reactions towards valuable applications in purification and energy conversion. Photocatalysis, in addition to pollutant degradation, also has useful applications in the areas of solar energy conversion and hydrogen evolution. Semiconductor materials, especially titanium dioxide (TiO_2), are increasingly utilized due to their stability, non-toxicity, and enhanced photocatalytic activity. In particular, TiO_2 is capable of catalyzing the splitting of water and the reduction of carbon dioxide to provide the transformation of solar power into chemical power. Composite materials, for instance, have been proven in recent studies in the testing of thin films of TiO_2 with gold and silver nanoparticles to increase photocatalytic activity. The optimal thickness of the films also enhances performance, provided certain conditions are met for effective water oxidation. These discoveries collectively point to the need for ongoing research to uncover efficiencies and applications that can ultimately harness solar energy in sustainable and cost-effective ways (Darr et al., 2011) (Figure 1.1).

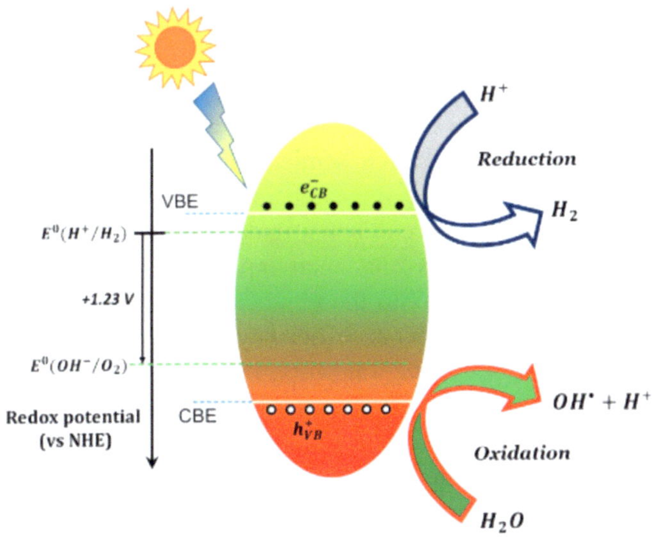

Figure 1.1. Elementary scheme of a basic photocatalytic process (water splitting) caused by illumination of a semiconductor photocatalyst. The scheme illustrates the reduction of H^+ ions through electrons photogenerated in the conduction band and the oxidation of hydroxyl ions, resulting in the formation of hydroxyl radicals through photogenerated holes in the valence band.

Source: https://encyclopedia.pub/entry/10109.

In addition, photocatalysis has implications for limiting the danger of antibiotic resistance, an emerging global health concern. Recent studies have demonstrated that traditional disinfectants, such as chlorine and ozone, are unable to cut across the transmission of antibiotic resistance genes. Photocatalytic reactions, however, have proven capable of breaking down plasmid DNA of resistant bacteria, thereby reducing the likelihood of resistance transfer. For instance, studies have indicated that photocatalytic treatment led to visible conformational changes in the DNA, thereby rendering the resistant strains unable to be transformed, an effect that chlorine treatment could not achieve. This remarkable ability of photocatalysis positions it not only in wastewater treatment but also as a technique in combating antibiotic resistance threats, making it a valuable tool in contemporary environmental and health research (Balcioglu et al., 2011).

Historical Background and Development

The origins of photocatalysis, a key area at the interface of chemistry and materials science, date back to the pioneering work on semiconductor photophysics in the early 20th century. Early works laid the groundwork by exploring the semiconductor-type activity of materials like titanium dioxide (TiO_2), which is currently well established for its photocatalytic activity. Photocatalysis emerged in the late 1970s, following revolutionary developments that unlocked the means by which semiconductors could harness sunlight to energize chemical processes. Researchers demonstrated that TiO_2 could catalyze the decomposition of pollutants under ultraviolet (UV) irradiation, showing it had potential for environmental applications. This groundwork not only spurred scientific inquiry into the photocatalytic process but also raised significant questions about energy efficiency and environmental cleanup, which set in motion an innovative trajectory that would revolutionize the field in the following decades. The 1990s were a critical time as researchers began developing photocatalysis for applications beyond environmental clean-up. Researchers investigated material usage in energy storage and conversion, primarily through novel routes of water splitting and hydrogen generation. The decade saw the emergence of photocatalytic materials tailored for specific reactions, indicating a growing understanding of molecular interactions and surface engineering to enhance catalytic performance. Doped and high-surface-area materials began to feature prominently in the literature, pushing the boundaries of theories established during the preceding decades. With the implementation of nanotechnology into photocatalysis, scientists accessed unique characteristics that were incredibly valuable in the conversion of solar energy. With increasing global concerns about climate change, this period fueled research into photocatalysis for sustainable energy applications, cementing its position in contemporary research and development (Figure 1.2).

Figure 1.2. Photons from a light source (out of frame on the right-hand side) are absorbed by the surface of the titanium dioxide (TiO_2) disc, exciting electrons within the material. These electrons then react with water molecules, splitting them into their constituents, hydrogen and oxygen. In this experiment, chemicals dissolved in the water prevent the formation of oxygen, which would otherwise recombine with the hydrogen.

Source: https://en.wikipedia.org/wiki/Photocatalysis#/media/File:Anatase_(titania,_TiO$_2$)_photocatalyst_producing_hydrogen.jpg.

Photocatalysis, in the 21st century, has evolved into a multidisciplinary field that encompasses numerous applications, spanning both environmental and energy sectors. Consequently, advancements have multiplied to encompass new photocatalytic material development and the design of sophisticated photoreactors that enable optimal reaction conditions. Today's focus is on integrating artificial intelligence and machine learning to predict the performance of photocatalytic systems, which ultimately drives innovation. Additionally, photocatalysis has been applied to antibacterial coatings, self-cleaning surfaces, and the elimination of CO2, in response to the growing demand for sustainability worldwide. As this rapidly advancing science continues to grow, it is essential that environmental scientists, chemists, and engineers collaborate to enhance photocatalytic processes and their consequences. With these developments, photocatalysis has the potential to make a significant contribution to future energy infrastructure and environmental management practices (Figure 1.3).

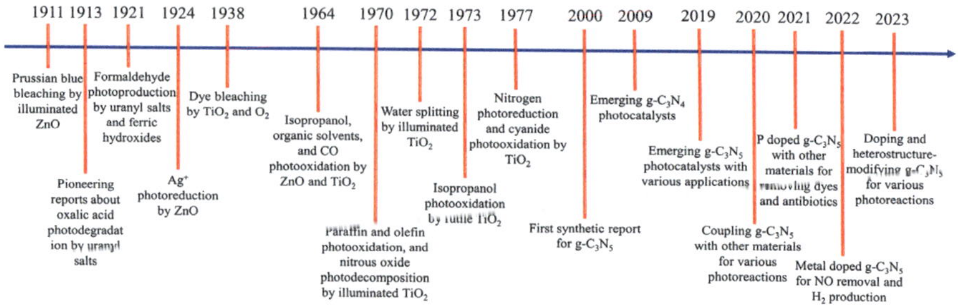

Figure 1.3. The evolution of photocatalytic studies throughout history.

Source: https://www.researchgate.net/figure/The-evolution-of-photocatalytic-studies-throughout-history-in-the-literature_fig1_371856147.

Importance in Modern Science and Technology

The importance of photocatalysis in modern science and technology lies in its ability to help address major environmental problems, particularly those related to water and air quality. The increased urbanization and development of industrial operations necessitate the development of new and sustainable wastewater treatment technologies. Photocatalysis provides a green process for degrading pollutants, purifying water supplies of contaminants through artificial photosynthesis-like processes. The shift to semiconductor-supported photocatalysts is an indication of the improvements in nanotechnology, especially formulated for optimum efficiency in pollutant elimination. Indeed, the technology is highly conducive to sustainable development principles, as it seeks to minimize ecological footprints while offering a cost-effective response to crucial environmental issues (Mohamed et al., 2023). The prospects for photocatalytic systems are promising, given that they offer extensive applications in various industries, including energy harvesting and waste treatment. The interfacing of photocatalysis with the development of nanomaterials also increases its significance in modern scientific research. By employing severe plastic deformation techniques, researchers have developed ultrafine-grained and nanostructured materials that exhibit enhanced photocatalytic properties. These advances in materials science have enabled the catalytic synthesis of more efficient and better-functioning materials, making them essential for applications in energy conversion and environmental remediation. For instance, photocatalytic reactions can enable hydrogen production through water splitting and the conversion of carbon dioxide into sustainable fuels, thereby contributing to renewable energy solutions. The constant advancement of photocatalytic materials, driven by progress in materials science and nanotechnology, focuses attention on the role of photocatalysis in addressing the sustainability and energy crisis challenges in the modern world (Edalati et al., 2022). Moreover, the integration of artificial intelligence (AI) and machine learning into the field of photocatalysis demonstrates the evolution of scientific approaches that support the process of material design. By applying computational models, researchers can predict the efficiency of newly synthesized photocatalysts, thereby drastically reducing the time required for experimental testing. This merging of AI and photocatalysis not only accelerates the discovery of more efficient catalysts but also opens doors for targeted solutions specifically engineered to address specific environmental challenges. With the battle to cope with global warming and resource deprivation, the ability to innovate and optimize photocatalytic systems at an unprecedented rate is an absolute necessity. This represents not only a quantum leap in the realm of technological progress but also a watershed moment in transitioning towards a more resource-efficient and sustainable future, where photocatalysis serves as the cornerstone of addressing environmental and energy issues (Mohamed et al., 2023; Edalati et al., 2022).

1.1. MECHANISM OF PHOTOCATALYSIS

The photocatalytic process largely relies on the application of photocatalysts, often semiconductors, to facilitate photochemical reactions on exposure to light. On colliding with photons that have sufficient energy, the photocatalyst surface induces excitation of electrons from the valence band to the conduction band and thus creates electron-hole pairs. This process is the basis of photocatalytic activity, with the created holes capable of oxidizing substrates, and the excited electrons capable of reducing other entities in the environment. Such a process might be highly dependent on the properties of the semiconductor material used, e.g., the band gap energy and morphology. Moreover, the surface stability and properties of the photocatalyst significantly influence its working life and efficiency, and therefore its functional applications in mitigating environmental issues (Králová et al., 2012). In response to the extensive applications of photocatalysis, it is worth noting that the mechanism is not only suitable for the breakdown of pollutants but also points towards energy production. For instance, photocatalysis has proved to be successfully employed for the degradation of organic pollutants from wastewater, where the oxidative capacity of the photogenerated holes breaks down large molecules into smaller non-toxic species. Recent research has also explored the use of photocatalysts in hybrid processes, including the cooperative degradation of pollutants and the production of hydrogen through water splitting. This two-pronged application is a testament to the versatility of photocatalytic systems, which makes them instrumental in environmental remediation and renewable energy technologies. With advancements in the material science field, the discovery of new photocatalytic materials will further expand their horizon of applications, making them even more critical (Yin et al., 2018). The introduction of innovative methods in the synthesis of photocatalysts also holds promise for significantly optimizing photocatalytic mechanisms. For example, the synthesis of all-inorganic perovskite nanocrystals has demonstrated promising photocatalytic performance, particularly under sunlight. These materials exhibit favorable band gap properties and maximum light absorption capabilities, which are crucial for achieving peak photocatalytic efficiency. With techniques such as sol-gel synthesis and chemical vapor deposition, researchers can tailor the surface morphology and crystallinity of photocatalysts to achieve optimal performance in specific reactions. As implied by recent studies, improving structural features can lead to enhanced charge separation and reduced recombination rates, thereby increasing photocatalytic efficiency. Continuous exploration and discovery in this field position photocatalysis at the forefront of green processes in energy conversion and environmental protection (Králová et al., 2012) (Figure 1.4).

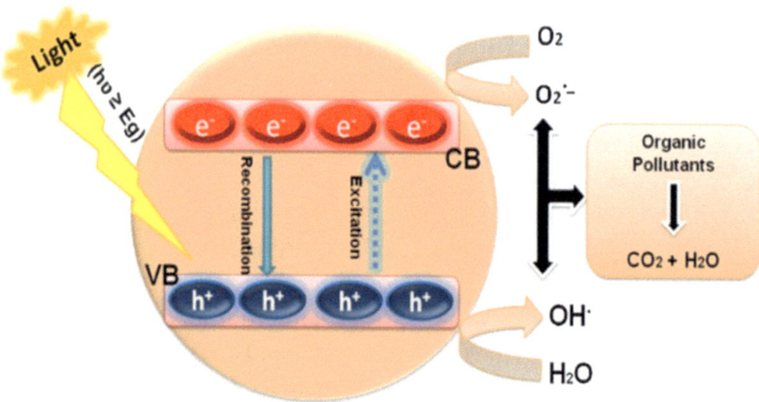

Figure 1.4. The basic mechanism of photocatalysis.

Source: https://www.researchgate.net/figure/The-Basic-mechanism-of-Photocatalysis-Reproduced-with-permission-17_fig1_336785403.

1.1.1. Role of Light in Photocatalytic Reactions

It is essential to grasp the role of light in photocatalytic reactions in understanding the mechanisms of such processes. Photocatalysis is essentially the process of photosorption of light, typically ultraviolet or visible, by a photocatalyst, which can be in the form of semiconductors such as titanium dioxide (TiO_2). When light reaches the semiconductor, it raises the electrons, which transfer from the valence band to the conduction band, thus creating electron-hole pairs. This particular property enables the electron and the hole to participate in redox reactions, effectively degrading the pollutants or forcing chemical conversions. The energy and the wavelength of the incoming light thus directly determine the selectivity and the efficiency of photocatalytic reactions and hence it is critical that optimized light sources are used in actual applications. The photocatalytic activity is also significantly enhanced by doping and structural optimization, which facilitates better light absorption. For instance, research shows that incorporating some dopants can stabilize multiple phases of TiO_2, such as anatase or rutile, which possess different photocatalytic activity. To this end, dopants such as fluorine, silicon, and iron have been discovered to control the bandgap energy of TiO2, thereby enhancing photocatalytic activity under visible light (Assadi et al., 2012). The ability to control the electronic structure of photocatalysts highlights the importance of both material chemistry and light chemistry in achieving maximum photocatalytic efficiency. As a result, the search for new materials and dopant technologies remains a current research area in the field. Beyond fundamental concepts, the applications of photocatalytic systems in technology increasingly involve the exploitation of the unique interactions between light and photocatalysts. Advances in microfluidic technology suggest the potential for integrating photocatalytic processes into confined spaces, as demonstrated by TiO_2 thin-film-based blood oxygenation machines (MNF et al., 2009). These devices not only demonstrate the versatility of photocatalysis but also underscore the central role of light in catalyzing a specific reaction under controlled conditions. The fabrication of these devices exemplifies the practical application of the principles of photocatalysis, demonstrating how light can be harnessed to drive vital biochemical reactions in an optimal manner. Hence, with increased research, the interplay of light functions in photocatalytic reactionresponses will be at the center of generating innovative and sustainable technologies in a variety of fields.

1.1.2. Types of Photocatalysts Used

The photocatalyst family is extensive, and some of the most extensively studied materials include titanium dioxide (TiO_2), which exhibits superior photocatalytic activity and stability. Anatase, rutile, and brookite phases of TiO_2 exhibit different efficiencies based on their crystallographic structures and surface areas. Advanced preparation techniques such as sol-gel and hydrothermal synthesis have significantly enhanced the photocatalytic activity of TiO_2. For instance, TiO_2 research supported on substrata like soda-lime glass showed that preparation pathways can influence properties such as morphology and crystallinity, which are directly related to photocatalytic activity in processes like the decomposition of organic pollutants (Králová et al., 2012). Tuning these properties on the nanoscale demonstrates the adaptability and inherent relevance of TiO_2 in photocatalytic processes. In addition to TiO_2, some other semiconductor materials are efficient photocatalysts with unique photochemical characteristics.

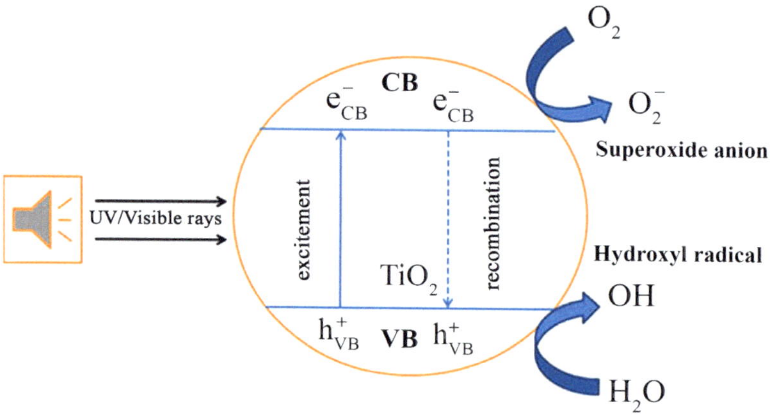

Figure 1.5. The photocatalytic activity of TiO_2 under UV irradiation.

Source: https://www.researchgate.net/figure/The-photocatalytic-activity-of-TiO2-under-UV-irradi-ation_fig3_258784932.

One of them is zinc oxide (ZnO), which has gained popularity due to its high exciton binding energy and intensive antibacterial activity. Similar to TiO_2, the morphology of ZnO can be controlled, and metals or non-metals can be doped to enhance its photocatalytic behavior. Moreover, materials such as graphitic carbon nitride (g-C_3N_4) and metal–organic frameworks (MOFs) have been explored for their ability to absorb visible light and catalyze a wide range of reactions. These novel materials expand the realm of photocatalysis beyond traditional practices, representing a crucial step toward sustainable and eco-friendly methods of environmental cleaning (Mart et al., 2019). The quest for these new photocatalysts is crucial in overcoming the limitations of TiO2, particularly in terms of visible light activity. The efficiency of photocatalysis depends not only on the materials but also on reaction conditions and reactor geometries. Heterogeneous photocatalysis, which is widely applied in suspension, enables efficient removal of contaminants, particularly in wastewater treatment processes. For example, suspension photocatalysts have a crucial role in achieving high decolorization and mineralization rates in studies on the photodegradation of azo dyes. The procedure is facilitated by *in situ* monitoring methods that track dye degradation without disturbing the reaction environment, thereby maintaining the photocatalytic process as continuous. These developments allow researchers to gain a greater understanding of the dynamics of photocatalytic reactions, which ultimately serves as a stepping stone for improved pollutant removal strategies (Mart et al., 2019). The multilateral nature of these studies also sets the stage for the necessity of fine-tuning catalyst composition and operating conditions in driving the study of photocatalysis forward (Figure 1.5).

1.1.3. Reaction Pathways and Processes Involved

In understanding the photocomplexity of photocatalysis, the reaction pathways and processes that catalyze reactants to products are of pivotal importance. Photocatalysis harnesses light energy to power chemical reactions, i.e., by exciting electrons from the valence band to the conduction band of a semiconductor material. Electron-hole pair generation is created through this mechanism, which plays a crucial role in subsequent redox reactions. The activity of the photocatalytic system is significantly influenced by the catalyst surface properties, which determine reactant adsorption and activation. For example, the processes of reactions differ considerably at the interfaces of solvent/catalyst and electrolyte/catalyst systems, impacting product distribution and reaction kinetics in general, and illustrating the complexity of photocatalytic processes in various environments (Alexandrova et al., 2020). One of the highlights of recent research has been the photoelectrocatalytic reduction of CO_2, a reaction process with the potential to address both energy storage and climate concerns. By employing photoelectrocatalytic reactions, CO_2 reduction to functional small molecules is attainable with higher efficiency than through traditional photocatalysis. DFT-based research has provided insights into how different interfaces affect electronic structure and reaction dynamics. The study defines typical catalytic cycles responsible for product variation, for example, the preferred production of methanol in the presence of a suitable electrolyte (Alexandrova et al., 2020). This emphasizes designing photocatalytic materials and operating conditions to optimize specific reaction routes to desired outcomes. Moreover, advancements in catalytic materials, such as bimetallic hydride clusters, are emerging trends in the field of photocatalysis. Mass spectrometric analysis has established the formation of intermediate species, $PdCuCO_2H_4^-$, by reaction of palladium-copper tetrahydride anions with CO_2. The identification of numerous isomers, particularly the higher-energy ones, highlights the complexity of the catalytic mechanism and its consequent functions in catalyzing reactions. This research exemplifies the critical importance of understanding reaction pathways and intermediates, as the presence of certain species can have a dramatic impact on catalytic efficiency and performance. As research in photocatalysis continues, future exploration of such reaction dynamics will be critical to the development of novel solutions for energy conversion and environmental sustainability.

1.2. APPLICATIONS OF PHOTOCATALYSIS

Photocatalysis extends a broad number of applications that contribute in an essential way to environmental remediation and energy alternatives on a sustainable basis. Among the most promising uses is water treatment, particularly in the oil and textile industries, which generate huge volumes of contaminated effluent.

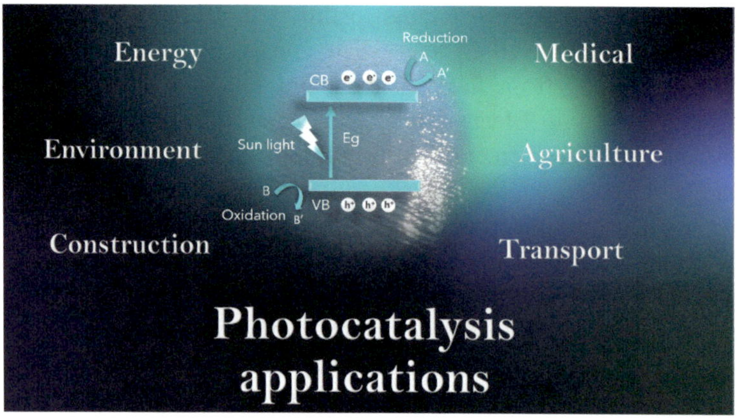

Figure 1.6. Applications of photocatalysis.

Source: https://external-content.duckduckgo.com/iu/?u=https%3A%2F%2Fcicenergigune.com%2Fmedia%2Fuploads%2Fmediacenter%2Fphotocatalysis-cicenergigune1.png&f=1&nofb=1&ipt=9a80f0c732c7c972a20b3bdb6d91292a998278 1d3278959ac2f8ae2e5ec48ace.

The issue is how to effectively annihilate poisonous organic compounds, which are likely to include carcinogenic contaminants. Recent advances in photocatalytic technology have shown improvements in this process. To illustrate, the use of titanium dioxide (TiO_2) combined with carbon nanotubes and magnetic iron oxide enables enhanced degradation of dispersed organic pollutants, highlighting a multidimensional approach to wastewater treatment (Mason et al., 2018). Besides ensuring the faster breakdown of poisonous substances, this new method highlights the necessity to further develop it in order to incorporate photocatalysis into more effective water treatment plants. In addition to wastewater treatment, photocatalysis is also utilized in air purification to destroy organic pollutants and volatile organic compounds (VOCs) found in both indoor and outdoor environments. TiO_2 is currently leading in terms of photodegradation using UV assistance due to its high activity. However, the limitations associated with TiO_2's photonic response have prompted researchers to seek out modifications that improve its activity under visible light. These techniques, such as noble metal deposition and cationic doping, have been employed to enhance the photocatalytic activity of TiO_2s and thereby their application in environmentally-friendly solutions (Cho et al., 2008) (Figure 1.6).

This advancement into visible light catalysis not only facilitates photocatalysis as a more viable option for application in home and industrial air cleaning systems but also complements the broader objective of reducing reliance on non-renewable energy sources and curtailing pollution. Furthermore, the emergence of photocatalysis as a method for generating clean energy through hydrogen production adds a further level of significance to it. Using the photocatalytic process of water splitting, researchers are investigating the ability of TiO_2 to play this role effectively, converting water into hydrogen fuel. Two-for-one benefits are inherent in this technology: providing energy while simultaneously advancing the use of cleaner technologies. Hybrid nanostructures and innovative design concepts, combined with photocatalytic systems, enhance efficiency and provide avenues for practical applications in renewable energy generation. With the maturity of such technologies, the synergy between photocatalysis and clean energy solutions will yield unprecedented significant breakthroughs in alleviating global energy challenges as well as environmental issues.

1.2.1. Environmental remediation

The demand for effective techniques of environmental cleanup has generated considerable interest in emerging technologies, one of which is photocatalysis. Photocatalysis applies sunlight to facilitate the degradation of contaminants and promote cleaner environments. Photocatalysis has shown immense capability in cleaning polluted water resources using semiconductors such as titanium dioxide (TiO_2). The process relies on the photoexcitation of electrons on irradiation with light, which subsequently reacts with water and oxygen to form reactive species capable of destroying various pollutants. The more pollution problems society faces, the more the use of photocatalytic techniques for environmental cleaning is not just novel but also imperative in guaranteeing sustainable development, and thus in harmony with the endeavor to combat water scarcity and enhance the quality of natural resources worldwide. One of the first studies into photocatalysis is the investigation of the ligand-to-metal charge transfer (LMCT) mechanism in TiO2 nanoparticles, which makes them more responsive to visible light. Historically, TiO_2 has been a poor performer under visible light, and researchers have been seeking alternative mechanisms to enhance its performance. The LMCT mechanism enables organic or inorganic adsorbates to interact with TiO2, resulting in additional absorption bands in the visible wavelength range. This approach extends the scope of applicability of photocatalytic systems for environmental cleaning since it allows for the degradation of a variety of organic pollutants via photochemical pathways (Figure 1.7).

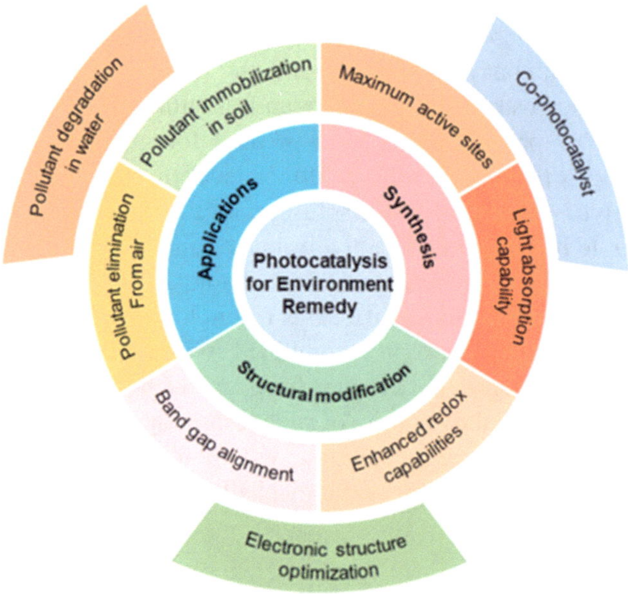

Figure 1.7. Photocatalysts for environmental remediation.

Source: https://ars.els-cdn.com/content/image/1-s2.0-S0013935122024112-ga1_lrg.jpg.

Thus, the LMCT-catalyzed methods exhibit significant versatility through the use of varying surface complexes that effectively stimulate the degradation of contaminants, thereby enhancing the variety of advanced oxidation processes that can be utilized in environmental remediation (Choi et al., 2025). Additionally, the successful integration of advanced oxidation processes (AOPs) within photocatalytic systems represents a significant paradigm shift in addressing wastewater treatment challenges. Research indicates that AOPs, especially when combined with biological treatment, cause the degradation of the majority of pollutants, like harmful chemicals from municipal and industrial sources of pollution (Agüera et al., 2019). By employing photocatalytic mechanisms that tap into available sunlight, AOPs can significantly reduce the environmental footprint of water treatment facilities. Countries such as China, Spain, and the USA are also at the forefront in this regard, cooperating to develop and enhance these technologies. The interplay of photocatalytic processes and AOPs is not only the key to an effective means of decontaminating polluted environments but also to a means of attaining sustainable water resource management in the face of intensifying worldwide demand for clean water.

1.2.2. Energy Production and Conversion

The field of energy conversion and generation has been closely aligned with the advancement of photocatalysis, particularly in terms of clean energy production methods that are sustainable. Photocatalysis is the acceleration of photoreactions in a photo-induced process, whereby sunlight can be converted into fuels that can be stored, such as hydrogen. It has the potential to be the solution to the global energy crisis and to limit greenhouse gas emissions. Previous works have concentrated on optimizing the solar-to-hydrogen (STH) efficiency by synthesizing photocatalysts that can utilize a broader spectrum of sunlight. Proper photocatalytic system design, such as that incorporating photothermal effects integration, plays a key role in maximizing reaction kinetics and ultimately optimizing the rates of hydrogen production, as seen in experiments achieving impressive rates of 603 mmol h^{-1} m^{-2}. By optimizing these processes, photocatalysis can contribute immensely to a cleaner energy future in the energy production and conversion paradigm. A critical direction in this energy paradigm is the use of novel materials, such as covalent organic frameworks (COFs). These crystalline porous polymers are not only surface area-rich but also possess tunable electronic properties, making them ideal photocatalysts for various uses, including energy storage and conversion. 2D conjugated COFs' synthesis and design, for instance, have been shown to possess the possibility of providing a massive boost to photocatalysis by virtue of their ability to enable effective charge separation and transport. Structural flexibility enables the incorporation of active sites that promote photocatalytic activity. The implementation of methods like skeleton engineering in COF design is aimed at enhancing the structure-to-efficiency ratio of energy conversion balance, thus enabling novel applications in sustainable energy storage and harvesting, which play a central role in transitioning to renewable energy systems (Figure 1.8).

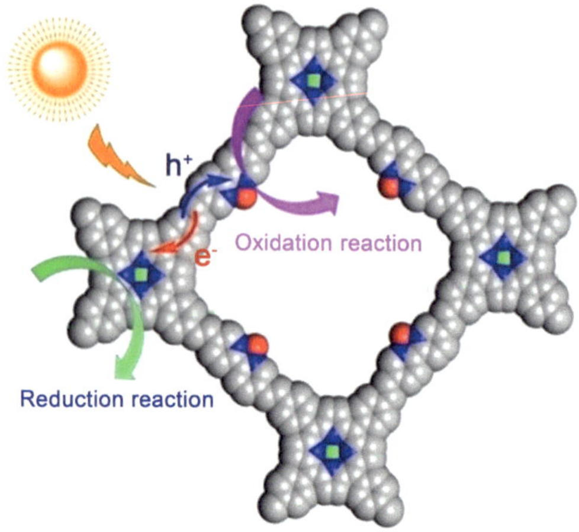

Figure 1.8. Covalent organic frameworks for energy conversion in photocatalysis.

Source: https://external-content.duckduckgo.com/iu/?u=https%3A%2F%2Fonlinelibrary.wiley.com%2Fcms%2Fasset%2F01d1eb0e-4cd2-4e40-b556-1c110f84f878%2Fanie202303086-toc-0001-m.jpg&f=1&nofb=1&ipt=e75d78b224a363589cd07cc39698ca2b0e38272019b55fd01ec0b0f10ccd92c4.

With the growth of photocatalysis, the point of convergence of material science and energy conversion holds in store revolutionary breakthroughs. Improvements in photocatalyst structure and performance enhance efficiency, thereby widening the horizons of energy conversion technologies to make them more viable for large-scale implementation. Researchers are interested in integrating approaches that combine light absorption with innovative substrate materials, thereby increasing operating spectra and overall activity. Such developments represent the key role that next-generation photocatalytic systems will play in revolutionizing how we generate and utilize energy in an increasingly sustainable future. The integration of these new technologies underscores the importance of interdisciplinarity in addressing energy challenges and developing a more secure energy system for the future.

1.2.3. Chemical Synthesis and Manufacturing

The coincidence of chemical manufacturing and production with photocatalysis has a high capability for enhancing industrial processes and mitigating environmental impacts. Photocatalysis can use light to activate chemical reactions, which is particularly beneficial in industries where toxic impurities are prevalent. For instance, in the nitration reactions of aromatic compounds, which are widely used in the manufacturing of pharmaceuticals and pesticides, significant amounts of toxic byproducts can be produced. Researchers have researched the efficiency of using titanium dioxide (TiO_2) as a photocatalyst to purify wastewater from such processes. A test on a lagoon's wastewater was able to identify the presence of poisonous chemicals, such as RDX and HMX, alongside herbicides like Dinoterb and Dinoseb. The employment of UV/TiO_2 treatments facilitated the extensive degradation of dinitrophenolic compounds, demonstrating the potential of photocatalysis in transforming hazardous industrial effluents into safer compounds through chemical synthesis and process development (Hafidi et al., 2009).

The limitations associated with treating phenolic compounds in wastewater further emphasize the applications of photocatalytic processes in chemical synthesis. Phenols are notorious for their persistence in the environment and toxicity, which make them hard to remove from industrial effluent. Factories that produce phenolic compounds rely on sophisticated and costly methods to overcome contamination. Photocatalytic technology has now demonstrated that TiO_2-based catalysts can effectively promote the degradation rates of these pollutants. Through research, it has been established that doping TiO_2 with nitrogen or copper, for instance, during its modification, can significantly increase the efficiency of degradation processes using a variety of light sources.

This innovative process not only diminishes the environmental footprint of chemical manufacturing but also concurs with industry-wide sustainability objectives and encourages greener synthesis processes (Steiner et al., 2017). The use of photocatalysis in chemical synthesis and manufacturing processes is a paradigm shift towards productive and environmentally friendly industrial processes. Through the utilization of photocatalytic processes, manufacturers can address complex wastewater issues while reducing operating costs associated with traditional waste treatment processes. The nature of catalysts like TiO_2 allows them to be applied to a wide range of manufacturing processes, including pharmaceutical and agrochemical synthesis. As the capabilities of tunable photocatalysts are continually unraveled by science, the vision of large-scale application becomes increasingly tangible. Ultimately, the synergy between chemical synthesis, manufacturing, and photocatalysis holds promise for the future pursuit of green industrial processes, leaving space for innovation that balances productivity and environmental stewardship equally (Hafidi et al., 2009; Steiner et al., 2017).

1.3. MATERIALS USED IN PHOTOCATALYSIS

In photocatalysis, the materials used are the key to the effectiveness and efficiency of the process. Semiconductors, particularly metal oxides, are the foremost photocatalytic materials due to their preferred electronic properties and light-exposure stability. Titanium dioxide (TiO_2) is the most extensively studied photocatalyst, primarily due to its high photocatalytic activity, non-toxicity, and chemical stability. TiO_2 is employed for the degradation of various organic contaminants and has extensive applications in water treatment processes. Novel developments include the surface modification of TiO2 to enhance its activity in the visible light region, making it more effective than employing ultraviolet light (Maury et al., 2009).

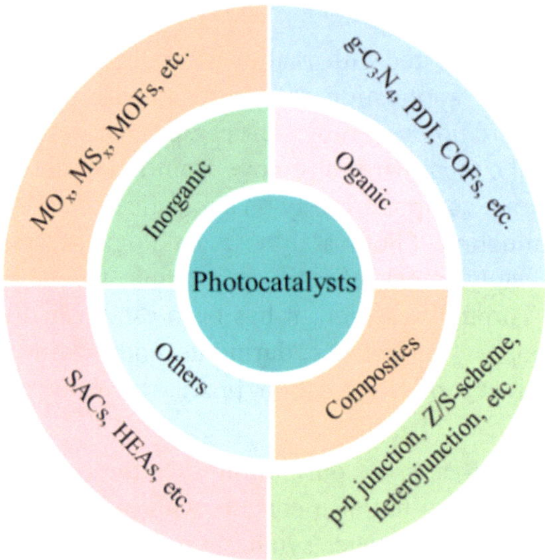

Figure 1.9. Types of materials explored for use as potential photocatalysts.

Source: https://www.mdpi.com/1420-3049/30/2/269.

Variations in microstructure resulting from different deposition techniques also influence its photocatalytic activity, promoting further research aimed at optimizing these properties. Besides titanium dioxide, other materials, such as zinc oxide (ZnO), tin oxide (SnO2), and carbon nanomaterials, are being increasingly researched for their photocatalytic applications. Zinc oxide is also reported to be very photocatalytic and stable, especially in the aqueous phase. Similar to TiO_2, though, its action is limited only to ultraviolet light. In response to this, researchers are developing composite materials in which these semiconductors are combined with metal nanoparticles or carbonaceous materials, allowing them to function with a wider range of light wavelengths. Such composites not only enhance photocatalysis but also augment the dynamics of the charge carrier in the system, leading to more effective removal of pollutants (Maury et al., 2009). The interaction between different materials can produce new features that significantly enhance the general efficiency of photocatalytic reactions. Furthermore, reactor design is a significant factor since it comes into direct contact with the characteristics of the materials used. Reactor design features, such as the illumination source and reactor shape, have a critical impact on the improvement of photocatalyst efficiency. For example, exterior illumination and some reactor designs, such as annular and round geometries, may dictate light penetration and photocatalyst-contaminant interaction. Quartz or some glasses can also be chosen based on transparency to maximize exposure to light (Lee et al., 2004). The future of photocatalytic technology is highly dependent on the innovative combination of world-class materials and reactor design, which will lead to more efficient and sustainable environmental solutions (Figure 1.9).

1.3.1. Semiconductor Materials

The significant role of semiconductor materials in photocatalysis cannot be overstated as they are the foundation for the majority of photocatalytic reactions. The photocatalysts often undergo electronic transitions upon absorbing light, enabling them to generate charge carriers—electrons and holes. Charge carriers participate in redox reactions that enable the degradation of organic pollutants and also contribute to energy conversion reactions. Various semiconductor compounds, such as titanium dioxide (TiO2), zinc sulfide (ZnS), and bismuth sulfide (Bi2S3), are typically investigated for their photocatalytic activity. TiO_2, for instance, has extensive applications due to its stability, non-toxicity, and high UV light-induced photocatalytic activity. However, the quest to expand the range of light absorption by semiconductors beyond UV wavelengths has led to innovative solutions, such as the development of composite materials that can enhance overall photocatalytic activity. Composite semiconductors have emerged as a focal point for maximizing the efficiency of photocatalysis. Nanocomposites of different semiconductor materials are capable of availing themselves of the specific advantage of each component and result in greater charge separation with reduced rates of recombination. For example, the synthesis of a nanocomposite material in which titanate nanofibers (TNF) are mixed with ZnS and Bi_2S_3 illustrates this trend. The functionalization of TNF with these semiconductor nanoparticles, as discussed in the hydrothermal synthesis route, renders various photocatalytic functions. Assays of such nanocomposites, as in the study where hydroxyl radical generation was measured, are characterized by excellent photocatalytic activity relevant to applications such as the degradation of organic dyes from wastewater. These advances underscore the significance of composite semiconductor structural engineering in optimizing photocatalytic activity. Research on the fabrication method of semiconductor materials is just as crucial in determining photocatalytic performance. Techniques such as chemical vapor deposition (CVD) and sol-gel synthesis have been found actually to influence the morphology and crystallinity of semiconductor thin films. For instance, thin titanium dioxide films prepared by sol-gel and hydrothermal techniques were tested for their photocatalytic activity against various organic compounds, such as 2,6-dichlorindophenol. The relative efficiency of the movie may vary under conditions like sol composition or curing conditions, as demonstrated in intensive studies analyzing the role of PEG additives and curing times. It is worth noting that the processing methods will be optimized to yield improved photocatalytic materials, thereby enhancing their prospects for environmental purification processes and renewable energy devices (Králová et al., 2012; Silvestre et al., 2015) (Figure 1.10).

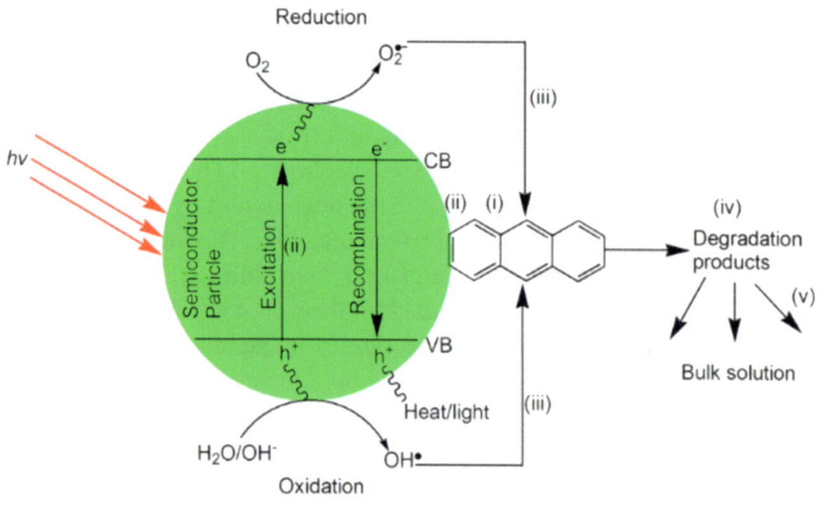

Figure 1.10. Key steps in semiconductor photocatalysis.

Source: https://www.researchgate.net/figure/Key-steps-in-semiconductor-photocatalysis_fig2_303627507.

1.3.2. Metal-Based Photocatalysts

Metal-based photocatalysts have made significant advancements in the field of photocatalysis, primarily due to their ability to drive chemical reactions upon exposure to light. Traditional inorganic photocatalysts, such as titanium dioxide (TiO2), have dominated the landscape to date; however, the need for efficiency and sustainability has opened doors to the study of semiconducting metal-organic frameworks (MOFs). These materials exhibit exceptional properties, including tunable band gaps, large surface areas, and the capacity to integrate multiple metal ions. This adaptability enables the creation of photocatalysts that can respond to various light sources and reaction conditions, making them an indispensable component in establishing the practice of green chemistry. Notably, the introduction of transition metals into the MOF framework can lead to enhanced charge separation and improved catalytic performance, thereby expanding the scope of visible light photocatalysis to non-traditional horizons. Notably, the discovery of red light-mediated photocatalysis has marked a paradigm shift in organic molecule synthesis, primarily by minimizing energy consumption and maximizing reaction selectivity (Figure 1.11).

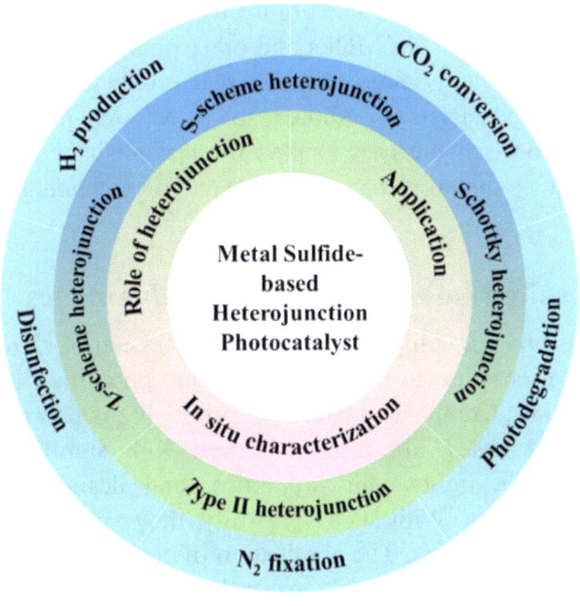

Figure 1.11. Metal sulfide-based heterojunction photocatalysts.

Source: https://external-content.duckduckgo.com/iu/?u=https%3A%2F%2Fscx2.b-cdn.net%2Fgfx%2Fnews%2F2023%2Fa-review-of-the-latest.jpg&f=1&nofb=1&ipt=477dfa80ac6607fe4fc3043f78845ae562bda732235bd24009632cdc485e00bc.

Tungsten (W) and molybdenum (Mo) are potential candidates for use in designing red-light-absorbing photocatalysts. Due to their ability to absorb low-energy light, they not only minimize potential damage arising from exposure to

UV light but also provide improved penetration depth within reaction media. Different organic transformations carried out using earth-abundant transition metal-based photosensitizers have been documented in recent studies, which function effectively in reactions such as C-C coupling and polymerization. Enhanced photocatalytic activity under red light can be attributed to mechanisms such as sensitized triplet-triplet annihilation, which enables efficient transformation of photonic energy into chemical energy, thereby opening up new avenues in organic synthesis (Sinha et al., 2025). S-scheme heterojunctions represent a promising approach for enhancing the performance of metal-based photocatalysts. By constructing such heterojunctions from complementary semiconductors, scientists have achieved significant advancements in charge separation and the utilization of sunlight's energy. The new architecture of S-scheme heterojunctions enables facile electron transfer, thereby enhancing the photocatalytic activity of materials in essential applications, such as hydrogen production and CO_2 reduction. The ability to construct such heterojunctions using various combinations of MOFs and other photocatalysts opens the way to resolving critical energy and environmental problems. While the demand for efficient photocatalysts continues to grow, integrating metal-based photocatalysts into S-scheme systems presents a promising horizon that has the potential to enhance the efficiency of photocatalytic technologies (Yuan et al., 2024).

1.3.3. Composite Materials and Their Advantages

Over the last few years, composite materials have been of significant interest in many applications due to their advantageous properties over individual, traditional materials. Composite materials, comprising two or more constituent materials in most cases, are often endowed with significantly improved mechanical strength, lightweight properties, and designable thermal and electrical conductivities. With such versatility, they are very valuable in the application of photocatalysis. The application of composite structures, such as titanium dioxide in a polymer matrix, can enhance photocatalytic performance and improve environmental stress stability. Material mixture optimization enables scientists to create new surfaces that capture the maximum amount of light, which is crucial for efficient photocatalytic reactions. The advanced composite fabrication processes also complement the need for effective and environmentally friendly materials in a wide range of industries, particularly in environmental remediation. The advantages of composite materials extend beyond the betterment of mechanical properties; they also extend to the justified improvement of photocatalytic activity. For instance, composite development can facilitate the dispersion of photocatalytic agents over a wide surface area, thereby increasing the interaction with light and pollutants to the maximum extent. In their practical applications, e.g., wastewater treatment, composites facilitated through techniques like sol-gel synthesis have exhibited high degradation rates of organic pollutants, including toxic compounds. Through the exploitation of the synergistic effect of composite materials, scientists can engineer more effective catalysts that operate under diverse environmental conditions. The synthesis method employed for the preparation of these composites, such as chemical vapor deposition elaborated in studies, allows scientists to have , reasonable control over the morphology and structure of the final material, which are essential for the photocatalytic activity. Finally, the use of composite materials in photocatalysis is a pioneering step towards the resolution of environmental issues. With the introduction of new composites, such as all-inorganic metal halide perovskites, there is broader utility beyond solar harvesting and toxic substance degradation. The stability and efficiency of these materials can have significant potential to enhance the overall performance of photocatalytic systems. Material pairing advancements, such as titanium dioxide and metal nanoparticles, push the boundaries of traditional photocatalytic reactions, offering a direction for stronger and more efficient solutions. As researchers continue to explore the new properties and applications of composite materials, it is essential to strike a balance between scientific and practical aspects in fields such as photocatalysis, making them valuable tools in achieving a sustainable world (Králová et al., 2012; Yin et al., 2018).

1.4. FACTORS INFLUENCING PHOTOCATALYTIC ACTIVITY

The efficiency of photocatalysis is significantly influenced by various environmental and material-related conditions affecting the photocatalytic activity of materials. It is critical to consider the intrinsic nature of the photocatalytic materials themselves, including their crystal structures and morphology. The crystalline phase and defect concentration can significantly impact the ability of photocatalysts to absorb light and generate electron-hole pairs. For instance, the use of proper band gap energies in materials facilitates light absorption and charge separation, which is paramount for the efficient degradation process. In organic pollutant research, various variables, such as the concentration of catalyst and initial dose of dye, have been identified as determinants of the degradation rates of pollutants like crystal violet directly. Therefore, optimizing these parameters has become necessary for enhanced photocatalytic activities (Shabna et al., 2024). The temperature and solution pH are also important determinants of the photocatalytic reactions. Adjusting the pH can control the charge on the surface of the photocatalyst, affecting its interaction with ionic dyes and altering the adsorption characteristic. For example, decreased or increased pH would promote or restrict the adsorption of specific pollutants and thus directly affect the efficiency of the photocatalytic process.

In addition, temperature plays a critical role in kinetic energy, which governs the reaction rate; high temperatures can accelerate the degradation process through enhanced molecular motion and increased reaction rates. Therefore, determining the optimal conditions for pH and temperature is crucial to optimize photocatalytic activity and ensure the complete mineralization of harmful substances (Shabna et al., 2024). In addition, incident light intensity is a governing parameter in photocatalytic activity and has a direct influence on photon absorption rate and electron excitation. Improvement of light intensity may increase the electron-hole pairs generated, improving the degradation efficiency of the photocatalyst.

The relationship between light intensity and photocatalytic activity requires careful optimization to avoid saturation effects in the utilization of certain materials. In current studies of photocatalytic hydrogen production through water splitting, the need for improved photocatalyst designs is now evident, and the details of the crystal properties, particularly regarding surface states and lattice defects, are therefore crucial. Development of the understanding of these coupled parameters not only enhances theoretical insight but also offers practical design criteria for achieving more effective photocatalytic systems (Figure 1.12).

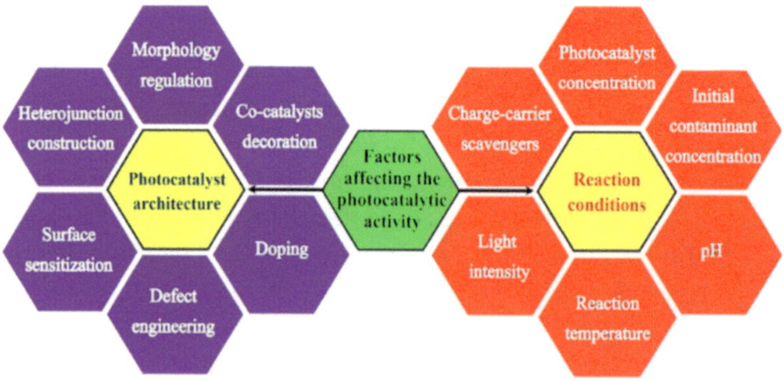

Figure 1.12. The most critical factors affecting activity of a photocatalytic reaction.

Source: https://www.researchgate.net/figure/The-most-critical-factors-affecting-activity-of-a-photocatalytic-reaction_fig7_372385929.

1.4.1. Light Intensity and Wavelength

The interaction between wavelength and the intensity of light plays a crucial role in photocatalysis, and both the efficacy and efficiency of catalytic reactions are impacted.

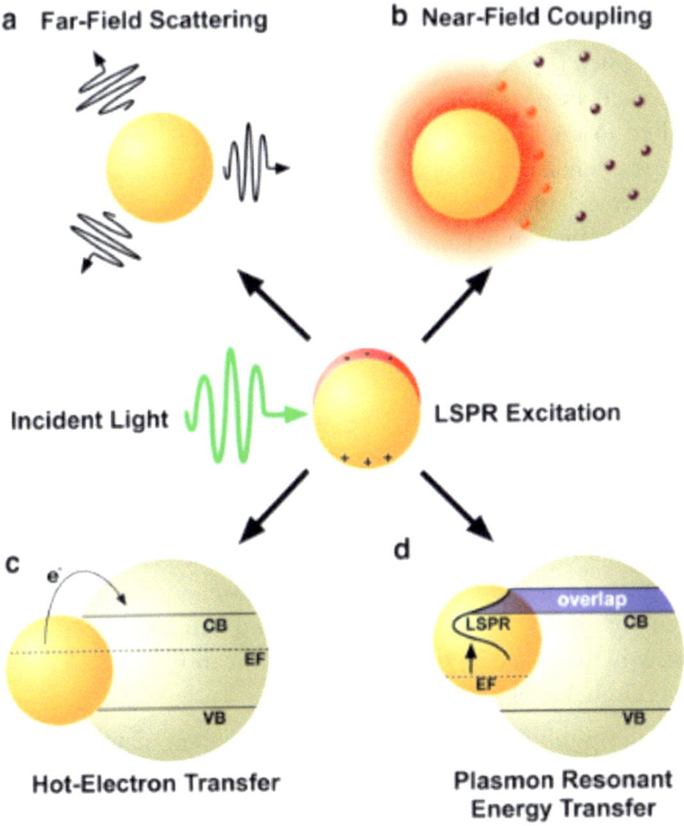

Figure 1.13. The excitation of localized surface plasmon resonance and different energy transfer mechanisms responsible for plasmon-driven performance enhancement.

Source: https://www.researchgate.net/figure/The-excitation-of-localized-surface-plasmon-resonance-and-different-energy-transfer_fig7_321910427.

Photocatalysts, typically semiconductor materials, utilize light to activate electrons from their valence band to the conduction band, thereby creating electron-hole pairs that lead to chemical transformations. The wavelength of incoming light determines the photon energy; hence, light must possess sufficient energy to cross the bandgap of the photocatalyst. For example, ultraviolet light may be more effective due to its shorter wavelength and higher energy compared to visible light. But the overall efficiency also relies on the intensity of light, which is proportional to the density of photons available for absorption by the catalyst, ultimately affecting the rate of the chemical reactions involved in the photocatalytic process. Furthermore, the significance of localized surface plasmon resonance (LSPR) becomes apparent in the context of photocatalysis when one considers how light interacts with nanostructured catalysts. LSPR can be applied to increase light absorption at particular wavelengths, thereby accelerating the generation of charge carriers utilized in catalytic reactions. The phenomenon is particularly pronounced when light intensity is increased, thereby enhancing the interaction between matter and light. Observations indicated that materials exhibiting significant LSPR effects are capable of boosting reaction rates under particular illumination, thus demonstrating a complex connection between the physical properties of lights and the efficiencies of photocatalysts. As hinted at in recent literature, this interaction offers avenues for further research on optimizing photocatalytic performance through the selection of suitable catalyst materials and optimal operating light parameters, thereby advancing LSPR to the forefront in the development of photocatalytic applications (Verma et al., 2024). Experimental studies also further affirm the importance of wavelength and light intensity for photocatalytic effectiveness. For example, incorporating titanium oxide colloids into supercapacitor frameworks revealed that wavelengths in the blue spectral range exhibited significant light-induced phenomena, which enhanced capacitance significantly compared to unaltered frameworks (Grebel et al., 2022) (Figure 1.13).

This result highlights the wavelength-dependent nature of photonicity and suggests the capability of judiciously adapting light parameters to yield significant increases in reaction rates and resultant catalytic activity. It follows that understanding the specific roles played by light intensity and wavelength becomes imperative to effectively exploit photocatalysis in practical applications, which may provide opportunities for innovations in using light more efficiently in schemes for energy conversion and environmental cleanup.

1.4.2. Surface Area and Morphology of Catalysts

The surface area and morphology of catalysts are critical factors that significantly influence their activity in photocatalytic reactions. The larger the surface area, the more interaction between the catalyst and the reactants, providing more active sites for the photoreaction to occur. This feature is critical in the case of semiconductor catalysts like titanium dioxide (TiO_2), where the nanostructured materials may exhibit enormous enhancements in photocatalytic activity due to the large surface area. Experiments using techniques such as N_2 adsorption have shown that modifying the fabrication processes, e.g., sol-gel treatments or co-doping with other elements, can directly influence the surface area and the final photocatalytic activity of the materials, thereby impacting their application in environmental clean-up technologies and energy conversion devices (Avci et al., 2012). Moreover, morphological characteristics of catalysts, including pore shape and size, play a key role in their efficiency. More specifically, accessibility of mesopores could provide for optimum diffusion of the reactants into the catalyst's active sites to improve reaction kinetics. The function of morphology is demonstrated in research with composite catalysts, in which gold or tungsten oxide materials added to TiO_2 forms new structures for enhanced light trapping and charge separation. It is through these heterojunctions that electron transfer dynamics can be significantly boosted. The structural integrity and material organization can be well-engineered to deliver optimal photocatalytic performance, highlighting the significance of both surface area and morphology in the design of effective photocatalysts (Parker et al., 2016). Regarding the impact of doping and synthesis protocols on surface area, morphology, and photocatalytic activity, the effect is more pronounced. For instance, doping TiO_2 with other metals, such as vanadium or tantalum, not only modifies its crystalline form but also its surface properties, thereby enhancing its optical and photocatalytic activity in visible light. This is particularly important in applications involving organic pollutants, as doped catalysts tend to exhibit higher degradation rates. The synthesis paths can create defects and alter the surface condition of the catalyst, which in turn impacts charge carrier dynamics and the rate of photocatalytic reactions. Ultimately, a comprehensive understanding of how surface area and morphology interact is crucial for the development of catalysts tailored to specific photocatalytic applications, thereby enhancing efficiency in energy and environmental processes (Avci et al., 2012; Parker et al., 2016).

1.4.3. Environmental Conditions (pH, Temperature)

The impact of environmental parameters such as pH and temperature is key to an understanding of photocatalysis and its efficiency. pH can have a dramatic on the surface properties of photocatalytic materials, which in turn affects performance and reactivity. For example, varying pH conditions can alter the semiconductor surface charge, facilitating or inhibiting the adsorption of reaction intermediates. This pH regulation of catalytic activity emphasizes the importance of pH in attaining maximum photocatalytic activity. Furthermore, by adjusting pH values, researchers can enhance the contact between the photocatalyst and the subject pollutant of interest, thereby promoting degradation rates in various applications, such as wastewater treatment and air purification. Clarification of the function of pH is thus fundamental to improving photocatalytic technology and developing solutions for specific environmental challenges. Temperature, which is another key ecological parameter, plays an essential role in controlling the kinetics of photocatalytic reactions. When the temperature increases, reaction rates typically rise due to increased molecular activity, which tends to result in improved photocatalytic performance. Optimum temperatures, however, need to be established, as very high temperatures can impede photocatalytic efficiency or cause undesirable side reactions. The investigation of the effect of temperature is not limited to the catalysts but extended to the substrates and the surroundings in which reactions occur. For instance, temperature fluctuations may affect the solubility of reactants and the stabilization of intermediates, which in turn dictate the overall route of the photocatalytic reaction. According to research findings, temperature control plays a crucial role in enhancing the efficiency of photocatalysis under realistic conditions. Finally, the combined action of pH and temperature can have a cumulative effect on photocatalytic systems. For instance, under certain temperature conditions, the effect of pH can become more pronounced, either encouraging or suppressing the degradation of contaminants depending on their chemical nature. These effects pose a challenge to the complexity of optimizing photocatalytic systems for practical applications. Clearly, understanding how these variables impact photocatalytic activity can lead to more effective experiments and more efficient processes. As shown in experiments with mycotoxin production, variations in environmental conditions typically result in drastic variations in outcomes, which means that the same principles apply to photocatalytic processes as well. Understanding these nuanced interactions can inform future scientific and technological advances in photocatalysis and its applications across various sectors (Matuska-Łyżwa et al., 2024; Bata-Vidács et al., 2025).

1.5. CHALLENGES IN PHOTOCATALYSIS

The field of photocatalysis, although highly promising across a broad spectrum of environmental and energy-related applications, is hindered by several challenges that slow its widespread deployment. Among the principal difficulties is the inefficacy of photocatalytic materials, particularly titanium dioxide (TiO_2), which has received much more attention and application due to its advantageous characteristics. Despite this, some of TiO_2's bulk properties remain poorly understood, particularly its electronic band gap and how structural defects, such as oxygen vacancies, influence it. Such shortcomings can have adverse effects on its photocatalytic efficiency and necessitate further research into how the optical properties of TiO_2s can be optimized through means such as doping and nanostructuring (Chiodo et al., 2010). The requirement for increased understanding is apparent, as increased photocatalytic activity is needed to optimize the use of such materials in processes like the degradation of pollutants and the conversion of solar energy. Scalability in the production of efficient photocatalysts is also a challenge in the area of photocatalysis. Although several fabrication methods, including sol-gel synthesis and chemical vapor deposition (CVD), have been developed to produce TiO2 thin films, the reproducibility and effectiveness of these processes can be highly unpredictable. For instance, the photocatalytic activity of TiO_2 films can be controlled by varying synthesis conditions, such as temperature, time, and precursor concentration. Experiments have demonstrated that the structural and morphological characteristics of TiO_2 can vary considerably depending on the method used, which impacts their photocatalytic activity (Králová et al., 2012).

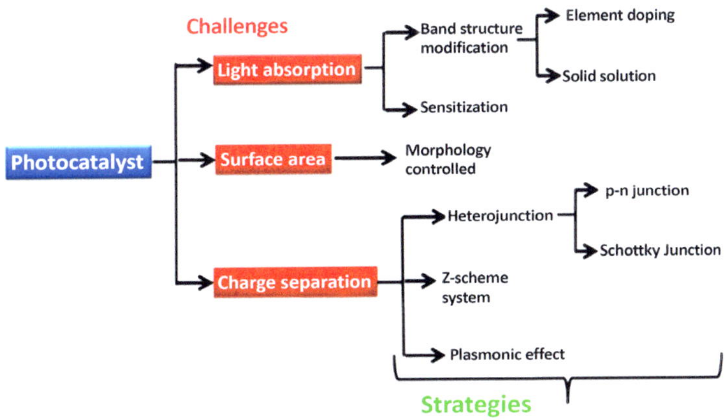

Figure 1.14. The field of photocatalysis, while promising for various environmental and energy applications, is fraught with multiple challenges that hinder its broader implementation.

Source: https://ars.els-cdn.com/content/image/3-s2.0-B9780128190494000118-f07-03-9780128190494.jpg.

These differences in production processes need to be addressed to scale up photocatalytic technologies from the lab to the factory floor and make them more suitable for use. Lastly, the stability and durability of photocatalytic materials are other concerns. Over time, material photocatalytic properties, such as those of TiO_2, may diminish through mechanisms such as photo-corrosion and depletion of passive surface sites. Degradation limits the life of photocatalysts, thereby increasing operational costs and reducing efficiency in extended operations. Environmental factors, such as moisture and impurities, can also adversely impact photocatalytic efficiency, leading to the need for robust materials capable of withstanding operational conditions that vary from optimal. Research efforts aimed at improving photocatalyst stability through new coatings, composite materials, and hybrid structures are crucial for addressing these challenges and ensuring the durability and reliability of photocatalytic processes in practical applications (Chiodo et al., 2010; Králová et al., 2012) (Figure 1.14).

1.5.1. Efficiency and Scalability Issues

The success of photocatalysis relies heavily on the choice of catalyst materials, which in turn affects the overall performance of photoreactors. Typical photocatalysts, such as TiO2 and ZnO, have achieved a mixed level of success; however, their low light absorption and charge carrier recombination restrict their practical applications. Recent advances, including the application of nanoparticles, have demonstrated potential for enhancing catalytic activity via the amplification of light absorption and charge separation phenomena. Hybrid materials also hold promise in addressing these efficiency issues, and through hybridized materials, pollutant degradation with higher efficiency is possible in various environments. However, the practical application remains limited by catalyst instability and recyclability, underscoring the need for solutions that not only exhibit superior photocatalytic performance but also demonstrate longevity and reliability in continuous operation (Qasim et al., 2025). Another significant issue affecting commercialization is scalability. The transition from laboratory-scale plants to industrial applications involves optimizing several factors, including reactor design and catalyst composition. Nanoparticle-enhanced photocatalysts have shown promise in laboratory studies, but scaling up production with similar efficiency at a lower cost remains a significant challenge. The environmental implications of large-scale photocatalyst applications are also a concern, particularly with respect to energy requirements and potential ecological damage. These scalability issues must be addressed by the establishment of standardized protocols for the synthesis, characterization, and utilization of photocatalytic systems for practical applications. Without modification of these parameters, the scope of photocatalysis in resolving environmental problems can be significantly limited (Qasim et al., 2025). New trends, such as the development of perovskite quantum dots (PQDs) and S-scheme heterojunctions, are providing solutions to efficiency and scalability problems in photocatalysis. The new materials not only enhance catalytic performance but also result in a cleaner approach to energy conversion and remediation. PQD-based structures reveal superior charge carrier motion and durability, which are helpful for their utilization in processes such as CO_2

photoreduction. However, their large-scale integration remains a significant challenge, specifically in terms of the long-term stability and economic viability of production technologies. These challenges need to be met through ongoing research and dialogue between material scientists and engineers to rationalize new photocatalytic structures and make innovations realistically deployable at the commercial scale. The potential of such technologies provides a rationale for developing scalable systems to effectively harness photocatalysis for addressing environmental and energy challenges.

1.5.2. Stability and Durability of Photocatalysts

The stability and longevity of photocatalysts are of utmost importance to extend their applicability to a wide range of applications, from energy conversion to environmental remediation. The photocatalyst materials undergo degradation or modification under prolonged irradiation and exposure to the atmosphere, resulting in a loss of efficiency. To optimize the use of these materials, methods that enhance their structural stability need to be developed. For instance, surface treatment by chemical or physical means can form a protective layer, hence strengthening the functional lifespan of photocatalysts. Evaluation of the long-term functionality of the same materials, particularly in real applications, requires thorough testing under various conditions to find out if they are rigid and stable enough over a period of time so that they can be suitable for industrial applications, particularly for water treatment and air purification. The current progress in the field has highlighted the means of promoting the stability of photocatalysts through new approaches, such as dopant doping and composite formation. Experiments have demonstrated how the incorporation of carbon dots (CDs) into traditional photocatalysts significantly enhances their stability and photocatalytic activity. Their photoluminescence properties and charge separation ability allow for the creation of more stable photocatalytic processes. This addition not only helps minimize electron-hole recombination but also enhances the material's photocatalytic activity in fields such as hydrogen production and pollutant degradation. With a focus on the synthesis and engineering of such composite materials, researchers can create idealized photocatalysts that significantly outperform their traditional counterparts, demonstrating the potential for useful, durable applications in sustainability and green energy. Additionally, the long-term durability of photocatalysts should not be overlooked when evaluating their durability and stability. With the growing need for green technology, the development of stable and efficient photocatalytic materials is more relevant than ever. New avenues for highly durable photocatalysts have opened up through the exploration of various synthesis methods, including sol-gel processing, hydrothermal treatment, and even biogenic routes. These processes not only endow materials with enhanced structural properties but also pave the way for renewable materials to be utilized as precursors, which is particularly apt for achieving sustainable goals. In this context, it is imperative to conduct extensive research on the long-term behavior of these materials to facilitate their efficient utilization in actual applications, thereby providing higher-value integrated solutions to address global environmental issues (Boonsith et al., 2020; Hakki et al., 2017).

1.5.3. Economic Considerations and Cost-Effectiveness

The use of photocatalysis in many industries is greatly influenced by economic considerations and cost-effectiveness, which are prime for its widespread application. Photocatalytic reactions require careful consideration of initial costs, maintenance expenditures, and return on investment. When industries consider the application of photocatalysis in fields such as pollutant degradation and energy conversion, the financial feasibility of these technologies is a crucial consideration. The development of novel catalysts that facilitate efficient reactions cannot only decrease costs but also enhance productivity. The design of catalysts, for example, to maximize both efficiency and selectivity in photocatalytic reactions has a significant overall economic impact. The scalability potential of photocatalytic technologies also influences their cost considerations, as successful integration into larger systems can lead to overall cost savings in the long run, potentially making them a preferable option for industries seeking sustainable solutions. The availability of resources and the environmental impact of sourcing and using materials also influence the cost-effectiveness of photocatalysis. The shift towards using substitute raw materials, such as waste products and biomass, holds high economic potential, realizing double dividends of energy generation and waste reduction. Innovations in catalyst fabrication and processing can even boost the viability of using these alternative resources. A two-stage processing regime, for instance, could involve generating fuels from supply gases derived from waste, yielding economic benefits (Ball et al., 2025). Furthermore, the development of chemically stable catalysts that retain their potency under harsh conditions makes long-term cost effectiveness possible through lower requirements for frequent replacement or maintenance, thereby providing maximum value for money. Sustainability in the environment is being progressively recognized as a core component of cost-effectiveness in photocatalysis, particularly in agricultural uses where the balancing act between productivity and environmental safety is most urgently needed. The application of nanotechnology to agricultural cultivation increases the demand to minimize harmful agents and increase crop output. As explained in recent studies, nanomaterials can be used to improve agricultural production systems as well as decontaminate polluted environments. The dual focus on productivity and environmental well-being is a reflection of a more sophisticated understanding of economic drivers in photocatalysis (Bacchini et al., 2013). Through the implementation of more efficient photocatalytic systems, agricultural stakeholders can achieve environmentally friendly farming practices that are beneficial to the environment as well as profitable, demonstrating the entwinement of economic sustainability and environmental responsibility in the new science of photocatalysis.

1.6. RECENT ADVANCES IN PHOTOCATALYSIS RESEARCH

Recent advancements in photocatalysis research have significantly enhanced our understanding of semiconductors, particularly titanium dioxide (TiO_2), in promoting photocatalytic reactions under visible light. The poor excitation of TiO2 generally plagues traditional photocatalytic pathways under the visible region. A novel approach with ligand-to-metal charge transfer (LMCT) has now been proposed, which can allow TiO_2 nanoparticles to utilize visible light effectively. This discovery has broad implications for environmental remediation because LMCT facilitates the coupling between TiO2 and common surface adsorbates, thereby forming new absorption bands that enhance photocatalytic activities. These developments not only expand the functional purposes of TiO2 but also shed light on the potential use of LMCT processes as a versatile method in solar energy engineering (Choi et al., 2025). Further, the marriage of nanotechnology with photocatalysis is yielding interesting results. As agriculture evolves to meet the enormous food production needs, increasing quantities of fertilizers and pesticides are being used. While they have their uses, uncontrolled release is giving rise to environmental concerns, especially about soil and groundwater pollution. Recent research indicates that nanomaterials can significantly enhance agricultural production and environmental health through targeted detection and removal of toxins. Nanotechnology in photocatalysis facilitates the degradation of harmful pollutants, which plays a crucial role in modern agro-environmental strategies. While toxicity cannot be neglected, research indicates that the majority of nanomaterials are found to have low adverse effects, thereby qualifying as an effective tool for sustainable uses (Bacchini et al., 2013). The future of photocatalysis research looks promising, with ongoing innovations consistently enhancing the efficiency and efficacy of photocatalysts. Improvements in photocatalytic material design and synthesis emphasize the need for designing photocatalysts that not only remain efficient under visible light but are also economical and environmentally friendly. For instance, surface property optimization of semiconductors and composite material design, which synergistically enhance photocatalytic performance, are promising directions. Furthermore, continued investigation into other mechanisms of activation, such as the aforementioned LMCT, suggests a trend toward more flexible and resilient photocatalytic systems. As research goes on, these breakthroughs have much potential to solve immediate ecological concerns, such as pollution and sustainability, and also to contribute to the world's shift towards energy (Choi et al., 2025).

1.6.1. Innovations in Photocatalyst Design

With the high-tech nature of the photocatalysis field, design innovations in photocatalysts play a crucial role in the efficacy and applicability of photocatalytic reactions. One of the promising avenues here is the creation of covalent organic frameworks (COFs), as these compounds exhibit marvelous structural tunability and surface characteristics. COFs are crystalline, porous solids with a high surface area and stability, making them superior candidates for light-induced chemical transformations. Their unique photoelectronic properties allow them to trap energy and catalyze upon visible light irradiation efficiently. Through the rational optimization of synthesis and functionalization procedures, researchers can tailor COFs for enhanced photocatalytic activity in specific applications, including organic conversions and environmental purification. Research into such cutting-edge materials helps fill the gaps in state-of-the-art photocatalytic technology, thereby enabling sustainable operations in the field of chemical engineering and providing environmentally friendly solutions in various areas (Alberto et al., 2022). Another interesting aspect of innovation in photocatalyst development is overcoming pressing environmental challenges through effective abatement of pollution. Microplastics, renowned for their persistence and bioaccumulation in ecosystems, necessitate new methods for degradation and removal. Recent advances in photocatalysis highlight its potential to catalyze the degradation of microplastics into harmless organic compounds, such as water and carbon dioxide. It not only aims for the eradication of microplastics but also a grander vision of conjoining waste management with photocatalytic technologies. The effectiveness of novel photocatalysts in degrading such contaminants underscores the need for interdisciplinarity in materials science and environmental engineering. Through the application of photocatalytic technology, society can evolve towards an integrated approach to prevent microplastic pollution, while leveraging the unique properties of new photocatalysts to achieve overall environmental benefits (Chen et al., 2022). Research on hybrid photocatalysts has also garnered significant attention as researchers strive to optimize photocatalytic activity and multiplicity. By bringing different classes of photocatalysts together or coupling them with another material, researchers can exploit synergies to enhance light harvesting, charge separation, and catalytic activity. Hybrid systems may exhibit better photocatalytic activity than their component parts by leveraging the strengths of individual materials. These advances not only optimize the utilization efficiency of solar energy but also extend the reach of targetable compounds for photocatalytic reactions. Thus, hybrid photocatalysts are at the forefront of research on designing efficient, sustainable processes for energy production and environmental purification. The ongoing research into the interactions and characteristics of hybrid photocatalysts is a promising indicator of further advancements in photocatalytic technologies and new solutions in photocatalysis (Alberto et al., 2022) (Figure 1.15).

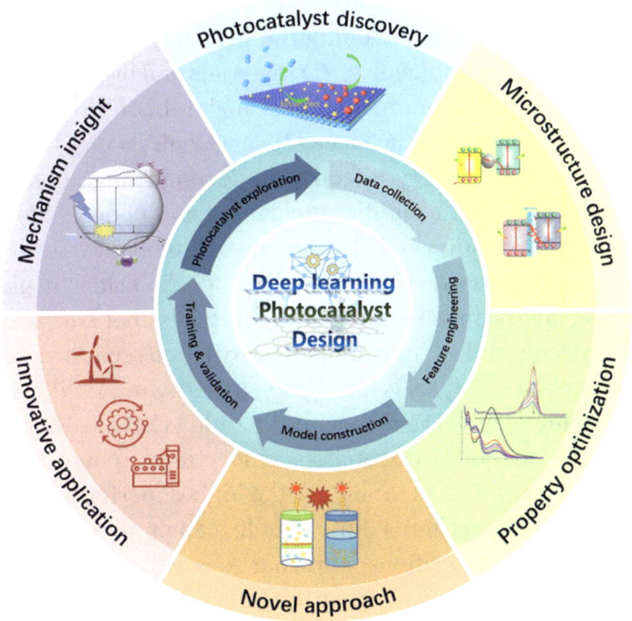

Figure 1.15. Transformative strategies in photocatalyst design.

Source: https://www.oaepublish.com/articles/jmi.2024.48.

1.6.2. Novel Applications and Emerging Technologies

The evolution of photocatalysis is closely tied to novel applications and promising technologies, extending its efficiency and scope. One area of combining photocatalytic processes with the latest surface-enhanced Raman spectroscopy (SERS) techniques, for instance, is gaining attention. Another method includes the use of a photo-induced photo-catalytic SERS (PI-PC

SERS) approach to the development of nanocomposites for ultrasensitive molecule detection. The process involves a two-step interaction, where pre-irradiation enhances the Raman signal and post-irradiation facilitates the removal of contaminants from the substrate, thereby restoring functionality. Such innovation not only increases sensitivity, as seen in detection limits of up to 1.02 × 10−14 M, but also extends the lifespan of the sensors through effective handling of the substrate. Consequently, the application of these advanced methodologies puts photocatalysis at the frontline of analytical chemistry, paving the way for more viable applications in environmental monitoring and diagnostics (Mai et al., 2025). Advanced technologies are practically revolutionizing the future of photocatalysis, particularly through the emergence of new materials. The quest for more photocatalytic activity has led to innovation in materials, including the investigation of two-dimensional (2D) materials and organic molecules. 2D materials and organic molecules have unique electronic and optical properties that make them a sought-after option for effective photocatalysts. For instance, 2D materials, such as transition metal dichalcogenides, have been very effective in decomposing contaminants with visible light irradiation. These abilities are essential in photocatalytic processes for environmental remediation and clean energy technologies. At the same time, research is ongoing to unravel the nanoscale mechanisms of photocatalytic reactions, driving innovation in composite materials that can absorb a wider range of light spectra and exhibit higher energy conversion rates (Katehi et al., 2025). While photocatalysis is emerging as a growing field, its intersection with artificial intelligence (AI) and machine learning (ML) holds high potential. By leveraging data analytics, researchers can optimize photocatalytic system design and operating conditions to achieve the highest performance at a reduced cost. Machine learning algorithms, for example, can predict the performance of various photocatalyst compositions, allowing scientists to iterate and optimize in the laboratory quickly. Such a synergistic interaction not only accelerates the discovery of new photocatalytic materials but also facilitates their deployment in real-world applications, such as wastewater treatment and solar energy harvesting. Furthermore, the movement towards innovative photocatalytic systems has the potential to establish a novel paradigm for environmental engineering, integrating sophisticated control mechanisms that can adapt to variable conditions for increased efficiency and sustainability. The meeting of photocatalysis with these emerging technologies promises a revolutionary age in the search for sustainable solutions to world challenges.

1.6.3. Future Trends and Potential Breakthroughs

The future potential of photocatalysis lies in its potential for dramatic breakthroughs, which will revolutionize environmental cleanup and energy conversion. Researchers are striving to develop more efficient photocatalysts, especially hybrid materials. By combining materials such as titanium dioxide, carbon nanotubes, and iron oxide, one can synthesize more effective photocatalysts for photochemical wastewater treatment. The blend of materials not only enhances photoactivity but also facilitates the effective removal of contaminants from complex wastewaters, addressing crucial environmental challenges posed by toxic chemicals prevalent in the oil and textile industries. In particular, the novel use of Pickering emulsions in reactor design can maximize contact between the photocatalyst and pollutants, making photocatalytic degradation a more viable option for large-scale implementation (Mason et al., 2018). Additionally, novel reactor designs are required to make photocatalysis applicable in real-world settings. Current limitations are often caused by inefficient reactor designs that do not provide maximal light exposure or catalyst contact. Future research will continue to advance these designs to incorporate continuous flow and higher mass transfer rates. As seen between geopolymer concrete and photocatalysis, the direction is towards integrating hybrid systems where building materials can be integrated with photocatalytic materials. This not only enhances the structural integrity of the concrete but also enables the material to act as a filter, cleansing nitrogen and other contaminants from water sources. Such applications can lead to dual-use materials with widespread environmental benefits (Aris et al., 2015). In addition to material and reactor design innovation, another revolutionary trend in photocatalysis is the application of artificial intelligence (AI) and machine learning. These technologies can aid in identifying and developing new photocatalytic materials by analyzing large datasets generated from experiments. By using predictive models, scientists can forecast which combinations of elements will yield the most effective photocatalysts, thereby accelerating the process. The union of AI-based simulations and experimental evidence can lead to discoveries, not only in photocatalytic performance but also in the scalability of the technologies towards industrially viable applications. In addressing the immediate global concern of water contamination and energy sustainability, the future of photocatalysis will undoubtedly emerge as a pillar of modern environmental technology (Aris et al., 2015; Mason et al., 2018).

SUMMARY

Photocatalysis research is indisputably at the core of developing sustainable solutions to genuine environmental problems. As this essay has emphasized, photocatalytic reactions utilize light energy to induce chemical transformations, thereby opening up new opportunities for air purification, water treatment, and energy conversion. Integration of materials, such as all-inorganic metal halide perovskites, has introduced novel paradigms into this field. These materials, especially $CsPbX_3$, possess excellent photophysics that enhance their utility for various photovoltaic and optoelectronic applications. By putting into perspective the limitations of past photocatalysts, especially stability, researchers have paved the way for future development with a focus on material innovation being the crucial factor for the success of photocatalysis. Hence, present research on novel photocatalytic systems is not only unavoidable but also critical to achieving pragmatic applications necessary for effectively addressing environmental issues. In addition, advances in two-dimensional materials over the last few years have expanded the horizon of photocatalysis by introducing new surface properties favorable for catalytic reactions. Phosphorene has been the forerunner, with its tendency to possess high chemical reactivity and suitable electrical properties, thus making it a game-changing candidate in photocatalytic applications. Synthesis and functionalization of phosphorene, including its interaction with metal particles, portend new dimensions in photocatalytic efficiency. The ability of these 2D materials to facilitate electron transfer supports the rate and selectivity of photochemical reactions, thereby further validating that material engineering is the foundation upon which photocatalytic research stands. As we progress through these advances, it is clear that the overlap of new materials and photocatalytic technology will be at the forefront of subsequent breakthroughs. In summary, the understanding and development of photocatalysis encompass material science as well as practical applications, and thus require interdisciplinary approaches. The knowledge of high-technology materials and their performance has led to the design and development of efficient photocatalysts. Research focused on all-inorganic perovskites and phosphorene is indicative of the rapid advancement of the field, where interactions between light and matter yield revolutionary outcomes. As the difficulties of stability and efficiency persist, the scientific community must unite to continue advancing these materials and methods. The future of photocatalysis thus bodes well, as continued research continues to harness the full potential of these next-generation materials in addressing global environmental concerns.

The discussion of photocatalysis highlights the central role materials, particularly titanium dioxide (TiO_2), have in the establishment of this technology. TiO_2 has drawn immense attention due to its multidisciplinary uses that range from catalysis to photovoltaic use. Despite the extensive research on this material, there are some intrinsic properties of TiO_2, such as the electronic band gap and the influence of oxygen vacancies, which are still controversial points. Advanced computational methods, such as density functional theory (DFT), provide a stable platform for investigating the electronic, structural, and optical characteristics of TiO_2 nanostructures. Advanced computational methods not only enhance our understanding of TiO_2 but also address issues such as narrowing the band gap, a crucial factor in improving photocatalytic functionality in real-world applications (Atwater et al., 2020). In addition, the creative application of nanomaterials, such as copper nanoparticles in photocathodes, is a perfect example of the potential to enhance selectivity during photocatalysis. By combining copper and p-type nickel oxide (p-NiO), scientists have discovered that optical excitation generates hot electrons, thereby sustaining reactions such as CO_2 reduction. The process is a good demonstration of the way plasmonic influences can alter catalytic activity and achieve increased selectivity for desired products with fewer unwanted byproducts such as hydrogen. These findings highlight the significance of material design and the interaction between the various components of photocatalytic systems, paving the way for the optimization of artificial photosynthesis and energy conversion technology (Atwater et al., 2020). Lastly, the synthesis of photocatalyst materials should focus on both the electronic and structural attributes of the materials. The integration of doping methods with TiO_2 nanostructures offers a means to control the electronic properties of the materials successfully. For instance, boron or nitrogen doping can lead to the creation of mid-gap states, which favor band gap narrowing and result in higher photocatalytic activity under solar illumination. These advancements not only improve basic research but also pose profound implications for the design of highly efficient photocatalysts. These complementary viewpoints highlight the interdisciplinary approach necessary to harness the full potential of photocatalysis in addressing global challenges, including energy conversion and environmental remediation (Chiodo et al., 2010).

REFERENCES

1. Garrido-Cardenas, J. A., Esteban-García, B., Agüera, A., Sánchez-Pérez, J. A., & Manzano-Agugliaro, F. (2019). Wastewater treatment by advanced oxidation process and their worldwide research trends. *International Journal of Environmental Research and Public Health*, *17*(1), 170. https://doi.org/10.3390/ijerph17010170Alexandrova, A. N., Blankenhorn, M., Bowen, K. H., Liu, et al. (2020). CO_2 hydrogenation to format and formic acid by bimetallic palladium-copper hydride clusters. https://core.ac.uk/download/323082880.pdf (accessed on 28th September 2025).
2. Alexandrova, A. N., Cronin, S. B., Gong, X. Q., Li, et al. (2020). (Photo) Electrocatalytic CO_2 reduction at the defective anatase TiO_2 (101) surface. https://core.ac.uk/download/323311157.pdf (accessed on 28th September 2025).
3. Aris, A., Berhanuddin, M. S., & Shamsudin, S. (2015). An overview of geopolymer materials-based zeolite as an antifouling layer in improving water quality. https://core.ac.uk/download/83533062.pdf (accessed on 28th September 2025).
4. Assadi, M. H. N., Hanaor, D. A. H., Li, S., et al. (2012). Ab initio study of phase stability in doped TiO_2. http://arxiv.org/abs/1210.7555 (accessed on 28th September 2025).
5. Atwater, H. A., Cheng, W. H., DuChene, J. S., et al. (2020). Optical excitation of a nanoparticle Cu/p-NiO photocathode improves reaction selectivity for CO_2 reduction in aqueous electrolytes. https://core.ac.uk/download/287633875.pdf (accessed on 28th September 2025).
6. Avci, N., Cimieri, I., Geens, J., et al. (2012). Sol-gel preparation of pure and doped TiO_2 films for the photocatalytic oxidation of ethanol in air. https://core.ac.uk/download/55755375.pdf (accessed on 28th September 2025).
7. Bacchini, F., Greppi, G., Mura, S., et al. (2013). Advances of nanotechnology in agro-environmental studies. https://core.ac.uk/download/16749559.pdf (accessed on 28th September 2025).
8. Balcioglu, I. A., Balcıoğlu, I. A., Menceloglu, Y. Z., et al. (2011). Comparison of the effectiveness of chlorine, ozone, and photocatalytic disinfection in reducing the risk of antibiotic resistance pollution. https://core.ac.uk/download/pdf/11742857.pdf (accessed on 28th September 2025).
9. Ball, D. W., Chianelli, R. R., Hepp, A. F., et al. (2025). Novel catalysts and processing technologies for production of aerospace fuels from non-petroleum raw materials. https://core.ac.uk/download/pdf/42722367.pdf (accessed on 28th September 2025).

10. Bata-Vidács, I., El Hathat, C., Inotai, K., Kosztik, J., Szekeres, A., Horváth-Szanics, E., Zalán, Z., et al. (2025). The effects of pH, temperature, and oxygen availability on the production of aflatoxin B1 by Aspergillus flavus Zt41 and sterigmatocystin production of Aspergillus creber 2663 isolates. Journal of Food Safety. https://www.semanticscholar.org/paper/7565f5642ca18556c7376723c1610d9a09672398 (accessed on 28th September 2025).
11. Benito, I. I., Bini, R., Bolognesi, M., et al. (2019). A perspective on recent advances in phosphorene functionalization and its application in devices. https://core.ac.uk/download/301577585.pdf (accessed on 28th September 2025).
12. Boonsith, S., Kim, D. C., Sakdaronnarong, C., et al. (2020). Recent developments in synthesis and photocatalytic applications of carbon dots. https://core.ac.uk/download/323383933.pdf (accessed on 28th September 2025).
13. Chen, J., Wu, J., Sherrell, P. C., Chen, J., Wang, H., Zhang, W. X., & Yang, J. (2022). How to build a microplastics-free environment: Strategies for microplastics degradation and plastics recycling. Advanced Science, 9. https://doi.org/10.1002/advs.202103764 (accessed on 28th September 2025).
14. Chiodo, L., García-Lastra, J. M., Iacomino, A., et al. (2010). Tailoring electronic and optical properties of TiO_2: Nano structuring, doping and molecular-oxide interactions. https://core.ac.uk/download/36054376.pdf (accessed on 28th September 2025).
15. Cho, D. L., Kim, J. B., Kim, J. H., et al. (2008). Visible light responsive titanium dioxide (TiO_2). https://opus.lib.uts.edu.au/bitstream/10453/9199/1/2008005324.pdf (accessed on 28th September 2025).
16. Choi, W., Kim, G., & Zhang, G. (2025). Visible light driven photocatalysis mediated via ligand-to-metal charge transfer (LMCT): An alternative approach to solar activation of titania. https://core.ac.uk/download/148797921.pdf (accessed on 28th September 2025).
17. Darr, J. A., Hyett, G., Kafizas, A., et al. (2011). An investigation into the effect of thickness of titanium dioxide and gold-silver nanoparticle titanium dioxide composite thin-films on photocatalytic activity and photo-induced oxygen production in a sacrificial system. https://core.ac.uk/download/9040748.pdf (accessed on 28th September 2025).
18. Ding, L., Li, K., Li, J., Lu, Q., Fang, F., Wang, T., & Chang, K. (2023). Integrated coupling utilization of the solar full spectrum for promoting

water splitting activity over a CIZS semiconductor. ACS Nano, 17, 11616–11625. https://doi.org/10.1021/acsnano.3c02029 (accessed on 28th September 2025).

19. Edalati, K., Bachmaier, A., Beloshenko, V. A., Beygelzimer, Y., Blank, V. D., Botta Filho, W. J., Bryła, K., et al. (2022). Nanomaterials by severe plastic deformation: Review of historical developments and recent advances. *Materials Research Letters, 10*, 163–256. https://doi.org/10.1080/21663831.2022.2029779 (accessed on 28th September 2025).

20. Grebel, H., & Chowdhury, T. S. (2022). Optically controlled TiO_2-embedded supercapacitors: The effects of colloidal size, light wavelength, and intensity on the cells' performance. Nanomaterials, 12. https://www.semanticscholar.org/paper/79611802794f0a8114784c5be097823055b79d4d (accessed on 28th September 2025).

21. Hafidi, M., Merlina, G., Perchet, G. T., et al. (2009). Evaluation of a TiO_2 photocatalysis treatment on nitrophenols and nitramines contaminated plant wastewaters by solid-phase extraction coupled with ESI HPLC–MS. https://core.ac.uk/download/12041818.pdf (accessed on 28th September 2025).

22. Hakki, A., MacPhee, D. E., Wang, F., et al. (2017). Photocatalyst efficiencies in concrete technology: The effect of photocatalyst placement. https://core.ac.uk/download/131024803.pdf (accessed on 28th September 2025).

23. Hassaan, M. A., El-Nemr, M. A., Elkatory, M. R., Ragab, S., Niculescu, V. C., & El Nemr, A. (2023). Principles of photocatalysts and their different applications: A review. Topics in Current Chemistry, 381. https://doi.org/10.1007/s41061-023-00444-7 (accessed on 28th September 2025).

24. Katehi, L., Yi, S., Lin, Y., Banerjee, S., Xia, Q., & Yang, J. J. (2025). Emerging nonvolatile memory technologies in the future of microelectronics. ACS Omega, 10, 28492–28498. https://www.semanticscholar.org/paper/63fd93548fabcd0ee314383f9014c328c5cc5ab3 (accessed on 28th September 2025).

25. Králová, M. (2012). Self-cleaning properties of thin printed layers of titanium dioxide. https://core.ac.uk/download/30310816.pdf (accessed on 28th September 2025).

26. Lee, S. K., & Mills, A. (2004). Detoxification of water by semiconductor photocatalysis. https://core.ac.uk/download/9013846.pdf (accessed on 28th September 2025).

27. López-Magano, A., Daliran, S., Oveisi, A. R., Mas-Ballesté, R., Dhakshinamoorthy, A., Alemán, J., García, H., et al. (2022). Recent advances in the use of covalent organic frameworks as heterogeneous

photocatalysts in organic synthesis. Advanced Materials, 35. https://doi.org/10.1002/adma.202209475 (accessed on 28th September 2025).

28. Mai, Q., Trang, D. T. H., Bach, T. N., Na, V. T. L., Pham, A. T., & Le, A. T. (2025). Synergizing PIERS and photocatalysis effects in a photo-responsive Ag/TiO_2 nanostructure for an ultrasensitive and renewable PI-PC SERS technique. RSC Advances, 15, 4149–4162. https://www.semanticscholar.org/paper/aefca298608bc94d7a704d4a547a48c38c43744f (accessed on 28th September 2025).

29. Mason, E. (2018). Synthesis and photocatalytic performance of TiO_2-CNT and magnetized Fe_3O_4-TiO_2-CNT multifunctional hybrids: A Pickering emulsion platform for organic degradation. https://core.ac.uk/download/211346833.pdf (accessed on 28th September 2025).

30. Matuska-Łyżwa, J., Duda, S., Nowak, D., & Kaca, W. (2024). Impact of abiotic and biotic environmental conditions on the development and infectivity of entomopathogenic nematodes in agricultural soils. Insects, 15. https://www.semanticscholar.org/paper/e80d551c42dba7d0f26627a9475fd3f6353d4602 (accessed on 28th September 2025).

31. Maury, F., & Mungkalasiri, J. (2009). Chemical vapor deposition of TiO_2 for photocatalytic applications and biocidal surfaces. https://core.ac.uk/download/12042017.pdf (accessed on 28th September 2025).

32. McDonnell, K. A., Percy, J., Anders, L., Brown, M. J., Richman, A. R., Deardorff, J., Ruiz, M. S., et al. (2025). Investing in the development of the next generation of MCH leaders. Frontiers in Public Health, 13. https://www.semanticscholar.org/paper/b8cb15f57546be8c1cf2db23359bf3e89b9a1eef (accessed on 28th September 2025).

33. Parker, D. M. (2016). Photocatalytic degradation of an organophosphorus compound by porous Au- and WO_3-modified TiO_2. https://core.ac.uk/download/215152772.pdf (accessed on 28th September 2025).

34. Qasim, R., Balqees, K., Nasir, A., Hamza, S. M., Nadeem, T., Parveen, A., Rasheed, R. M., et al. (2025). Nanoparticle-enhanced photocatalysis for advanced wastewater treatment: Mechanistic insights, performance evaluation, and sustainable applications. Scholars Journal of Engineering and Technology. https://www.semanticscholar.org/paper/723a1da6eef45349475f60c5f39d22de9fad42b1 (accessed on 28th September 2025).

35. Romeral Martínez, J. L., Tapia Tlatelpa, T., & Trull Silvestre, J. F. (2019). In situ decolorization monitoring of textile dyes for an optimized UV-

LED/TiO_2 reactor. https://core.ac.uk/download/294829789.pdf (accessed on 28th September 2025).

36. Rueda-Márquez, J. J., Levchuk, I., Fernández-Ibáñez, P., & Sillanpää, M. (2020). A critical review on application of photocatalysis for toxicity reduction of real wastewaters. Journal of Cleaner Production, 258, 120694–120694. https://doi.org/10.1016/j.jclepro.2020.120694 (accessed on 28th September 2025).
37. Shabna, S., Singh, C. J. C., Dhas, S. D. S. J., Jeyakumar, S. C., & Biju, C. S. (2024). An overview of prominent factors influencing the photocatalytic degradation of cationic crystal violet dye employing diverse nanostructured materials. Journal of Chemical Technology & Biotechnology. https://www.semanticscholar.org/paper/6a7fc811638eccca3835a1a02e00eaf68a210fef (accessed on 28th September 2025).
38. Silvestre, A. J., Augugliaro, B., et al. (2015). Synthesis of titanate nanofibers co-sensitized with ZnS and Bi_2S_3 nano crystallites and their application on pollutants removal. http://arxiv.org/abs/1503.08693 (accessed on 28th September 2025).
39. Sinha, N., Kumar, N., Sharma, T., Thakur, N., & Jain, R. (2025). Abundant transition metal based photocatalysts for red light driven photocatalysis. Chemistry. https://www.semanticscholar.org/paper/60cbe96d8d5335b27cd8e56514cd36fc70a213ad (accessed on 28th September 2025).
40. Soni, V., Rana, P., Chauhan, R., Singh, P., Singh, A., Chaudhary, V., Khan, A. A. P., et al. (2025). Recent trends in perovskite quantum dots-based S-scheme photocatalysis for CO_2 photoreduction. ChemNanoMat. https://www.semanticscholar.org/paper/7823bb3952cb6b1a8b78859ea71c62f9fcdf7efc (accessed on 28th September 2025).
41. Steiner, M. G. (2017). Photocatalytic decomposition of phenol under visible and UV light utilizing titanium dioxide-based catalysts. https://core.ac.uk/download/84124263.pdf (accessed on 28th September 2025).
42. Verma, R., Sharma, G., & Polshettiwar, V. (2024). The paradox of thermal vs. non-thermal effects in plasmonic photocatalysis. Nature Communications, 15. https://www.semanticscholar.org/paper/2e58a2f91f3e51d738e7db60abe0c36c6d0537db (accessed on 28th September 2025).
43. Wexler, R., Knezek, P., Rudnick, G. H., Zellner, N., Eckert, K. D., McBride, J., Patterson, M., et al. (2025). Recommended actions for the American Astronomical Society: CSWA's perspective on steps for a more inclusive astronomy–I. Background and methods. Bulletin of the AAS. https://www.semanticscholar.org/paper/

c7b6847cda2d380738e07663f1f7b4c3052da551 (accessed on 28th September 2025).

44. Yamada, M., Yamada, T., Yanagisawa, K., Tipplook, M., Hayashi, F., & Teshima, K. (2024). Data-driven analysis of contributing factors in spectra toward quantitative prediction of photocatalytic activity of $BaTaO_2N$. ECS Meeting Abstracts. https://www.semanticscholar.org/paper/b114937913f0199b6ae0e62f4db8cfad7b67e5b5 (accessed on 28th September 2025).

45. Yin, Y., & Zhang, Q. (2018). All-inorganic metal halide perovskite nanocrystals: Opportunities and challenges. https://core.ac.uk/download/323067543.pdf (accessed on 28th September 2025).

46. Yuan, L., Du, P., Yin, L., Yao, J., Wang, J., & Liu, C. (2024). Metal-organic framework-based S-scheme heterojunction photocatalysts. Nanoscale. https://www.semanticscholar.org/paper/944089d4b49aca141fba73f67e4365e0dcff5eb5 (accessed on 28th September 2025).

47. Yue, M., Lambert, H., Pahon, E., Roche, R., Jemeï, S., & Hissel, D. (2021). Hydrogen energy systems: A critical review of technologies, applications, trends and challenges. Renewable and Sustainable Energy Reviews, 146, 111180–111180. https://doi.org/10.1016/j.rser.2021.111180 (accessed on 28th September 2025).

48. Zhang, T., Zhang, G., & Chen, L. (2022). 2D conjugated covalent organic frameworks: Defined synthesis and tailor-made functions. Accounts of Chemical Research, 55, 795–808. https://doi.org/10.1021/acs.accounts.1c00693 (accessed on 28th September 2025).

CHAPTER 2

FUNDAMENTALS OF PHOTOCHEMISTRY

Contents

INTRODUCTION

Photochemistry is a significant branch of physical chemistry that fundamentally studies the interaction of light and matter. The discipline explores how photons, or particles of light, can initiate chemical reactions to convert reactants into products. Theoretical photochemistry relies on quantum theory and examines how electronic transitions induced by light absorption can result in diverse chemical changes. This understanding is crucial for applications ranging from plant photosynthesis to the production of photovoltaic cells. Through continuous investigation of such light-matter interactions by researchers, there are continued advances in catalytic processes, particularly in green applications. One example is the fact that photocatalytic processes can break down toxic substances within wastewater, a reaction to a general societal need for more sustainable practices (Mason et al., 2018). Understanding the basic principles of photochemistry not only enriches our knowledge of natural processes but also helps direct state-of-the-art technologies. The practical application of photochemistry encompasses not only environmental detoxification but also extends to material science, biochemistry, and pharmacology. The ability to control chemical reactions with light provides precise control in the synthesis of new chemicals or drugs. For example, the use of photochemical reactions enables the synthesis of sophisticated organic compounds with superb efficiency, which is priceless in drug production. Moreover, photocatalyst engineering now stands at the forefront of research activities with the aim of improving efficiency and selectivity for target reactions. Progress in material properties, as seen with the use of titanium dioxide as a photocatalyst, demonstrates the intertwined nature of fundamental research and applications. This interplay of theory and application underscores the importance of problem-solving in the field, such as enhancing photocatalytic efficiency and longevity for industrial applications, ultimately contributing to a sustainable chemical economy (Mason et al., 2018). Besides its environmental and synthetic importance, photochemistry is also of first-order importance in excited-state research and its dynamics. Time-dependent density functional theory (TD-DFT) provides a model that describes the alteration of electronic states in response to perturbations, allowing researchers to simulate subsequent processes following photon excitation. The theoretical model elaborates on the complexity of excited-state behavior, which is a characteristic of most photochemical reactions. Using linear response theory, scientists can derive details about excited states, underscoring the nuanced interactions in energy transfer and reaction pathways. Along with it, hybrid functional construction in TD-DFT delivers better accuracy in these theoretical studies, while it can potentially uncover more about nonlocal interactions in time-dependent systems. Along with the improvement in research, the addition of advanced computational methods to experimental findings will significantly enhance our understanding of photochemistry and its vast applications in contemporary science.

Definition and Scope of Photochemistry

Photochemistry is essentially the chemistry of chemical reactions and processes brought about or influenced by light. Electronic excitation, the focus of most photochemical processes, occurs as a result of interaction between matter and light. Photochemistry, as an emerging field of study, harnesses the dictates of quantum mechanics to describe how photons influence molecules to have their electronic state changed, thereby bringing about chemical change. The essence of photochemistry lies not just in the knowledge of these reactions, but in applying such knowledge to various fields of science, including biology, environmental science, and materials science, thereby demonstrating its universality of application and central importance to the evolution of many sciences. Photochemistry has applications in the study of various phenomena resulting from matter-light interactions, including photophysical and photobiological processes. The heart of this discourse lies in considering electronically excited molecular states as the foundation for a broad range of applications, spanning from photosynthesis to the development of solar energy technology and phototherapeutic procedures in medicine. The contemporary-day image of photochemistry increasingly finds a place for robust computational methods like machine learning, designed to improve simulation and prediction of excited-state dynamics. These developments represent a paradigm shift in the approach scientists use to address photochemical problems, aiming to optimize efficiency and enhance innovation in the sector (Westermayr et al., 2020) (Figure 2.1).

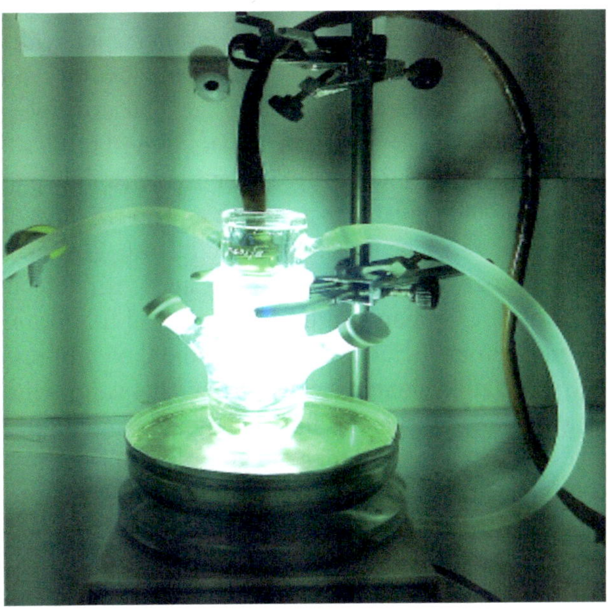

Figure 2.1. Photochemistry.

Source: https://chemistrytalk.org/photochemistry/.

In addition, photochemistry's applications reach into the solution of today's challenges, making it vital in green chemistry and sustainability. Utilizing light to drive chemical processes provides scientists with an avenue to reduce energy use and byproducts that are detrimental to the environment, aligning with sustainable practices. The application of photochemistry in areas like materials science has opened doors to new solutions in the development of materials with special properties that can be switched on by light. In addition, continuous developments in plasma technologies demonstrate how photochemistry converges with other scientific fields, underscoring its value across multiple disciplines and applications. This cross-disciplinary process is essential in addressing world problems, highlighting the revolutionary power of photochemistry in forging a greener future (Adamovich et al., 2022).

Historical Background and Key Discoveries

The history of photochemistry is traceable to the first experiments of pioneers who began to create a connection between light and chemical change. Thomas Young and Joseph Priestley were among the scientists during the 18th century who laid down the early foundations by exploring the relationship between chemical reactions and light, but scientific understanding of the process was primitive. The late 19th century saw a significant turning point, with the invention of spectroscopy enabling scientists to study the interaction of substances and light. This was a time of building on the fundamental principles of quantum mechanics, which ultimately enabled researchers like Max Planck and Albert Einstein to describe the composition of light as both a wave and a particle, thereby driving the study of photochemistry as a science and effectively revolutionizing the study of molecular interactions. Throughout the 20th century, several influential findings progressively advanced the study of photochemistry. The groundbreaking work of Grotian and others facilitated the formulation of the principles of photochemical reactions, including the theory of excitation states and their roles in facilitating chemical change. Such breakthroughs were central to understanding how energy changes initiated by light could drive phenomena like photosynthesis and radiation-induced synthesis. During this early period of development, scientists began to establish a more precise concept of excited molecular states, which became central to explaining the mechanisms of chemical light-induced reactions. This evolving knowledge found expression in applications ranging from environmental science to material design, illustrating the wide-ranging significance of photochemistry to technology and society. New challenges have emerged in recent research that compel scientists to rethink traditional models of photochemical behavior. Evidence for the presence of alternative nitrogen cycling pathways in marine ecosystems, for instance, highlights the complexity of photochemistry in nature, as denitrification reactions become essential for the global nitrogen cycle (Aller et al., 2005). Concurrent progress in fields such as space science, as represented by work under initiatives funded by NASA, also demonstrates the ongoing applicability of photochemical principles (N/A, 2025). Together, these advances suggest that the historical context of photochemistry not only accounts for past successes but also informs future and current avenues of research. With the continued advancement of knowledge, the interface between photochemistry and other sciences continues to evolve, innovate, and address issues at the world level.

Importance of Photochemistry in Modern Science

The significance of photochemistry in modern science is underscored by its application in the development of energy-efficient processes and clean energy resources. Photochemistry involves the interactions between matter and light, leading to a variety of chemical reactions that form the basis of solar energy conversion technologies. The utilization of sunlight by photochemical reactions, e.g., photosynthesis by plants, serves as a paradigm for artificial systems to convert solar energy into chemical energy. Advances in photochemical processes have also driven research towards artificial photovoltaics and solar fuels, which will help alleviate the current energy crisis. As the demand for sustainable energy sources rises, the discovery and investigation of photochemical mechanisms will be crucial in shaping future energy strategies and reducing our reliance on fossil fuels, thereby increasing environmental sustainability. Apart from its applications in energy, photochemistry also plays a pivotal role in various fields, including medicine and environmental science. Drugs often rely on photochemical reactions in drug development, as light is capable of initiating certain chemical transformations that lead to the creation of complex molecules. Photodynamic therapy, for instance, has been employed to employ light-activated compounds to kill cancerous cells, showing photochemistry's therapeutic potential selectively. Furthermore, in the field of environmental science, photochemical reactions play a crucial role in understanding atmospheric processes and the decomposition of pollutants. Photochemical reactions in the atmosphere may lead to the production of secondary pollutants, such as ozone, with global implications for air quality and climate change. With these mechanisms explained, photochemists can contribute to the development of effective measures for controlling environmental pollution and monitoring climate change. Theoretical underpinning of photochemistry, particularly as formulated through density-functional theory (DFT) and its time-dependent extension (TD-DFT), provides a powerful means of understanding complex molecular systems. DFT, as envisioned by Hohenberg, Kohn, and Sham, replaces complicated multi-electron wavefunctions with simple electron density functions, effectively removing computational complexity from the study of photochemical reactions. While traditional DFT is limited in handling excited states and time-dependent systems, TD-DFT is a viable method that gives helpful information about electronic transitions and reaction dynamics under perturbation. According to linear response theory, TD-DFT has become indispensable for calculating excited-state properties, allowing scientists to simulate absorption spectra and other photochemical phenomena of interest. These advances enhance our insight into molecular action in photochemical reactions and pave the way for state-of-the-art applications in a broad range of scientific disciplines (Casida et al., 2011, 2015).

2.1. BASIC PRINCIPLES OF PHOTOCHEMISTRY

The fundamental principles of photochemistry involve the interaction between matter and light, as well as the various processes triggered upon the absorption of photons by a molecule. Through this interaction, numerous photochemical reactions occur, primarily due to energy transfer processes. Photons, as quantized energy vectors, cause molecular electrons to be promoted to higher energy states, thereby forming excited species. Its molecular structure and environmental conditions fundamentally control the absorption spectrum of a substance, and how they affect the wavelengths capable of absorption. An understanding of these is crucial, as it controls the reaction dynamics of natural processes, such as photosynthesis, where light energy is converted to chemical energy, and synthetic processes, like photopolymerization in materials science (Bergqvist et al., 2005). Additionally, the various detection techniques employed in photochemistry significantly enhance our understanding of reaction mechanisms and rates. The use of photoacoustic spectroscopy to measure energy transfer and kinetic processes by creating sound waves generated from thermal expansion following photon absorption is one technique that can be employed. Such a method shows the overall relevance of photochemical principles, conveying their application not only in research work but also in fields about environmental monitoring and industrial processes. For this reason, the inclusion of advanced analytical techniques elevates the field to an extent that enables precise control and understanding of photochemical processes with extensive effects on atmospheric chemistry and pollution sensing (Berthelot et al., 1989). Finally, recent research has revealed the intricacy of photochemical systems, with a particular emphasis on the interplay between the reactive intermediates formed upon excitation. The intermediates often determine the reaction path and efficiency and are the major players in natural and synthetic processes. For example, the role of excited states in the process of emitting and triplet oxygen has drawn significant attention due to its importance for atmospheric chemistry and photodynamic therapy in medicine. Thus, a correct appreciation of these basic principles not only improves scholarship but also fosters innovation in photochemical processes across different sectors by highlighting the essential role played by light in triggering chemical changes (Berthelot et al., 1989; Bergqvist et al., 2005) (Figure 2.2).

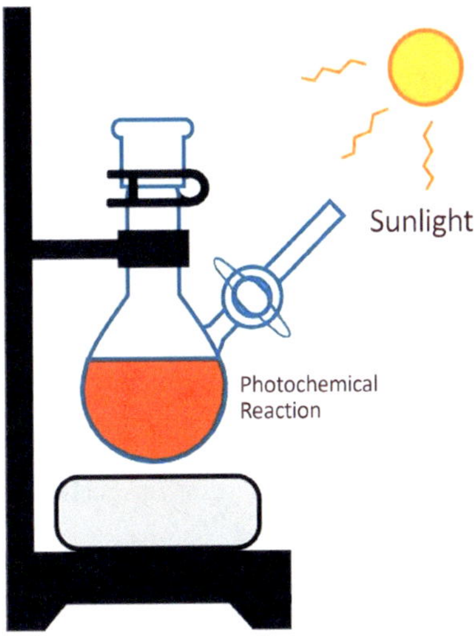

Figure 2.2. Basics of photochemistry.

Source: https://chemistrytalk.org/photochemistry/.

2.1.1. Interaction of Light with Matter

The influence of light on matter is the core of photochemistry as it determines processes that enable chemical changes through photonic energy. When matter interacts with light, it is either absorbed, reflected, or transmitted, thereby exciting electrons in molecules. Electron excitation results in the movement of electrons from lower to upper energy levels, basically altering the chemical nature of the substances involved. It is a multidimensional aspect with many factors at play, including the light wavelength, the character of the matter, and the chemical bonds in question. These interactions serve as a gateway to understanding a variety of applications, ranging from photosynthesis to photonic reactions utilized in materials science and nanotechnology. Advances in photochemical processes are generally catalyzed by meticulous surface alterations that utilize light-matter interaction at the molecular level. Advances in the design and construction of photochemical surface delivery vehicles have led to significant advancements in chemical control of functionality on a broad array of substrates. Specifically, surface modification protocols enable localized chemical surface alteration, thereby enabling patterned functionalization through exposure to photons. These responses not only enhance the efficiency of chemical reactions but also adhere to green chemistry principles, as they use photons as the sole reagent in most applications. This breakthrough illustrates the radical potential of light-mediated surface reactions across various disciplines, including biotechnology, nanotechnology, and environmental science (Carroll et al., 2009) (Figure 2.3).

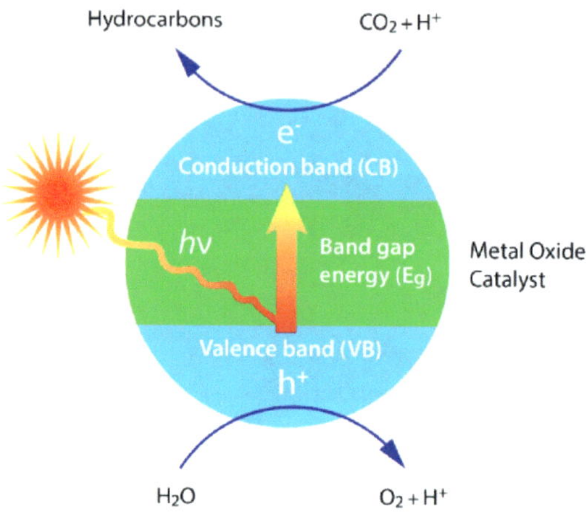

Figure 2.3. Principle of the photochemical reduction of CO_2 by water on a photocatalyst.

Source: https://www.researchgate.net/figure/Principle-of-the-photochemical-reduction-of-CO-2-by-water-on-a-photocatalyst_fig4_319214069.

Moreover, the time-dependent density functional theory (TD-DFT) theoretical framework has also emerged as a crucial tool in studies on the interaction of light with matter, particularly in the examination of electronic excitations responsible for photochemical processes. TD-DFT enables the examination of excited electronic states, which are highly valuable in the study of absorption spectra and other light-induced properties of medium- and large-sized molecules. This method provides bonds with insight into bond making and breaking, activities that are inherently reliant on electronic excitations triggered by photon interactions. Being good predictive tools, TD-DFTs are nonetheless plagued by the shortsightedness of functionals used in more complex systems.

Further efforts are ongoing to advance these theories, making them more widely applicable and useful in the description of photochemical dynamics (Casida et al., 2011).

2.1.2. Quantum Mechanics in Photochemical Processes

The overlap between photochemical processes and quantum mechanics is fundamental to understanding light interacting with matter. The photochemistry is based on photon absorption and the subsequent series of electronic transitions in the molecule. Quantum mechanics provides the tools to outline electronic states and the dynamics of what happens when a molecule absorbs energy. The wavefunctions play a crucial role, regulating the probability of finding electrons in specific states and arrangements. With the aid of tools such as time-dependent density-functional theory (TD-DFT), chemists can reliably predict the absorption spectrum and excited states of molecules, thereby obtaining information on energy transfer processes. This theoretical framework, with the rapid rise in computational power, enables scientists to investigate complex systems where classical approaches would fall short, embodying the requirements of quantum mechanics in photochemical studies. From theory to practice, the application of quantum mechanical principles in photochemistry entails describing reaction mechanisms and predicting how excited states participate in chemical reactions. Because photochemical responses are characterized by bond breaking and formation, the precision offered by quantum mechanics is indispensable. A case in point is the application of linear response theory within TD-DFT, which enables the calculation of excited state properties and the prediction of effects such as fluorescence and photochemical reactivity. Besides that, challenges such as avoiding the adiabatic approximation in TD-DFT calculations have driven the development of new approaches to treating inadequately described excited states in complex systems. By overcoming these obstacles, scientists can utilize quantum mechanics to gain a deeper understanding of the behavior of molecules interacting with light, shedding light on the mechanisms that control fundamental processes such as photosynthesis and photoredox reactions, which underlie both synthetic and natural processes. Furthermore, the incorporation of many-body corrections into quantum mechanical calculations enhances the description of non-additive interactions in large molecular systems. Quantum mechanics reveals that many-body effects and electron correlation are crucial factors in determining the properties and dynamics of excited states. Tracing back through history, relying on approximate treatments like the adiabatic approximation presents difficulties in predicting behavior in a vast number of photochemical scenarios. More recent work has focused on hybrid approaches that combine the efficiency of DFT with the qualitative insights of wavefunction theory, thereby improving the description of excited states. This evolving situation of quantum mechanical applications in photochemistry emphasizes the necessity of ongoing model improvement at the theoretical level, thereby further linking complex quantum systems to practical photochemical applications in materials science and organic chemistry, ultimately leading to the advancement of energy conversion technologies (Casida et al., 2011, 2015).

2.1.3. Energy Transfer Mechanisms

Energy transfer mechanisms play a crucial role in photochemical reactions, dictating how light energy is absorbed and distributed within a system. They have been extensively categorized as radiative and non-radiative processes. Radiative energy transfer is the mechanism by which photons are absorbed and emitted, allowing for the emission of energy in the form of light. On the other hand, non-radiative transfer is a method by which energy transfer happens without emission of photons, i.e., through vibrational relaxation or thermal conduction. Understanding these channels is essential for understanding the efficiency of photochemical energy conversion processes in photosynthesis or solar energy collection, where the transfer of energy must be optimized. For example, the efficiency of light-harvesting complexes relies on their ability to transfer excitation energy effectively among pigment molecules, thereby facilitating efficient photon utilization (Euclid et al., 2025). The dynamics of energy transfer mechanisms are strongly reliant on both molecular structure and the surrounding environment. Intramolecular interactions, such as electronic state-vibrational mode coupling, can significantly alter energy dissipation pathways. Additionally, such mechanisms rely on the environmental presence of surrounding media, particularly in complex systems such as biological membranes or artificial photonic devices. Quantum coherence intervenes when energy is transferred through multiple routes simultaneously, thereby increasing the probability of successful energy capture and transfer. This quantum mechanical framework provides a key understanding of energy transfer within systems exposed to stochastic environments, where the probability of fluctuations dictates the efficiencies and rates of transfer processes. The implications of this work extend to synthetic photochemistry, where constructed systems can efficiently utilize energy after natural processes (Euclid et al., 2025). Advances in computational simulations and machine learning algorithms have broadened our understanding of energy transfer processes by enabling the simulation of complex interactions from large data sets. Synthetic data for advanced radiation transfer models, for instance, provides scientists the ability to analyze the distribution of energy within various photochemical systems and uncover the complex interplay between molecular parameters and environmental conditions.

A study using high-resolution imaging and advanced computational methods has provided data on visible phenomena such as mass, metallicity, and stellar population profiles to uncover driving energy transfer machinery in galaxies (Euclid et al., 2025). The ability to track physical parameters in photochemical systems with increased accuracy enables a deeper understanding of underlying processes, such as the manner in which energy is absorbed, converted, and utilized in both natural and artificial environments. This intersection of advanced technology and photochemistry represents a significant step toward realizing efficient energy systems.

2.2. PHOTOCHEMICAL REACTIONS

Photochemical reactions are a crucial point of encounter between light and chemical change, in which photon absorption leads to a range of energetic processes. Photochemical reactions are central to many natural phenomena, most notably the process of photosynthesis, which enables plants to harness solar energy and store it in the form of chemical energy. Light-activated electronic processes in molecules lead to the formation of excited states, which drive subsequent chemical reactions, such as intramolecular electron transfer and bond rearrangements. An understanding of these processes is necessary for the construction of synthetic strategies in organic chemistry. The study by Oksdath et al. explores the photochemistry of N-(selenomethyl)alkyl-phthalimides, revealing that the reactions are accompanied by intramolecular electron transfer and subsequently followed by proton transfer, demonstrating the intricate behavior of reactive intermediates involved in photochemical reactions (Argüello et al., 2015). The effectiveness of photochemical reactions relies heavily on the character of the reaction medium, temperature, and the presence of solvents or additives. Modern applications of photochemistry have utilized these variations to enhance the conditions and yields of reactions. For example, microphotoreactors have revolutionized photochemical synthesis by enabling highly controlled light exposure and reaction times. In a microphotoreactor, the reaction volume has a significant influence on the mass and energy transfer processes compared to batch reactors. An experiment confirmed that the transformation from a batch setup to a microphotoreactor allowed for better chemical conversion and reduced irradiation times, with the process holding promise for more efficient photochemical applications (Aillet et al., 2013). The transformation is crucial to the development of synthetic methods in academic research and industrial applications. Moreover, not only do photochemical reactions influence simple synthetic processes, but they also extend into areas such as materials science and environmental chemistry. The ability to control light and chemical reactivity has led to the synthesis of new materials, such as photoresponsive polymers and light-controlled catalysts, which can profoundly change their physical properties upon exposure to specific wavelengths. Furthermore, an understanding of the mechanism of photochemical reactions allows the solution of environmental issues, including the photodegradation of pollutants and the optimization of solar energy conversion devices. As these systems become increasingly elaborate, sustained research is essential for the more effective and sustainable utilization of photochemical processes. Through continued exploration of both underlying principles and applied applications, photochemistry remains on the threshold of great leaps and scientific contributions in a broad variety of disciplines (Figure 2.4).

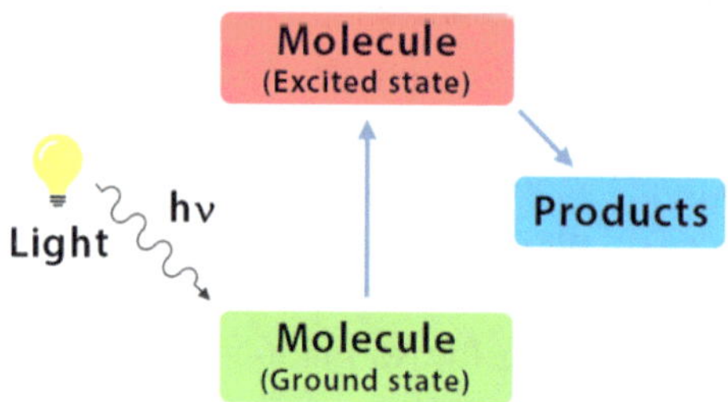

Figure 2.4. Photochemical reactions.

Source: https://external-content.duckduckgo.com/iu/?u=https%3A%2F%2Fwww.chemistrylearner.com%2Fwp-content%2Fuploads%2F2020%2F09%2FPhotochemical-Reaction.jpg&f=1&nofb=1&ipt=0f7aa2044f1ffceebd2632ceb6a3411894e35ee5ee9c8e517b3bcafe649bec53.

2.2.1. Types of Photochemical Reactions

Photochemical reactions are a diverse array of processes that are initiated primarily by light energy, particularly in the ultraviolet and visible ranges. They can be categorized into two broad forms: those involving molecular dissociation and others involving molecular rearrangement. Molecular dissociation, or reactions involving bond cleavage, occurs when photons are absorbed, causing disruption of chemical bonds. Conversely, molecular rearrangements involve readjustments of atoms without bond rupture, leading to the formation of isomers or new compounds. Familiarity with such forms is essential because they form the basis for various uses, ranging from synthetic chemistry to environmental science. The distinction between these types of reactions also helps clarify the complexities of photochemistry in more complex systems, including those found in natural conditions and constructed environments. In addition, intramolecular and intermolecular photochemical transformations are special classes of the interaction between molecular species. Intramolecular reactions describe a process involving one molecule, where the absorption of light may be employed to induce a process, such as cyclization or fragmentation, that directly involves the atoms of the same molecule. This differs from intermolecular reactions, which involve more than one molecule interacting through light absorption. This can be described by numerous mechanisms, such as energy transfer or the formation of photoadducts, for example, the dimerization of alkenes through photochemical processes. The identification of these reactions within the broader context of photochemistry is crucial in fields such as organic synthesis, where the achievement of effective transformations is highly desirable. These categories and their corresponding mechanisms outline the inherent character of light-activated processes in chemical reactivity (Figure 2.5).

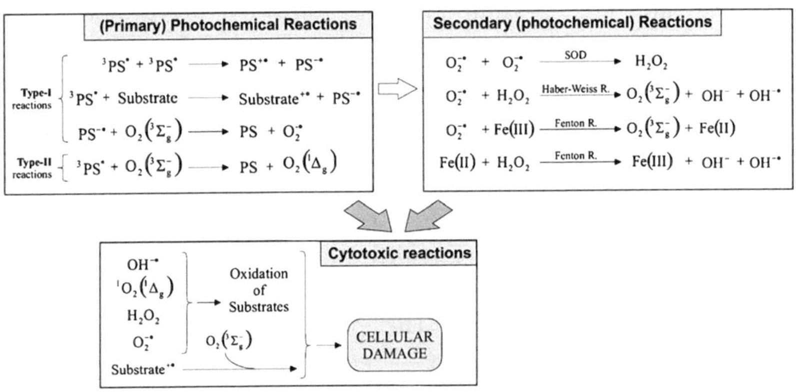

Figure 2.5. *Overview of photochemical reactions during PDT. Several types of primary and secondary photochemical reactions cause the production of reactive oxygen species and dose-dependent cellular damage.*

Source: https://www.researchgate.net/figure/Overview-of-photochemical-reactions-during-PDT-Several-types-of-primary-and-secondary_fig5_5605298.

The kinetics and dynamics of photochemical reactions are controlled by a range of factors, including solvent effects, temperature, and catalysts. Solvent polarity, for instance, can have a quite dramatic effect on the rate and mechanism of photochemical processes through the stabilization or destabilization of transition states. Moreover, photochemical reactions of transition metal complexes have been studied, providing a basis for identifying new photophysical and photochemical phenomena, such as long-lived excited states and charge transfer reactions. Current studies reveal the relevance of the responses in atmospheric chemistry, where they contribute to the formation and breakdown of the ozone layer (Albercook et al., 1998). In synthetic organic chemistry, the investigation of unusual photochemical routes, such as those involving selenium-substituted compounds, elucidates the complexities of how varied molecular structures determine reaction outcomes (Argüello et al., 2015). Such factors not only enhance the fundamental knowledge on photochemical reactions but also spur developments in theoretical and practical chemistry.

2.2.2. Reaction Mechanisms and Pathways

Understanding reaction paths and mechanisms in photochemistry is essential because it determines the path or course of light-induced reactions at the molecular level. Mechanisms involve a sequence of steps that affect the absorption of photons, electronic excitation, and eventual reorganization of molecular structures. To illustrate, the formation of specific molecular products can be directed by the energy landscape governing these transitions. Conditions such as the availability of catalysts, solvating environments, and temperature play a crucial role in determining the effectiveness and outcome of photochemical reactions. Unraveling these pathways not only enhances our understanding of fundamental chemical behavior but also lays the groundwork for applications in energy conversion systems, such as solar cells and artificial photosynthesis. Moreover, reaction pathway complexity becomes increasingly evident while investigating astrochemical environments, as in the case of investigating protoplanetary disks. The existence of various isotopologues, such as deuterated species, requires a more rigorous investigation of their formation mechanisms. As an example, the varying distribution patterns between DCN and DCO+ revealed in the disk of TW Hya are evidence of their distinct chemical formation processes. Specifically, the central peaking of DCN suggests that it is formed at higher temperatures and under changed initial conditions than DCO+. This point highlights that multiple reaction paths must be synthesized in an integrated manner to account for the deuterium fractions of Solar System bodies, underscoring the complexity of molecule formation in diverse environments (Aikawa et al., 2012). Additionally, the examination of the chemistry of nitrogen-containing compounds in planetary atmospheres, such as that of Jupiter, further exemplifies the significance of understanding reaction paths in photochemical systems. As revealed from the studies, the unpredictable abundance of hydrogen cyanide (HCN) is partially explained by thermochemical and photochemical constraints. Within Jupiter's troposphere, physical separation between photolytic regions for methane (CH_4) and ammonia (NH_3) inhibits effective HCN formation, illustrating the impact that physical separation has on chemical interconversion. The integration of thermochemical and photochemical models provides a more accurate representation of the pathways through which these nitrogen compounds are formed and converted. Thus, a comprehensive understanding of reaction mechanisms is crucial for interpreting the sophisticated chemistry that dominates planetary atmospheres, as well as the larger universe (Achterberg et al., 2010) (Figure 2.6).

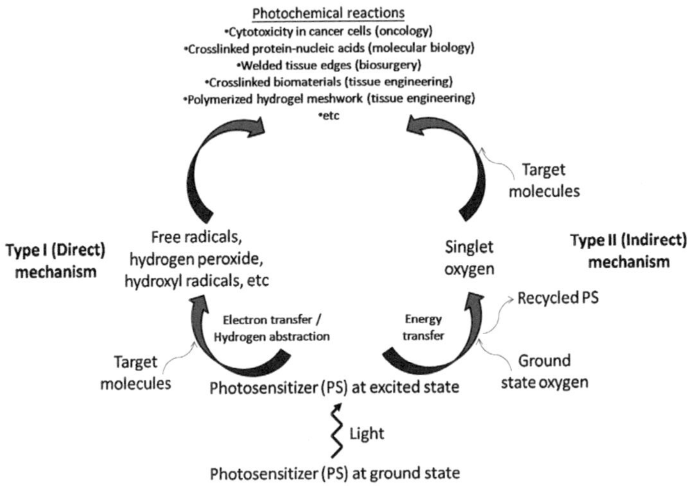

Figure 2.6. Type I and type II mechanisms of photochemical reactions.

Source: https://www.researchgate.net/figure/Type-I-and-type-II-mechanisms-of-photochemical-reactions_fig3_43180479.

2.2.3. Factors Influencing Photochemical Reactions

Photochemical reactions are also greatly reliant on a variety of exterior and interior conditions that contribute to their efficiency and the yield of the product. Among these determinants is the wavelength and intensity of the incident light, which must be compatible with the energy levels of the reactants involved. For instance, the absorption of specific wavelengths may cause electronic transitions in molecules to produce excited states that are instrumental to the photochemical reaction. Furthermore, the immediate surroundings, such as temperature and solvent polarity, play a crucial role in stabilizing the excited states and thus influencing reaction kinetics. For instance, lower temperatures may restrict molecular motion, which can slow down reactions that rely on the dynamic interaction between radical species. This intricate relationship between light and environmental factors is a manifestation of the multi-aspectual nature of photochemical systems, suggesting that a multi-aspectual approach is necessary in model representation and prediction of reaction behavior under different conditions (Brasseur et al., 1999). The molecular structure of the reactants and their chemical properties also determine the pathways of photochemical reactions. Molecular conformation changes, such as stereochemistry and the presence of functional groups, can alter the excited state energetics and the likelihood of radical formation. For example, the photolysis of different ketones may yield distinct product distributions based on their adsorption properties in complex matrices, such as zeolite structures, which alter the accessibility and reactivity of the reactive species. Turro et al. (1987) considers how radical variations influence recombination and final product formation in adsorption, a process that is central to the phenomenon of selectivity in chemical processes. Molecular design and strategic control of structural elements can therefore be valuable tools for optimizing photochemical processes, enhancing selectivity, and even guiding the synthesis of desired chemical products. Additionally, the solvent medium in which the photochemical reaction occurs should not be overlooked. Solvent effects, for instance, can dictate the reaction mechanism by stabilizing specific intermediates or transition states over others. Polar solvents can enhance specific interactions through solvation effects, resulting in differences in reactivity among isomeric molecules. The influence of the presence of other molecules, either via cooperative effects or competitive adsorption, may also play an essential role in the reaction dynamics by changing the neighborhood of the reactive molecules. These interactions are even more relevant in mixtures, as they may lead to unexpected reaction channels or the formation of side products, rendering the chemical scenario more complex overall. It is essential to understand these multifaceted impacts to encourage photochemical processes and develop more efficient reaction systems based on light as the driving force for chemical reactions (Brasseur et al., 1999).

2.3. APPLICATIONS OF PHOTOCHEMISTRY

Photochemistry, or the examination of light-activated chemical processes, is a central aspect in many applications across various disciplines, particularly in materials science. A rather significant application is the use of photochemical processes in the synthesis of complex organic molecules. Microphotoreactors, for example, are now efficient tools for photochemical processes. A microphotoreactor can enhance reaction kinetics and reduce irradiation time, allowing for higher chemical conversions compared to a standard batch photoreactor. This change is crucial to maximize photochemical synthesis, making the process industrially viable. The precise control of reaction conditions in a microphotoreactor, such as temperature and light exposure, presents a gigantic advantage in the development of photochemical approaches that can maximize synthetic pathways and potentially reduce waste in chemical manufacture (Aillet et al., 2013). Another wide application of photochemistry is in environmental remediation, which can be utilized in wastewater treatment. Industrial processes, such as those in the oil and textile industries, produce huge amounts of wastewater that is typically rich in toxic chemicals. Photocatalytic degradation has been proposed as a potential solution to this issue. Scientists have begun to improve the efficiency of photochemical processes in breaking down organic pollutants by utilizing photocatalysts such as titanium dioxide (TiO_2). This involves the creation of new reactor systems that permit the degradation of toxic compounds, as well as the simplicity with which photocatalysts can be recovered from treated water. The integration of novel materials, such as carbon nanotubes (CNTs), with photocatalysts enables increased photoactivity, making photocatalysis a competitive option for large-scale wastewater treatment (Mason et al., 2018). In energy production, photochemistry has also gained significant application in the production of photovoltaic devices and photovoltaic solar energy technologies. Solar cells, for instance, utilize photochemical principles to convert sunlight directly into electricity. Efficiency in these cells is fundamentally dependent on the materials used and their capacity to absorb and convert light energy. Recent advancements in photochemistry have led to the development of organic photovoltaics and dye-sensitized solar cells, which hold promise for lower production costs and increased application versatility. Further research in this field aims to enhance the stability and energy conversion efficiency of these devices, ultimately contributing to a more sustainable future. With growing global energy needs, the role of photochemistry in harnessing and utilizing solar energy is now greater than ever before, and it thus receives special attention in research and development (Aillet et al., 2013).

2.3.1. Role in Photosynthesis

The role of photosynthesis in life support on Earth cannot be overstated, as it forms the basis for energy transfer within ecosystems. Green plants, algae, and certain bacteria convert light energy into chemical energy through the process of photosynthesis, producing glucose from carbon dioxide and water. Trapping of light is primarily accomplished through pigments like chlorophyll, which absorb best at red and blue wavelengths. While chlorophyll effectively captures energy from such wavelengths, it reflects green light, a phenomenon that has intrigued scientists for centuries. This discrimination is not merely a sign of evolutionary adaptability, but also adheres to the laws of photochemistry, i.e., maximizing the utilization of energy while minimizing waste, highlighting the delicate relationship between light and life. The absorption characteristics of photosynthetic pigments suggest an intricate relationship between light capture and the requirement for maximum energy conversion during photosynthesis (Bebout et al., 2007). Environmental parameters that control light quantity and quality also play a role in determining photosynthetic efficiency. Natural selection has favored the development of various photosynthetic organisms with a degree of adaptability to their environment, enabling them to harness energy from diverse light conditions. For instance, some plants are triggered to adopt longer wavelengths during periods of limited light to enhance their ability to capture energy. This robustness stems from photochemical events that allow for photon uptake even during scattering or attenuation. The structure of light-harvesting systems exhibits a subtle order regulated by the forces of evolution, which facilitates the comprehension of photon energy harvesting optimization and ecological niches. Knowledge of such dynamics underscores the pivotal role of photosynthesis in maintaining both ecological and environmental balance, as well as the broader context of energy transfer in the biosphere (Hadden et al., 2013) (Figure 2.7).

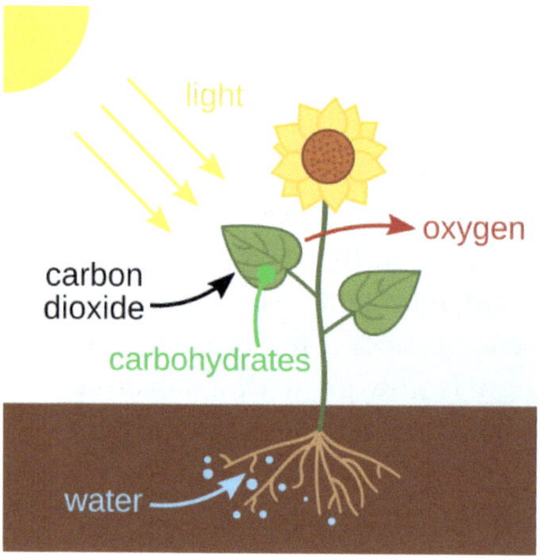

Figure 2.7. Process of photochemistry.

Source: https://chemistrytalk.org/photochemistry/.

The repercussions of photosynthesis go beyond ecological interactions, participating in vital biogeochemical cycles that determine global climate and atmospheric conditions. Through the process of converting carbon dioxide into oxygen during carbohydrate synthesis, photosynthetic organisms become key regulators of atmospheric gases that are fundamental to sustaining life. This process is not a local phenomenon; rather, it has far-reaching implications for climate control, symbolizing the interconnectedness of plant processes and global systems. By studying the spectral signatures of photosynthetic life, researchers can gain insight into how variations in light properties affect photosynthetic efficiency and, subsequently, carbon fixation. This understanding requires increased research into the manner in which photosynthetic activity can be influenced by climate change, as well as the need for ongoing photochemistry studies that range from basic science to ecological utility (Bebout et al., 2007).

2.3.2. Use in Solar Energy Conversion

In the field of solar energy conversion, identifying new materials has been at the heart of overcoming the intrinsic limitations of conventional semiconductors. Most conventional photovoltaic technologies, such as silicon solar cells, are afflicted by an intrinsic inefficiency due to the loss of excess photon energy as heat, particularly for energies exceeding the band gap. This loss lowers the net conversion efficiency, which is bounded by the detailed-balance limit of 33%.

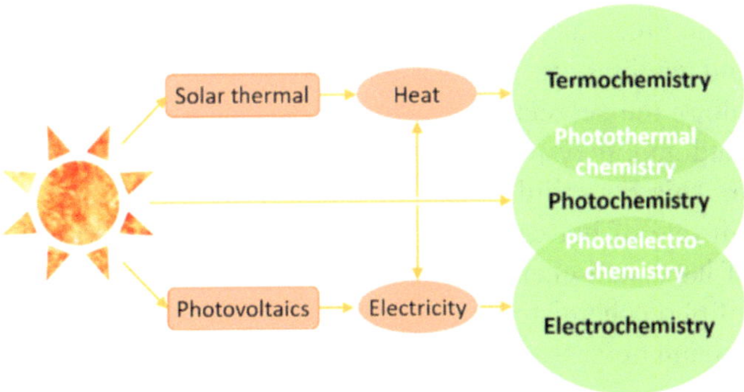

Figure 2.8. Schematic representation of the direct and indirect uses of solar energy to drive chemical reactions.

Source: https://www.researchgate.net/figure/Schematic-representation-of-the-direct-and-indirect-uses-of-solar-energy-to-drive_fig1_371162739.

Recent advances, however, have been built upon hot-carrier devices that promise to effectively harvest this lost energy. With the use of ultrathin two-dimensional (2D) semiconductors, such as monolayer MoS2 and WSe2, researchers have discovered new photophysical properties that can potentially facilitate energy harvesting. The importance of these materials lies in their ability to prevent energy loss and enhance efficiency in solar energy, thereby marking a paradigm shift in photovoltaic technology. Fundamental to the success of such hot-carrier systems is the ability to understand charge transfer processes in the 2D semiconductor structure. The research bridges a strategic knowledge gap in the interaction between energy levels and applied potential in materials such as monolayer MoS_2. A key factor here is the driving energy for charge transfer ($\Delta G^{\circ\prime}$), required to determine the hot-carrier extraction rate (kET). By specifying how $\Delta G^{\circ\prime}$ varies under illumination and with applied voltages, scientists can begin optimizing extraction rates in conjunction with cooling. The consequences of this understanding extend beyond simple improvement in energy conversion efficiencies; they also extend to solar fuels production, which holds a possible key to sustainable energy supply. Such research works testify to the central role that photochemistry is playing in transforming solar technology (Figure 2.8).

Experimental methods employed to study such materials further confirm their potential for converting sunlight to energy. Techniques such as photocurrent spectroscopy and femtosecond transient absorption spectroscopy enable researchers to elucidate the complex interplay between charge carriers and excitons in these 2D layers. The data obtained from such studies provide insights into the timescales associated with key processes, such as hot-carrier formation and cooling, as well as exciton formation and annihilation. As scientists unlock the secrets of exciton binding energies and changes to the electronic band gap, they pave the way for novel approaches to performance optimization. This bountiful source of information forms the foundation for the maximum rate of hot-carrier extraction, ultimately leading to maximized utilization of harvested energy. In short, these findings not only enhance our knowledge of photochemistry but also pave the way for new solutions in solar energy conversion processes and foretell an era of sustainable energy.

2.3.3. Applications in Pharmaceuticals and Materials Science

Photochemistry is equally significant in pharmaceuticals and materials science, where it can be employed to generate sophisticated solutions for complex issues. In pharmaceuticals, photochemical reactions are utilized in the synthesis of medicines that possess sophisticated therapeutic properties. It is thus possible to transform simple organic compounds into biologically active and more complex compounds by such reactions. Moreover, photochemistry is a component of photodynamic therapy (PDT) design, where light-initiated photosensitizing agents are employed to treat several cancers and other conditions. By using light energy, such agents catalyze cytotoxic reactions solely in the target tissues while minimizing damage to nearby healthy cells. The integration of photochemistry into drugs not only facilitates drug synthesis but also opens doors to targeted modalities of treatment, significantly improving patients' outcomes. In material science, photochemistry also finds significant application in the synthesis of novel materials with unique properties and functions. For instance, photopolymerization processes are widely used in the synthesis of high-performance polymers employed in coatings, adhesives, and 3D printing applications. The method involves the cross-linking of monomers in liquid form to form solid polymers using ultraviolet (UV) radiation, which enables a quick curing time and the desired material properties. Moreover, incorporating photochemical processes into manufacturing can facilitate environmentally friendly production by reducing demands on toxic solvents. Even newer advancements have demonstrated that photoresponsive materials can be utilized in smart coatings that respond to light stimulation by altering their properties, showcasing the potential for applications in novel fields such as energy harvesting and environmental sensing. This highlights the vast potential of photochemistry for material innovation. The and ecological effects of materials and drugs have raised concerns, triggering an increased focus on the role of photochemistry in these industries. Xenobiotic degradation, for example, pharmaceutical impurities in wastewaters, highlights the importance of photochemical reactions in environmental remediation. Photochemical processes are effective in reducing levels of such contaminants, which are notorious for their recalcitrance towards traditional biological treatment. Techniques such as heterogeneous photocatalysis have emerged as promising alternatives, using photocatalysts to promote the photolysis of toxic substances in wastewater. Combined technologies that integrate photochemical treatment with other emerging wastewater treatment technologies have also been more efficient in eliminating recalcitrant pollutants from water sources, thereby addressing the dual challenge of pollution and public health (Klementova et al., 2018). These advancements underscore the critical convergence of photochemistry with water resource purification and environmental sustainability as human civilization strives to mitigate the effects of pollution (Abou-Ghazala et al., 2014).

2.4. PHOTOCHEMICAL TECHNIQUES

Photochemical processes are key to the evolution of photochemistry, enabling new applications in biochemistry and materials science. The processes utilize the energy of photons to induce chemical reactions, thereby enabling the controlled conversion of molecules. One of the advantages of photochemical processes is that they enable reaction selectivity through selective activation, thereby minimizing the production of byproducts. Furthermore, the combination of photochemical processes with surface functionalization has provided a basis for the synthesis of next-generation functional materials. For instance, recent research has shown that surface chemical architectures can be regulated at the molecular level to create photosensitive surfaces that display useful photochemical reactions, thereby presenting prospects for new sensors and catalytic applications (Carroll et al., 2009). Second, the universality of photochemical methods stems from their ability to be utilized in various systems, ranging from solid to soft matter. One example of this variability is the use of photochemical surface delivery vehicles, whereby complex photoactive molecules are prepared to create stable surface monolayers. The cars are capable of delivering selective chemical functionalities to surfaces upon illumination, and it is possible to create complex patterns and tailored material properties. These systems underscore the need to design powerful photochemical reactions that are preprogrammed within surface delivery vehicles, enabling the development of sustainable and efficient processes that qualify as green chemistry. The potential of such methods is vast, ranging from laboratory research to practical application in industries (Carroll et al., 2009) (Figure 2.9).

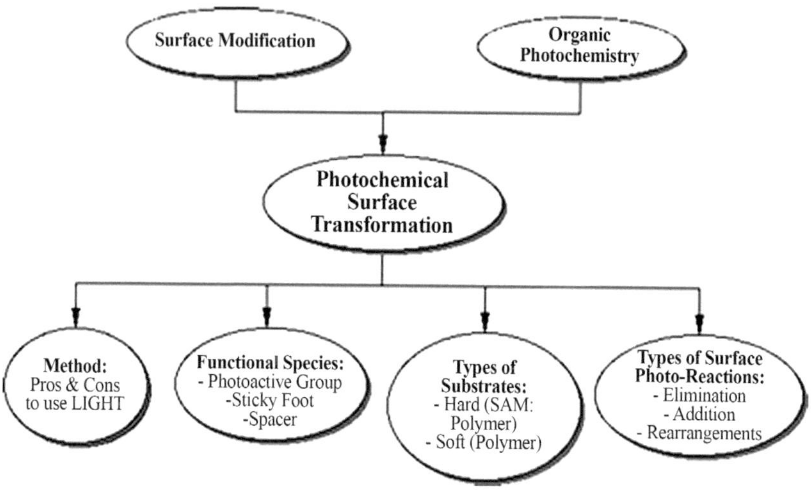

Figure 2.9. Photochemical methods of surface modification.

Source: https://www.researchgate.net/figure/An-outline-of-this-introduction-to-photochemical-methods-of-surface-modification_fig1_236883746.

The simple understanding of photon interaction with matter also motivates the development of theoretical frameworks, such as time-dependent density-functional theory (TD-DFT). This method enables researchers to model and predict molecular response to photonic perturbation exactly. By quantifying the evolution of electron density with time, which results from illumination by light, TD-DFT provides data that can be critical towards optimizing photochemical reactions. The continuous development of hybrid functional methods enables the inclusion of progressively more complex, nonlocal effects present in photochemical situations, driving research toward greater precision and applicability in determining outcomes. Thus, the interplay of theoretical models with experimental techniques in photochemistry presents an integrated perspective, allowing for deeper probing of the promise contained within photochemical systems.

2.4.1. Spectroscopy Methods in Photochemistry

The meeting ground of photochemistry and spectroscopic techniques is at the heart of demystifying molecular interactions on illumination. Spectroscopy, literally the science of the interaction between matter and electromagnetic radiation, holds crucial details of the electronic, vibrational, and rotational molecular states. Techniques such as UV-V sspectroscopy allow one to probe electronic transitions, allowing chemists to differentiate between the behavior of excited states and the resulting photochemical reactions that they will undergo. By quantifying the absorption of light at specific wavelengths, scientists can infer information about molecular structure and energy levels, thereby enhancing our knowledge of photochemical processes.

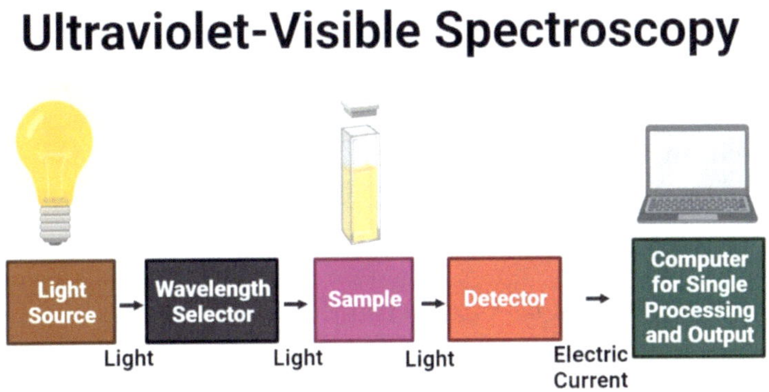

Figure 2.10. UV-Vis spectroscopy.

Source: https://external-content.duckduckgo.com/iu/?u=https%3A%2F%2Fmicrobenotes.com%2Fwp-content%2Fuploads%2F2023%2F11%2FUV-Vis-Spectros-copy-1024x536.jpeg&f=1&nofb=1&ipt=e6571acdde3e163732248cf105a024907c3a176dc1f8c0b40b6e1e877c41b373.

These methods extend beyond detection and assist in deciphering mechanisms and reaction kinetics, the foundation for photochemistry. Advances in time-resolved spectroscopy have revolutionized the study of fast photochemical reactions, enabling researchers to track molecular transformations in real time. Femtosecond and picosecond spectroscopy, for example, have enabled scientists to investigate ultrafast behavior in excited-state processes, such as electron transfer and bond formation. For instance,

N-(selenomethyl)alkyl-phthalimide research illustrates how quickly reactant-to-intermediate conversions may be monitored using time-resolved methods. Notably, the detection of intramolecular electron transfer reactions highlights the intricate interplay between chemistry and energy. These findings demonstrate how subtle variations in the molecular framework and solvent environment can influence reaction pathways, underscoring the key contribution of advanced spectroscopy to the broader discipline of photochemical science, as exemplified by new findings (Argüello et al., 2015) (Figure 2.10).

Furthermore, the demands of examining the photochemical activity of complex biological molecules necessitate new spectroscopic approaches. The DNA-ultraviolet light interaction is a paradigm example of the complex photochemical processes that could occur in biological systems. Understanding these interactions is crucial for gaining insights into the health effects of radiation and mutagenesis. Methods such as gas-phase spectroscopy and analysis of smaller biomolecular subunits are a systematic approach to deconstructing such complicated phenomena. Systematically increasing molecular complexity, researchers can trace the progression of photochemical processes and gain a deeper understanding of how large biomolecules respond when irradiated with light. With significant advancements, complexity in large structures continues to be a problem, which reinforces the need for continuous innovation in experimental methods to fully grasp the principles of photochemistry (Hadden et al., 2013).

2.4.2. Kinetic Studies of Photochemical Reactions

The kinetic study of the photochemical reaction offers a central paradigm for the description of light-induced dynamics of molecular transformations. The backbone of this topic is the electron transfer theory, which describes the processes often triggered by light absorption that form a sequence of intermediate reactive species, which may ultimately lead to the final products. For example, in *N*-(selenomethyl)alkyl-phthalimide research, researchers showed that the reaction mechanisms may vary significantly based on the structural properties of the reactants. Steady-state photolysis and time-resolved laser spectroscopy experiments revealed that such reactions proceed through intramolecular electron transfer between the selenium and phthalimide moieties. Such insight is vital to the uncovering of reaction mechanisms at play, wherein information regarding the ability of alterations in molecular shape to govern the efficacy and selectivity of the resultant photoproducts (Argüello et al., 2015) is communicated. Kinetic studies in photochemistry not only assist with mechanistic interpretation but also inform real-world applications in a variety of fields, from materials science to pharmaceuticals. The competition among different reaction pathways emphasizes the role of molecular interactions and steric effects in regulating outcomes. Experiments reveal that variations in the length of alkyl spacer chains between reaction sites influence proton transfer reactions, which in turn determine whether cyclization products are endo or exo biradicals. That such a delicate balance of factors contributes to the reaction behavior underscores the necessity of precise kinetic modeling to predict reaction behavior under various conditions. Additionally, a full understanding of such kinetic parameters can maximize photochemical reactions in industrial applications, as well as increase the yield and selectivity of target products (Argüello et al., 2015). The outcomes of kinetic studies also have far-reaching implications beyond synthetic routes to environmental science, where knowledge of the fate of photochemical agents is crucial. The reaction dynamics between atmospheric species, for instance, those that act toward the mitigation of pollutants, are significant determinants of environmental impacts. For instance, despite the prime focus of research on the effect of airborne fluoride, knowledge of the interaction between light and such species in the atmosphere can reveal mechanisms of transmogrification and elimination. By parallelizing field measurements with laboratory kinetic experiments, researchers can develop predictive models of photochemically active species in natural settings. Kinetic experiments thus form a pivotal component of establishing effective surveillance strategies and policy initiatives to check the harmful effects of photochemical reactions on industrial effluents.

2.4.3. Computational Methods in Photochemical Research

In photochemical investigation, computational methods have radically transformed the way researchers study molecular interaction and reaction mechanisms. Through advanced simulations, researchers can quantify the energetics and dynamics of excited states, which are at the core of describing the photochemical processes that form the basis of solar energy conversion and photodynamic therapy. Specifically, computational techniques such as density functional theory (DFT) and molecular dynamics simulations enable the sampling of potential energy surfaces and vibrational motion of molecules. Such a computational framework uncovers information that is usually out of reach by experiment alone, and transition states can be located, as well as reaction pathways predicted. As a result, these models enable the construction of novel photochemical reaction systems with enhanced efficiency and selectivity for various applications, including the synthesis of new materials and optimization of light-irradiation-induced chemical reactions. Another key aspect of computational methods in photochemistry is that they can correct the complexities in electronic structure calculations. Quantum chemical calculations of high level may be essential for explaining photochemical reaction mechanisms, particularly those involving numerous electronic states. Products designed to assist in analyzing and visualizing electronic transitions and facilitating these calculations are extremely valuable. For example, program packages can employ algorithms to blend multiple levels of theory, enhancing accuracy without excessive use of computer resources. The increased availability of computational resources with high calculation power helps to further simplify this endeavor, allowing scientists to tackle larger molecular aggregates and more complex photochemical scenarios. This aspect of computational photochemistry emphasizes precision and velocity in simulation, that predictions harmonize with experimental reality, and that results are capable of informing future research direction usefully. The marriage of computational methods with experimental verification is central to the development of photochemistry as a science. Techniques such as time-resolved spectroscopy can be combined with computational predictions to validate theoretical models, thereby yielding a more comprehensive understanding of photochemical processes. For instance, model building aimed at modeling the influence of solvent environments on photochemical reactivity can inform reaction dynamics. Likewise, with improved algorithms, specific methodologies, such as those introduced in the DICES framework, can enhance the analysis of data for spectral observations within photochemistry. As scientists continue to develop these computational techniques, the interconnection between synthetic methods, empirical findings, and theoretical computations will intensify, ultimately enhancing the credibility of photochemical research. These collaborative investigations highlight the paradigm-shifting potential of computational methodologies, promising new avenues for applications in energy conversion, environmental remediation, and the development of therapeutic drugs.

2.5. ENVIRONMENTAL IMPACT OF PHOTOCHEMISTRY

The environmental impact of photochemistry is multifaceted, most notably in the formation and modification of ozone in the atmosphere. Photochemistry is a significant contributor to the complex chemical processes that regulate ozone levels, thereby influencing air quality and climate. For example, studies conducted within the North Atlantic Regional Experiment (NARE) found that photochemical ozone generation above the ocean can reach as high as 3.6 ppbv d–1, indicating the significant role played by nitrogen oxides and solar radiation in photochemistry. The presence of contaminated continental plumes was revealed to impact background photochemistry, demonstrating that even remote areas can experience tremendous changes in air composition induced by regional emissions. The results highlight the intricate interplay between natural and anthropogenic processes of photochemistry, as well as the local sensitivity of atmospheric chemistry (Albercook et al., 1998). Though its role in ozone generation, photochemistry also plays a crucial role in the breakdown of pollutants, thus establishing its dual nature of both aiding and alleviating pollution problems. For instance, photochemical breakdown of volatile organic compounds (VOCs) can be a significant factor in reducing air pollution in cities. However, this response also forms secondary pollutants, such as ozone, which aggravates respiratory diseases and contributes to climate change. The equilibrium between these two opposing effects leaves no doubt that a general model encompassing various atmospheric conditions and chemical reactions is necessary. These models, which were designed to simulate photochemical processes, can provide policymakers with critical information to inform effective strategies for managing air quality that consider the benefits and drawbacks of photochemical reactions in the environment (Albercook et al., 1998) (Figure 2.11).

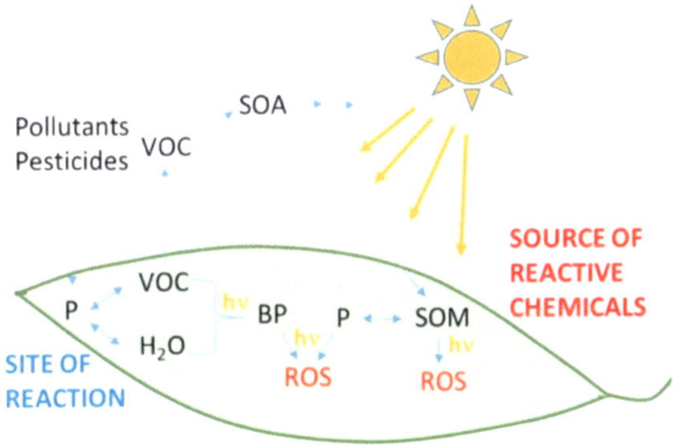

Figure 2.11. Environmental photochemistry.

Source: https://link.springer.com/article/10.1007/s43630-022-00228-w.

Furthermore, photochemical activity is priceless in quantifying the impact of fluorinated compounds, especially regarding plant health. Studies on the impact of fluorides on plants reveal the involvement of both dry and wet deposition, with precipitation playing a significant role in fluoride deposition on plant leaf surfaces. The interaction of moisture and pollutant accumulation shows a complex web of environmental responses promoted by photochemistry. Of particular importance is the capacity of fluorocarbon photodegradation to release inorganic fluoride, with serious consequences for plant life, necessitating a comparative analysis with established industrial discharges. Understanding of such effects lends insight into the broader implications of photochemical reactions to the environment and growing interest in the necessity for ongoing study and regulation of discharge and its environmental effects.

2.5.1. Photochemical Smog and Air Quality

The photochemical smog phenomenon represents a vital area of overlap between urban growth and environmental chemistry, posing a serious threat to air quality. It is a multicomponent mixture primarily produced in sunny, warm climates, where sunlight serves as a catalyst for the reaction between volatile organic compounds (VOCs) and nitrogen oxides (NO_x). These photochemical reactions produce ground-level ozone, a major component of smog that is both harmful to the environment and detrimental to human health. Photochemical smog is found in most significant quantities in large cities with high traffic density and industrial activities, where car and factory emissions are the source of the precursor pollutants responsible for the high density of the contaminants. With increased urbanization, there is a growing need to understand the dynamics of photochemical reactions in order to devise effective management strategies for successful air quality control. In addition to the effect on human health, photochemical smog can have far-reaching impacts on ecological systems too. The elevated levels of ground-level ozone weaken the health of plants by breaking down leaf tissue, reducing photosynthesis, and ultimately leading to reduced yields from crops and altered ecosystems. The exchange between smog ingredients and vegetation, particularly in relation to the deposition of toxic chemicals, highlights the need for a greater understanding of these environmental processes (Figure 2.12).

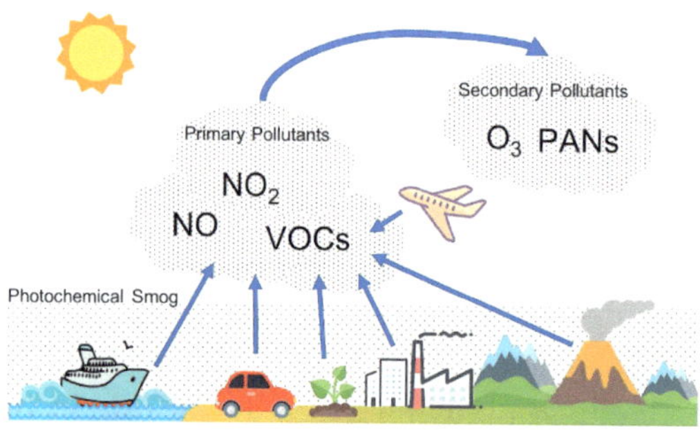

Figure 2.12. Photochemical smog.

Source: https://external-content.duckduckgo.com/iu/?u=https%3A%2F%2Fwww.worldatlas.com%2Fr%2Fw2000-h1125-q90%2Fupload%2F52%2Fbe%2F75%2Fphotochemical-smog-formation.png&f=1&nofb=1&ipt=629eb97ae64eab9cc968712bc35e0a5698b9a4be3c1ea58f6c50dbf9cf9d0eaf.

As hinted in fluoride effects on vegetation research, one realizes that pollutants like ground-level ozone are not unique in their behavior; they tend to accumulate and cause stress to plants through dry and wet deposition processes. The consequence is that the impact of photochemical smog is not limited to direct air quality problems. Still, it has a deeper incursion into ecological and agricultural environments, hence the need for an integrated environmental science. Investigations of nitrogen oxides (NO_x), one of the key precursors to photochemical smog formation, demonstrate the sophistication of urban air chemistry. As outlined in various studies, these emissions undergo a cascade of photochemical reactions to form a range of oxides and acids, which will in turn react and alter within the atmosphere. Modeling diurnal behaviors of NO_x reveals significant interactions with other atmospheric constituents, such as on urban scales of the South Coast Air Basin in southern California. The formation of nitric acid and its interaction with ammonia to produce ammonium nitrate illustrates how concentrated emissions can lead to degraded air quality. Understanding these intricate interactions is crucial for effective pollution control practices, highlighting that the fight against photochemical smog is a matter of thoroughly understanding the chemical processes involved and their broader environmental impacts, thereby laying the basis for sustainable urban growth and policy development.

2.5.2. Role in Atmospheric Chemistry

The role of photochemistry in atmospheric chemistry is significant, as it governs the reactions in the Earth's atmosphere under the control of solar radiation. When sunlight interacts with atmospheric compounds such as gases and aerosols, it initiates a chain of chemical reactions that can lead to the creation and breaking down of various substances. For instance, photochemical reactions influence greenhouse gas concentration, and consequently, climate change is influenced. Such processes also give rise to secondary pollutants such as ozone, which has a significant influence in regulating solar radiation reaching the Earth's surface. Therefore, knowledge of photochemical processes is essential for predicting atmospheric behavior and the consequent environmental impacts. Formaldehyde (CH_2O), a byproduct of various natural and anthropogenic processes, is a significant indicator of photochemical activity in the atmosphere. Its mixing ratios, as obtained during the NASA Pacific Exploratory Mission (PEM)-Tropics B mission, demonstrated how such compounds are integral to regional and global atmospheric chemistry. Median CH_2O concentrations, for instance, showed significant variation with altitude, highlighting the compound's significance in tropospheric chemistry. These findings demonstrate that with increasing altitude, levels of CH2O decrease drastically, thereby highlighting the impact of atmospheric layers on photochemical processes. The evidence suggests that photochemical processes regulating these levels have significant effects on air quality and climate models, thereby identifying formaldehyde as a critical area in atmospheric studies (Blake et al., 2001) (Figure 2.13).

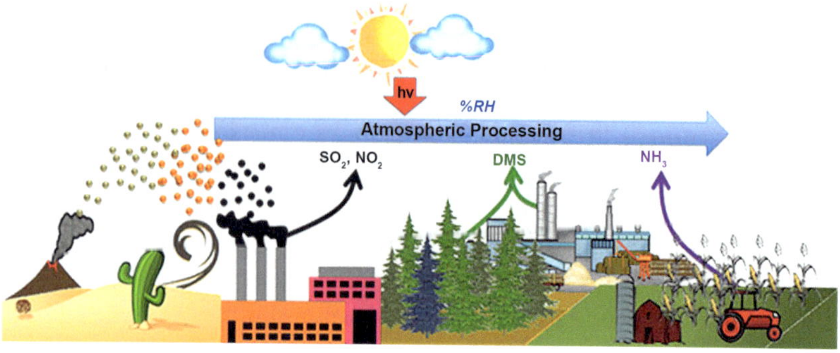

Figure 2.13. Photochemistry in atmospheric chemistry.

Source: https://gayannmt.us/atmospheric-chemistry-and-photochemistry.html.

Furthermore, observing ozone generation above oceanic backgrounds requires considering the interactions between different atmospheric constituents. Observations conducted during the North Atlantic Regional Experiment (NARE) demonstrated that ozone production rates differed with the introduction of nitrogen oxides and non-methane hydrocarbons, which are typically introduced by polluted continental plumes. The conclusions suggest that local meteorological regimes and cloudiness have a controlling influence on the photochemical processes within the region. It is thus desirable to understand these dynamics in order to accurately model atmospheric chemistry and estimate the potential for ozone formation under varying environmental conditions. This knowledge not only enhances our understanding of air quality, but it also has significant implications for climate change policy (Albercook et al., 1998).

2.5.3. Photodegradation of Pollutants

Photodegradation of pollutants is a significant process in environmental chemistry, as it utilizes sunlight to break down harmful substances into harmless products. The process is particularly relevant in industrial wastewater treatment, such as that of oil and textile industries, which release vast quantities of toxic chemicals. The pollutants are likely to be complex mixtures, making it harder to remove them. Sophisticated photocatalytic reactions, particularly those involving materials like titanium dioxide (TiO2), have shown great potential in enhancing the efficiency of photodegradation. Integration of sophisticated reactor designs into these processes for their improvement, for instance, by the application of Pickering emulsions with added magnetic iron oxide, has the potential to improve pollutant degradation significantly as well as simplify recovery of the used photocatalysts, thus improving the sustainability and efficiency of the treatment process (Mason et al., 2018). Varying operating parameters also influence photocatalytic activity in the removal of pollutants. For instance, the initial concentration of the pollutant, the amount of the photocatalyst, and illumination all play a crucial role in determining the rate of degradation. Some studies have investigated the photocatalytic degradation of phenylurea herbicides, revealing that the kinetics of degradation obey pseudo–first-order reaction rates, meaning that the concentration of the herbicides has an inverse impact on the efficacy of their degradation (David et al., 2007). In addition, the findings highlighted that increased illumination levels to an optimal degree enhance the rates of degradation; however, excessive light leads to electron-hole recombination, reducing efficiency. Such mechanistic information in photocatalytic degradation is crucial for optimizing these processes in any environmental remedy. The future growth of photochemistry as a discipline lies at the heart of creating sustainable methodologies of pollutant degradation. The combination of nanostructured materials with photocatalysis is an underexplored field with the potential to offer gains in reaction rate and overall efficiency. The fabrication of hybrid materials, wherein the properties of TiO2 are combined with those of carbon nanotubes and magnetite, can lead to innovative solutions that offer optimum photodegradation efficiency coupled with reduced waste generation (Mason et al., 2018). In addition, the pH influence on rate of degradation can be utilized to guide treatment system design based on site-specific conditions. This multifaceted approach is indicative of the complexity of integrating photochemical methodology into real-world applications for pollution control, and it highlights the need for continued research on the effectiveness of photodegradation processes.

SUMMARY

Photochemistry provides a window into the intricate interactions between light and chemical reactions, profoundly influencing atmospheric science, organic synthesis, and materials science. Underlying principles such as photon absorption and consequent energy transfer mechanisms are of prime importance in initiating reactions with consequences ranging from the formation of complex organic compounds to the degradation of pollutants in environmental processes. As noted throughout the essay, advancements in photochemical processes, including the transition from batch to microphotoreactor systems, signify a significant improvement in selectivity and efficiency, highlighting the potential of these technologies for future research and applications. Such innovations not only enrich our fundamental understanding but also underscore the need for optimization to achieve a desired outcome, thereby reiterating the importance of photochemistry in addressing contemporary problems. Furthermore, the uses of photochemical reactions extend beyond the laboratory to real-world applications such as atmospheric modeling and pollution control. A case in point is the research on atmospheric formaldehyde concentration, which provides information on the applicability of photochemical reaction knowledge at different altitudes and geographical locations. Variation in mixing ratios of CH_2O, as demonstrated in observations obtained during NASA Pacific Exploratory Mission, underscores the interconnected nature of chemical reactions that occur within the atmosphere and their environmental impacts as a whole. The findings highlight the necessity of incorporating photochemical principles into curricula for environmental ecological science to equip future scientists with the skills to address climate change and pollution. The synergy of knowledge from different branches of science expands the potential for innovative solutions, which are crucial in today's rapidly changing world. In short, the subjects discussed in this essay highlight the multidisciplinary aspects of photochemistry and its significant contributions to both theoretical and applied sciences. With the advent of new technologies, the future applications of photochemical methods are expected to grow, paving the way for novel syntheses and green processes that align with global sustainability programs. The close relationship between photochemical processes and natural environmental processes necessitates ongoing study and interdisciplinary collaboration in areas of chemistry, climatology, and engineering. Using knowledge from recent studies, such as comparative batch versus microphotoreactor studies, researchers can formulate informed strategies for maximizing chemical conversions and minimizing environmental impacts, ultimately contributing to the sustainable utilization of the world's resources.

The study of photochemistry is fundamentally based on the interaction between chemicals and light, which serves as the foundation of various applications across numerous fields of science. One of the essential aspects is the understanding of excited electronic states, which results from molecules' absorption of photons. The development of time-dependent density functional theory (TD-DFT) has significantly advanced an understanding of this aspect. Spurred by groundbreaking work by Hohenberg, Kohn, and Sham, TD-DFT offers a solid basis for the calculation of excited states, thus enabling the calculation of spectra associated with photochemical processes. Linear response theory has since confirmed the utility of TD-DFTs, positioning it increasingly at the forefront of the analysis of medium- and large-sized molecules. This emerging paradigm not only aids in the interpretation of absorption spectra but also highlights the ongoing need for improvement in the approximate functionals used in TD-DFT calculations, particularly regarding the treatment of doubly excited states (Casida et al., 2011). In the broader context of atmospheric chemistry, the principles of photochemistry have significant applicability in describing planetary atmospheres. The behavior of solar radiation and atmospheric components is closely studied to determine the intricate dynamics of greenhouse gases and clouds. Current observations from spacecraft have presented additional pressures on laboratory experiments to confirm results, ultimately driving an extension of our understanding of the chemical processes governing atmospheric action. This overlap between experimental and observational astrophysics highlights the need for comprehensive models that account for both molecular interactions and macroscopic atmospheric phenomena. These combined approaches not only refine the theoretical basis of photochemistry but also contribute to informing applied applications, such as climate models and the study of extraterrestrial atmospheres (Fox et al., 2025). With every step in the development of research, the recognition of limitations of existing methods, particularly in TD-DFT, is essential in an effort to catalyze innovation. Doubly excited states, which are particularly implicated in bond-breaking and bond-forming processes, further underscore the need to advance computational methodologies beyond the standard adiabatic approximations of TD-DFT. This pursuit of precision is essential in applications from introductory physical chemistry to practical applications such as solar energy trapping and photochemical synthesis. Overall, the study of these salient features through the fundamentals of photochemistry not only offers a complete foundation for researchers and students but also sheds light on the elaborate bouquet of light-matter interactions responsible for numerous chemical occurrences. Every step forward brings greater understanding into both theoretical underpinnings and empirical studies, driving the field towards new fields of understanding.

REFERENCES

1. Tijani, J. O., Fatoba, O. O., Madzivire, G., & Petrik, L. F. (2014). A Review of Combined Advanced Oxidation Technologies for the Removal of Organic Pollutants from Water. *Water Air & Soil Pollution*, *225*(9). https://doi.org/10.1007/s11270-014-2102-yAchterberg, A., Allen, M., Allen, J., Atkinson, D., Atkinson, T., Atkinson, R., Atreya, S., et al. (2010). On the abundance of non-cometary HCN on Jupiter. http://arxiv.org/abs/1006.3508 (accessed on 28th September 2025).
2. Adamovich, I., Agarwal, S., Ahedo, E., Alves, L. L., Baalrud, S., Babaeva, N. Y., Bogaerts, A., et al. (2022). The 2022 plasma roadmap: Low temperature plasma science and technology. Journal of Physics D: Applied Physics, 55, 373001–373001. https://doi.org/10.1088/1361-6463/ac5e1c (accessed on 28th September 2025).
3. Aikawa, Y., Asvany, O., Brown, J. M., Qi, C., Wilner, D. J., Hogerheijde, M., Öberg, K. I., et al. (2012). Evidence for multiple pathways to deuterium enhancements in protoplanetary disks. http://arxiv.org/abs/1202.3992 (accessed on 28th September 2025).
4. Aillet, T., Dechy-Cabaret, O., Loubière, K., Prat, L., et al. (2013). Photochemical synthesis of a "cage" compound in a microreactor: Rigorous comparison with a batch photoreactor. https://core.ac.uk/download/18540116.pdf (accessed on 28th September 2025).
5. Albercook, G. M., Blake, D. R., Blake, N. J., Carroll, M. A., et al. (1998). Photochemical production and loss rates of ozone at Sable Island, Nova Scotia during the North Atlantic Regional Experiment (NARE) 1993 summer intensive. https://core.ac.uk/download/215538921.pdf (accessed on 28th September 2025).
6. Alberti, M., Frank, A., & Kleidon, A. (2017). Earth as a hybrid planet - The Anthropocene in an evolutionary Astro biological context. http://arxiv.org/abs/1708.08121 (accessed on 28th September 2025).
7. Aller, R. C., Canfield, D. E., Dalsgaard, T., Engström, P., et al. (2005). Nitrogen removal in marine environments: Recent findings and future research challenges. https://core.ac.uk/download/12043452.pdf (accessed on 28th September 2025).
8. Argüello, J. E., Heredia, A. A., Oksdath Mansilla, G., Peñeñory, A. B., et al. (2015). Photochemistry of N-(selenoalkyl)-phthalimides. Formation of N, Se-heterocyclic systems. https://core.ac.uk/download/159300430.pdf (accessed on 28th September 2025).
9. Barham, J. P., & König, B. (2019). Synthetic photo electrochemistry.

Angewandte Chemie International Edition, 59, 11732–11747. https://doi.org/10.1002/anie.201913767 (accessed on 28th September 2025).

10. Bebout, B. M., Des Marais, D. J., Ehrenreich, A., Govindjee, Siefert, J., Joliot, P., et al. (2007). Spectral signatures of photosynthesis I: Review of Earth organisms. http://arxiv.org/abs/astro-ph/0701382 (accessed on 28th September 2025).
11. Bergqvist, A., Frantzich, H., Hasselrot, K., & Ingason, H. (2005). Fire and rescue operations in tunnel fires: A discussion of some practical issues. https://www.semanticscholar.org/paper/a5c8e5995d78ce72f06e97af09a5fba19091d6c3 (accessed on 28th September 2025).
12. Berthelot, Y., & Hess, P. (1989). Photoacoustics, photothermal and photochemical processes in gases. Journal of the Acoustical Society of America, 88, 1665–1666. https://www.semanticscholar.org/paper/f3e17c699f9a3769c20008a0f3c2419df2e8300d (accessed on 28th September 2025).
13. Bertram, A. K., Wahner, A., Carlton, A. G., Picquet-Varrault, B., Percival, C. J., Weschler, C. J., George, C., et al. (2017). The essential role for laboratory studies in atmospheric chemistry. https://core.ac.uk/download/78902283.pdf (accessed on 28th September 2025).
14. Blake, D., Blake, N., Chen, G., Crawford, J., et al. (2001). Formaldehyde over the central Pacific during PEM-Tropics B. https://core.ac.uk/download/287622899.pdf (accessed on 28th September 2025).
15. Bonomo, M., Cavallo, C., Di Pascasio, F., Dini, D., et al. (2017). Nanostructured semiconductor materials for dye-sensitized solar cells. https://core.ac.uk/download/54538567.pdf (accessed on 28th September 2025).
16. Brasseur, G., Orlando, J., & Tyndall, G. (1999). Atmospheric Chemistry and Global Change. https://www.semanticscholar.org/paper/c1a6dec947fe0e953ec0211c1c8d7a123630af72 (accessed on 28th September 2025).
17. Cambié, D., Bottecchia, C., Straathof, N. J. W., Hessel, V., & Noël, T. (2016). Applications of continuous-flow photochemistry in organic synthesis, material science, and water treatment. Chemical Reviews, 116, 10276–10341. https://doi.org/10.1021/acs.chemrev.5b00707 (accessed on 28th September 2025).
18. Carroll, G. T., Koberstein, J. T., Park, E. J., Turro, N. J., et al. (2009). Shedding light on surfaces: Using photons to transform and pattern material surfaces. https://core.ac.uk/download/161434895.pdf (accessed on 28th September 2025).

19. Casida, M. E., & Huix-Rotllant, M. (2015). Many-body perturbation theory (MBPT) and time-dependent density-functional theory (TD-DFT): MBPT insights about what is missing in, and corrections to, the TD-DFT adiabatic approximation. http://arxiv.org/abs/1008.1478 (accessed on 28th September 2025).
20. Casida, M. E., Frenkel, J., Gatti, M., Li, Z., Huix-Rotllant, M., Casida, M. E. (2011). Progress in time-dependent density-functional theory. http://arxiv.org/abs/1108.0611 (accessed on 28th September 2025).
21. David, B., Guittonneau, S., Zertal, A., Zouaghi, M., et al. (2007). Photocatalytic degradation of monolinuron and linuron in an aqueous suspension of titanium dioxide under simulated solar irradiation. https://core.ac.uk/download/59264088.pdf (accessed on 28th September 2025).
22. Euclid Collaboration, Kovavčić, I., Baes, M., Nersesian, A., Andreadis, N., Nemani, L., Abdurro'uf, Bisigello, L., et al. (2025). Euclid preparation. LXVIII. Extracting physical parameters from galaxies with machine learning. Astronomy & Astrophysics. https://www.semanticscholar.org/paper/176e166bbcf980d02f5422da1ff1a33eba742f3b (accessed on 28th September 2025).
23. Euclid Collaboration, Naidoo, K., Ruiz-Zapatero, J., Tessore, N., Joachimi, B., Loureiro, A., Aghanim, N., Altieri, B., et al. (2025). Euclid preparation. Accurate and precise data-driven angular power spectrum covariances. https://www.semanticscholar.org/paper/ac75b95cfb06fdf798ea54351079fd88f4492ab2 (accessed on 28th September 2025).
24. Euclid Collaboration, Scognamiglio, D., Schrabback, T., Tewes, M., Gillis, B., Hoekstra, H., Huff, E., Marggraf, O., et al. (2025). Euclid preparation LX. The use of HST images as input for weak-lensing image simulations. https://www.semanticscholar.org/paper/aab36fa0857315e56921ad1b7b8be1a93af05981 (accessed on 28th September 2025).
25. Euclid Collaboration, Zalesky, L., Weaver, J., McPartland, C., Murphree, G., Valdes, I., Jespersen, C. K., Taamoli, S., et al. (2025). Euclid preparation: TBD. Cosmic Dawn Survey: Evolution of the galaxy stellar mass function across 0.2. https://www.semanticscholar.org/paper/1ce3ba056d60a3860fec6404d0a3173755118dc4 (accessed on 28th September 2025).
26. Fox, K., Hornstein, J., & Mumma, M. J. (2025). Vibrational-Rotational Spectroscopy for Planetary Atmospheres (Vol. 1). https://core.ac.uk/download/pdf/42856431.pdf (accessed on 28th September 2025).
27. Hadden, D. J., Lodish, H., Rossle, S., Satzger, H., Scoles, G., Vastola,

F. J., & Yang, X. (2013). Recent advances in experimental techniques to probe fast excited-state dynamics in biological molecules in the gas phase: Dynamics in nucleotides, amino acids and beyond. https://core.ac.uk/download/18327700.pdf (accessed on 28th September 2025).

28. Klementova, S. (2018). Photochemical degradation of organic xenobiotics in natural waters. https://core.ac.uk/download/322436094.pdf (accessed on 28th September 2025).
29. Mason, E. (2018). Synthesis and photocatalytic performance of TiO_2-CNT and magnetized Fe_3O_4-TiO_2-CNT multifunctional hybrids: A Pickering emulsion platform for organic degradation. https://core.ac.uk/download/211346833.pdf (accessed on 28th September 2025).
30. McCune, D. C., & Weinstein, L. H. (2025). Assessment of effects on vegetation of degradation products from alternative fluorocarbons. https://core.ac.uk/download/pdf/42814572.pdf (accessed on 28th September 2025).
31. McRae, G. J., & Russell, A. G. (1982). Dry deposition of nitrogen containing species. https://core.ac.uk/download/33113716.pdf (accessed on 28th September 2025).
32. N/A. (2025). Resident research associateships. Postdoctoral and senior research awards: Opportunities for research at the Jet Propulsion Laboratory. https://core.ac.uk/download/pdf/42846348.pdf (accessed on 28th September 2025).
33. Rahimi, M., Yaghoobzadeh, Y., Daliri, M. R., et al. (2025). Explanations of large language models explain language representations in the brain. https://www.semanticscholar.org/paper/847238f439aeb46284f72dc17ef2645181c924ee (accessed on 28th September 2025).
34. Sambur, J. B. (2023). (Invited) Energy level alignment and hot carrier extraction in monolayer semiconductor photoelectrochemical cells. ECS Meeting Abstracts. https://www.semanticscholar.org/paper/9fecc24f159faaaf31b58d3ca122923aaf022247 (accessed on 28th September 2025).
35. Sambur, J. B., Austin, R., Almaraz, R., Krummel, A., Montoya-Castillo, A., Sayer, T., & Toole, J. (2024). (Invited) Photo electrochemistry of monolayer 2D semiconductors: Quantifying band gap renormalization effects and hot carrier extraction. ECS Meeting Abstracts. https://www.semanticscholar.org/paper/791bc978f286031826f740829ba278d76520384c (accessed on 28th September 2025).
36. Strelko, O., Pylypchuk, O., & Berdnychenko, Y. A. (2025). PREFACE. History of Science and Technology. https://www.semanticscholar.org/paper/15412810e923c7c081953712c15df3f47899fea1 (accessed on 28th September 2025).

37. Turro, N., Cheng, C., Abrams, L., & Corbin, D. (1987). Size, shape, and site selectivity's in the photochemical reactions of molecules adsorbed on pentasil zeolites. Effects of coadsorbed water. Journal of the American Chemical Society, 109, 2449–2456. https://www.semanticscholar.org/paper/b37cc11dd2c9089e2ca255577791bcc3e084e457 (accessed on 28th September 2025).

38. Westermayr, J., & Marquetand, P. (2020). Machine learning for electronically excited states of molecules. Chemical Reviews, 121, 9873–9926. https://doi.org/10.1021/acs.chemrev.0c00749 (accessed on 28th September 2025).

CHAPTER 3

TYPES OF PHOTOCATALYSTS

Contents

INTRODUCTION

Photocatalysis has emerged as a significant area of research in recent years, with a particular focus on environmental remediation and energy conversion. Photocatalysis refers to the process by which a photoreaction is catalyzed by a catalyst under illumination, and light-induced chemical reactions are initiated. Various materials exhibit photocatalytic activity; however, the material's photocatalytic activity is dependent on light absorption, charge carrier generation, and redox reaction promotion. The types of photocatalysts can be broadly classified into semiconductor materials, metal catalysts, and emerging composites. All these types possess different strengths and limitations with respect to light absorption, stability, and activity under varying light intensities, which structurally dictate their uses in applications ranging from water purification to air cleaning and fuel production. Semiconductor photocatalysts, particularly titanium dioxide (TiO_2), have been extensively investigated and utilized due to their favorable features, including low cost, non-toxicity, and excellent chemical stability. TiO_2 having a band gap of around 3.2 eV is predominantly active under UV radiation, which creates issues in the utilization of solar energy. Recent progress has tried to enhance its photocatalytic activity using various strategies, such as doping with a transition metal or hybridization with carbonaceous systems. For instance, the introduction of graphene into TiO_2 matrices has been promising for expanding the visible light-absorbing spectrum and improving the charge-separation ability. These developments are significant for enhancing the potential applications of TiO2 photocatalysts in addressing environmental problems, such as wastewater treatment, as emphasized by research on synthetic dye degradation using advanced composite materials (D'Amato et al., 2017). Aside from semiconductor- and metallic-based photocatalysts, new methods have paved the way for the synthesis of hybrid photocatalysts, which can leverage the unique properties of various materials. The use of metal-organic frameworks (MOFs) in photocatalytic systems and their synthesis is a notable trend. By combining the large surface area and tunable character of MOFs with robust semiconductors like UiO-66, researchers have created highly tunable platforms for photochemical reactions. These hybrid materials were found to facilitate charge transfer processes due to their thermal stability and electronic conductivity, thereby enhancing photocatalytic activity over traditional materials. This synergy strategy enhances understanding of electronic conversions in doped systems, as manifested by empirical models based on absolute intensity measurements (Vos et al., 2018). Collectively, these developments represent a promising trend in photocatalyst development, advancing universal applications in sustainable technologies.

Importance of Photocatalysis in Environmental Applications

The role of photocatalysis in environmental processes cannot be overstated, particularly in the prevention and mitigation of pollution, as well as in waste management issues. Photocatalysts, such as titanium dioxide (TiO2), have been recognized for their ability to induce degradation reactions upon illumination. This is particularly important in cleaning up toxic pollutants in wastewater, such as industrial dyes and organic compounds generated by industrial processes. These toxins are highly toxic to the environment, contaminating water supplies and retarding the natural environment. The application of photocatalytic technology presents a highly effective way to mitigate these risks, highlighting the role that advanced materials can play in environmental protection procedures and enhancing the ecological sustainability of industrial processes. Procedures that maximize the photocatalytic performance of materials, such as TiO2, are critical, providing increased efficiency for practical applications. The synthesis of photocatalyst materials has introduced new means of addressing environmental contamination. Sol-gel and hydrothermal synthesis procedures have proven effective in modifying the structural properties of TiO_2 for high photocatalytic activity. Immobilization of TiO_2 on various substrates has, for instance, been considered in order to promote its usability in practical applications optimally. Varying variables such as particle size, morphology, and crystallinity have been shown through experiments to have a and significant impact on the rate of degradation of contaminants such as formic acid and 2,6-dichlorophenol. This means that enhancing the photocatalytic activity through tailored synthesis techniques not only improves the effectiveness of contaminant removal but also paves the way for cost-effective technology in the purification of industrial wastewater, a giant leap towards environmental remediation technology. In addition, the quest for expanding the spectrum of light available to photocatalysts has increasingly been on the agenda, particularly with the inclusion of carbon-based materials. The combination of TiO_2 with graphene composites has been a promising solution for enhancing photocatalysis under visible light, thereby overcoming one of the limitations of conventional TiO_2 photocatalysts. These composites leverage the high electron mobility and surface area of graphene to facilitate electron transfer and adsorption of organic pollutants. In this integration, not only is the efficiency of photocatalysis improved in degrading harmful substances, but the application range of this technology is also widened in various industries, including wastewater treatment and air purifiers. The developments in the field of photocatalysts represent a crucial intersection of materials science and environmental sustainability, underscoring the transformative power of innovation in addressing pressing ecological concerns.

3.1. TYPES OF PHOTOCATALYSTS

The development of photocatalysts has taken an unprecedented pace, particularly in the fields of environmental sustainability and energy conversion. Among various forms of photocatalysts, titanium dioxide (TiO2) has been a thoroughly studied material due to its beneficial characteristics, such as chemical stability, non-toxicity, and outstanding catalytic activity. TiO_2-based photocatalysts are mainly employed for purposes ranging from water purification to air treatment, effectively decomposing organic contaminants under ultraviolet light. The drawback, though, is that their activity remains low under visible light, which constitutes the majority of the sun's light. To counter this, recent studies have focused on the synthesis of new TiO_2 composites and hybrid materials that enhance their photocatalytic activity under visible light, a giant step towards making photocatalytic processes more efficient (Jang et al., 2015). Besides titanium dioxide, other photocatalysts have been targeted of late. For instance, metal sulfides such as cadmium sulfide (CdS) and zinc sulfide (ZnS) have been found to create electron-hole pairs efficiently with visible light. These have been shown to be promising for catalyzing photocatalytic reactions for hydrogen evolution and pollutant degradation. Nevertheless, similar to TiO_2, metal sulfides too are plagued by issues of instability and self-aggregation. To address these flaws, researchers are exploring methods to enhance the structural stability and photocatalytic efficiency of their materials through the introduction of support materials or composite photocatalysts. This ongoing research highlights the dynamic nature of photocatalyst development, underscoring the need for innovative methods to enhance efficiency and utilization (Jang et al., 2015). Another important class of photocatalysts is metal-organic frameworks (MOFs) and graphitic carbon nitrides (g-C_3N_4). These have been appreciated for their tunability and large surface area, which enhance light absorption and catalysis. MOFs, in particular, are versatile as they can be functionalized using various metal sites, which improve photocatalytic activity. Graphitic carbon nitrides have also gained momentum due to their fascinating electronic properties and stability, making them suitable for visible-light-driven photocatalytic applications. Their potential to be combined with other materials to form heterojunctions is an interesting direction, as these structures can facilitate effective charge separation as well as photocatalytic reactions. The continued advancement of photocatalytic technologies, exemplified by the development of novel materials, marks a consistent endeavor to achieve more efficient and eco-friendly solutions to global energy and environmental challenges (Afzal et al., 2014).

3.1.1. Semiconductor Photocatalysts

Semiconductor photocatalysts are currently key materials in photocatalysis, primarily due to their ability to use solar light for chemical conversions. They typically consist of semiconducting compounds such as titanium dioxide, zinc oxide, or cadmium sulfide, and can be effective in using ultraviolet or visible light irradiation. The bandgap of the semiconductor determines its photocatalytic action in creating electron-hole pairs that catalyze redox processes in environmental applications. The ability of semiconductor photocatalysts to be applied towards degrading organic pollutants and reducing carbon dioxide into valuable chemicals positions them at the forefront in efforts towards reducing environmental concerns and guaranteeing sustainability. One of the most significant advancements in semiconductor photocatalysts is the introduction of new materials and structural forms that optimize their photocatalytic performance. Heterojunction photocatalysts, for example, combine different semiconductor materials to leverage their dissimilar bandgap properties, thereby enabling improved charge separation and enhanced photocatalytic activity. Furthermore, morphological changes, such as the formation of nanoscale structures or porous materials, have shown exceptional effects on surface area and light absorption capacity. Developments in such advances can improve reaction rates and overall efficiency in photocatalytic applications, such as organic synthesis and wastewater treatment (Antonietti et al., 2020) (Figure 3.1).

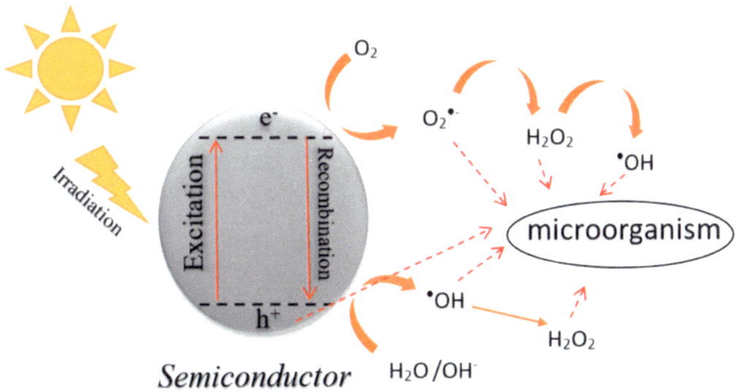

Figure 3.1. Semiconductor photocatalysts.

Source: https://www.mdpi.com/catalysts/catalysts-11-01498/article_deploy/html/images/catalysts-11-01498-g001.png.

The continued exploration of such advanced semiconductor platforms is key to realizing their full potential both in the laboratory and in practical applications. The optimization of the preparation methods for semiconductor photocatalysts also optimizes their performance. Sol-gel synthesis, hydrothermal treatments, and chemical vapor deposition have been explored to create thin films and powders with defined properties. For instance, sol-gel processing offers the precision necessary for controlling the morphology and crystallinity of end products, which is crucial for achieving maximum photocatalytic performance (Králová et al., 2012). Furthermore, dopant additions can be achieved that generate notable improvements of visible light activity, enabling sunlight to be used for photocatalytic applications. The continuous research and improvement of these synthesis routes are instrumental in advancing semiconductor photocatalysts to become more effective and widely applicable materials for environmental cleaning and energy conversion applications.

3.1.2. Metal-Based Photocatalysts

Metal-based photocatalysts play a pivotal role in the development of photocatalytic applications due to their unique electronic properties and enhanced stability under working conditions. The photocatalysts are typically composed of metals such as titanium, silver, and gold and have greater light absorption and charge carrier generation compared to their non-metal counterparts. The incorporation of metal nanoparticles into metal oxide matrices has been successful in enhancing photocatalytic efficiency.

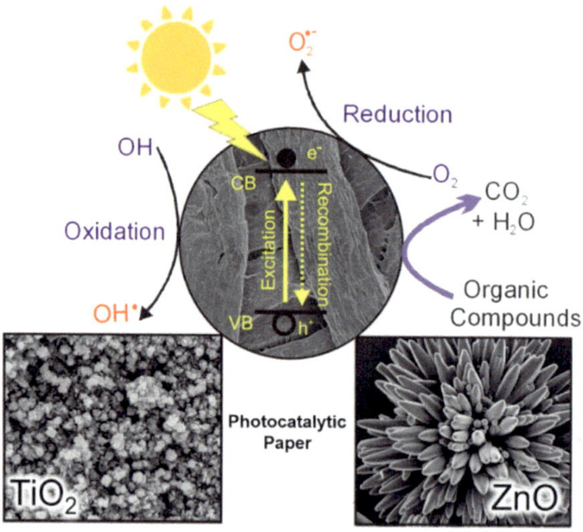

Figure 3.2. Metal-based photocatalysts are pivotal in advancing photocatalytic applications.

Source: https://pub.mdpi-res.com/catalysts/catalysts-11-00504/article_deploy/html/images/catalysts-11-00504-ag.png?1618906404.

Titanium dioxide (TiO_2), for instance, when doped with noble metals, exhibits enhanced surface plasmon resonance, thereby increasing its visible light activity. This change is of utmost relevance in fields ranging from environmental cleaning to energy conversion because it renders sunlight energetically helpful in enabling chemical reactions. In metal-organic frameworks (MOFs), new structural changes are transforming their photocatalytic efficiency. A classic case in point is UiO-66, whose photocatalytic performance can be significantly enhanced by introducing defects into its structure. These defects enhance ligand-to-metal charge transfer, which enables a more efficient electron transfer process essential for photocatalysis. Defect structures with appropriate properties are since they have a significant impact on the material's electronic behavior. Implementing a classification scheme for assessing these defects enables researchers to select the most efficient photocatalysts for specific applications. This paradigm shift in design emphasizes the importance of calibrating electronic structures to achieve the highest photocatalytic activity, beyond classical approaches that include linker tuning and metal replacement, highlighting the versatility of metal-based catalysts (Figure 3.2).

Their physical properties, such as film thickness and composite structures, are the same factors that determine the functionality of metal-based photocatalysts. Research indicates that the thickness of TiO_2 films greatly influences the activity of the photocatalyst, especially during water oxidation reactions. The 200 nm thickness is best for maximum performance when TiO_2 is pure or when it is supported with nanoparticles. These tuning holds promise to further enhance the generation of reactive oxygen species, which are crucial in most photocatalytic reactions. As demonstrated through work on nanoparticles embedded in TiO2 films, the composite materials exhibit significantly enhanced activity towards oxygen evolution compared to pure TiO_2. Through this approach, physical property control, alongside electronic property control, offers new means of optimizing the efficiency of metal-based photocatalysts for environmental and energy applications.

3.1.3. Carbon-Based Photocatalysts

Carbon-based photocatalysts represent a significant advancement in the field of photocatalysis, owing to their unique structural features and functional properties. Comprising various allotropes of carbon, such as graphene, carbon nanotubes, and graphene oxide, the materials possess good electron mobility and large surface areas. These properties significantly enhance their photocatalytic activity, particularly in energy conversion reactions such as water splitting and CO_2 reduction. Additionally, carbon photocatalysts can be combined with other components, such as metal oxides, to form composite materials with enhanced efficiency in visible light, thereby expanding their action spectrum beyond the typical ultraviolet spectrum. Hybrid materials utilize the benefits of each component, thereby fully realizing the potential of carbon as a catalyst. One of the prominent characteristics of carbon-based photocatalysts is their remarkable stability and biocompatibility upon application in various photocatalytic reactions. Compared to the traditional semiconductor photocatalysts, which are most often burdened with poor stability, carbon materials have been found to exhibit resilience to harsh working conditions and, therefore, are perfect materials for long-term applications. The stability is advantageous for environmental pollution remediation, as evidenced by studies comparing their performance in breaking down various organic pollutants. Their ability to maintain performance over extended periods under a range of conditions is a testament to their potential for applied use in sustainable environmental remediation technology. Pioneering research into this field addresses not only the operating advantages but also the scale of applications in real-world settings. Additionally, the functionalizability and tunability of carbon-based photocatalysts highlight their versatility. With the aid of various synthesis methods, including sol-gel processes and hydrothermal routes, the performance of carbon materials can be tuned to become highly photocatalytic. By adjusting aspects such as the degree of reduction or structural morphology, photocatalytic activity can be significantly enhanced. Surprisingly, introduction of carbon-based photocatalysts into the hybrid systems typically results in synergetic effects, and these further improve their photocatalytic activities. The continuous developments in this area not only have the potential to enhance existing photocatalytic technologies but also unveil new directions of pioneering applications across diverse industries, especially in energy and environmental management. The potential of carbon-based photocatalysts is an exciting area in the larger category of photocatalytic materials.

3.2. COMMON SEMICONDUCTOR PHOTOCATALYSTS

Among the most widely used semiconductor photocatalysts, titanium dioxide (TiO_2) stands out due to its excellent photocatalytic activity and stability under UV light. Its versatility arises from multiple preparation methodologies, such as sol-gel and hydrothermal synthesis, allowing for it to be highly tailored in morphology and crystalline structure. Research has shown that when immobilized on substrate materials like glassy carbon or soda-lime glass, TiO_2 exhibits dramatically improved efficiency when utilized in photocatalytic processes. The photocatalytic behavior of TiO_2 is evidently exhibited during the degradation of organic pollutants, as observed from a study in which the compound 2,6-dichloroindophenol degradation rate was investigated. By adjusting parameters such as sol-loading and incorporating anticracking agents, including PEG, the resulting TiO_2 thin films possess the capability to achieve efficiency levels comparable to those produced through more conventional chemical pathways, making them a benchmark within photocatalysis (Králová et al., 2012). Yet another essential semiconductor photocatalyst is potassium poly(heptazine imide) (K-PHI), which has been of interest because it can form organic radicals under visible light irradiation. The compound represents a departure from traditional UV-active photocatalysts towards those that respond to the visible spectrum, thereby enabling a wider range of applications. K-PHI has been employed by researchers to create dichloromethyl radicals from chloroform, opening the door to novel approaches of organic synthesis. The efficiency of the processes is optimized in a quasi-homogeneous flow photoreactor, resulting in a 19-fold increase in productivity compared to batch processing. The high yields of γ,γ-dichloroketones from the reaction demonstrate the promise of K-PHI as a bifunctional building block, with its potential not only as a photocatalyst but also for value addition to organic compounds, such as substituted furans and pyrroles (Antonietti et al., 2020). Complementing the well-studied TiO_2 and K-PHI, other semiconductor photocatalysts have been explored, each possessing unique strengths that depend on its particular structural and electronic properties. Materials such as zinc oxide (ZnO) and graphitic carbon nitride (g-C_3N_4) are also of considerable application in photocatalysis due to their appropriately positive band gap energies and ability to absorb visible light. ZnO has exhibited potent antibacterial activity coupled with its use in organic pollutant degradation, and it is thus attractive for environmental remediation. g-C_3N_4, on the other hand, has emerged as a strong competitor due to its inherent photostability and tunable electronic character, making it attractive for various chemical conversions. With continued investigation into and optimization of the efficiencies of such materials, the field of semiconductor photocatalysts will be expanded, offering innovative solutions to some of the most fundamental environmental challenges and promoting sustainable processes in chemical synthesis and energy conversion.

3.2.1. Titanium Dioxide (TiO_2)

Among the many photocatalysts examined in recent years, titanium dioxide (TiO_2) is a champion material due to its excellent photocatalytic activity, stability, and innocuity. This semiconductor material exhibits unique photoinduced activity that can effectively decompose organic pollutants and facilitate a vast range of chemical reactions under ultraviolet light. TiO_2 also occurs in multiple crystalline forms structurally, i.e., anatase, rutile, and brookite, each with different photocatalytic efficiencies. The anatase phase is particularly favored in photocatalytic reactions due to its increased surface area and enhanced electron mobility, which are central to making it more reactive and effective in environmental and energy processes. Hence, most of the research has been focused on enhancing TiO_2s properties through various synthesis and modification techniques with the final goal of optimizing its photocatalytic activity. The method of preparation dictates the properties and efficacy of titanium dioxide photocatalysts. Techniques such as sol-gel synthesis and hydrothermal methods are commonly utilized for the fabrication of TiO_2 architectures with specific morphologies and surface properties, thereby enhancing their catalytic performance. For instance, sol-gel enables the precise control of material morphology, which includes the introduction of additives that can enhance performance, such as polyethylene glycol (PEG), thereby preventing cracking during the film formation process. Furthermore, hydrothermal synthesis can regulate the size and phase of the crystals, thereby influencing the overall photocatalytic activity of the materials. Manipulation of such variables enables researchers to create TiO_2-based photocatalysts with precise functionalities for specific applications, such as water purification and air purification, which are essential in addressing worldwide environmental issues (Králová et al., 2012) (Figure 3.3).

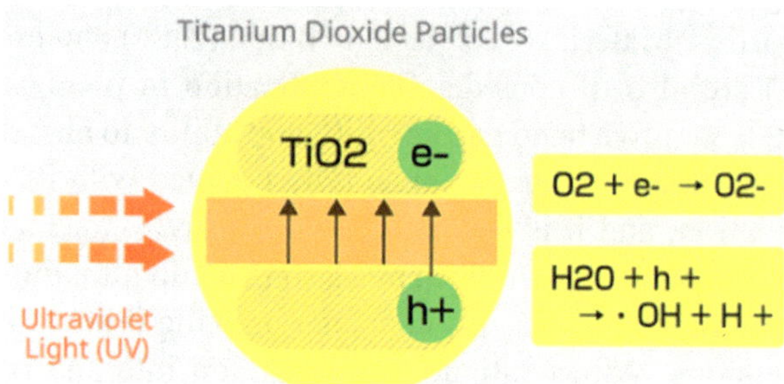

Figure 3.3. Photocatalyst.

Source: https://external-content.duckduckgo.com/iu/?u=http%3A%2F%2Fwww.biomimic-c.com%2Fen%2Fassets%2Fimages%2Fservice%2Fbiomimic_coat%2Fphotocatalyst%2Ffunction_img01-en.png&f=1&nofb=1&ipt=fd156d5767504833a54e4717c99c6a88e545f377829b0b7367e3af39d0d2b8e0.

The second significant advantage of titanium dioxide as a photocatalyst is its photoinduced superhydrophilicity potential, which enhances its usability in self-cleaning applications. TiO_2 surface modification is responsible for this phenomenon, where the water contact angle drops significantly, thereby enhancing water spreadability and facilitating the removal of dirt. This aspect is particularly beneficial in the development of self-cleaning surfaces and coatings, which significantly reduce maintenance costs and promote sustainability. Research on various deposition techniques for TiO2, including chemical vapor deposition (CVD) and material printing, has yielded promising results in enhancing the photocatalytic performance and superhydrophilicity of TiO_2 films. Comparative studies have revealed that modifying the power of incident light during deposition can further enhance the efficiency of TiO2 films, highlighting the intricate relationship between synthesis routes and photocatalytic performance (Abdullah et al., 2018).

3.2.2. Zinc Oxide (ZnO)

Zinc oxide (ZnO) is a better photocatalyst, particularly in the context of environmental pollution and water purification. Its good nature, such as high catalytic activity, non-polluting abundance, and low toxicity, renders ZnO a better alternative to traditional catalysts like titanium dioxide (TiO_2). ZnO has a direct bandgap of approximately 3.3 eV, which enables it to absorb UV light effectively and generate electron-hole pairs, a process of utmost significance for photocatalytic reactions.

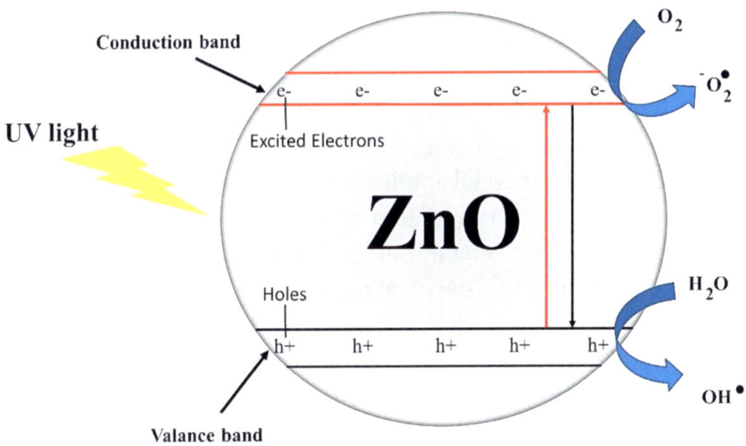

Figure 3.4. Photocatalytic activity of zinc oxide (ZnO).

Source: https://external-content.duckduckgo.com/iu/?u=https%3A%2F%2Fpub.mdpi-res.com%2Fpolymers%2Fpolymers-13-01227%2Farticle_deploy%2Fhtml%2Fimages%2Fpolymers-13-01227-g009.png%3F1628113447&f=1&nofb=1&ipt=ef33aca0d4c2af5558f814dfd6011a93f2a32c339fc76b385370d2a94461c8de.

The versatility of ZnO renders it applicable in a wide range of applications, from self-cleaning surfaces to removing air and water pollutants. In addition, its excellent stability under varied environmental conditions enhances its practical utility, thus making its photocatalytic processes and efficiencies more viable for probing. These characteristics underscore the importance of ZnO as a suitable photocatalyst, particularly in light of the growing demand for environmentally friendly approaches to wastewater treatment and environmental remediation. The photocatalytic activity of ZnO can be further understood with reference to its interaction with various reactive species in photocatalytic processes. Through exposure to light, the electrons generated can interact with oxygen to produce superoxide radicals, which, in turn, can create highly reactive hydroxyl radicals. These reactive species have been found to play an important role in degrading organic pollutants, and as such, ZnO is a valuable agent in contaminant degradation in industrial effluents. Moreover, the photocatalytic activity of ZnO-based photocatalysts can also be enhanced by changing their morphology, composition, and surface properties. Composite materials that include ZnO with other semiconductors or carbonaceous materials, for example, have exhibited superior photocatalytic activity. Such flexibility in design allows researchers to tailor ZnO photocatalysts for specific applications, essentially optimizing their potential in environmental cleanup processes (Figure 3.4).

Despite the advantages of ZnO, its potential environmental and health concerns must be addressed when it is applied. Like other nanomaterials, the life cycle of ZnO must be studied in depth to assess its toxicity on ecological systems and human health. Future studies recognize an emerging need to evaluate the impact of ZnO on various organisms, both aquatic and bacterial, due to the increasing commercial applications of such materials. In addition, the inclusion of nanotoxicology in research work is crucial for achieving safe and effective ZnO photocatalysts that minimize detrimental impacts on ecosystems. Increased research into ZnO's photocatalytic properties, along with rigorously performed assessments of its environmental footprint, will ultimately enable us to more fully appreciate its role in the overall scheme of sustainable water purification and pollution mitigation technologies (Bahnemann et al., 2011; Faria et al., 2014).

3.2.3. Cadmium Sulfide (CdS)

Cadmium sulfide (CdS) has emerged as a conventional photocatalyst due to its unique semiconducting character, which enables it to absorb visible light efficiently. Its ability to harvest solar energy and induce chemical transformations, primarily through the generation of electron-hole pairs, renders CdS a valuable material for various environmental and energy-related applications. The structural characteristics and photocatalytic activity of CdS largely depend on its synthesis routes. Techniques such as solvothermal and hydrothermal methods have been extensively utilized to synthesize CdS nanostructures with predetermined sizes and morphologies, thereby expanding their absorption spectrum. The evolution of the of synthesis makes it easier to control the atomic structure and surface area, therefore perfecting the photocatalytic activity of CdS in degradation and conversion processes, making it a subject of research in sustainable energy applications (Klochikhin et al., 2013). The doping of metal dopants into CdS also serves to enhance its photocatalytic activity. Doping with noble metals, such as platinum, creates heterostructures that significantly inhibit the recombination rates of electron-hole pairs, which are vital in sustaining photocatalytic activities. In recent literature, Pt-doped CdS/TiO_2

heterostructures have exhibited extraordinary efficiency for the photocatalytic reduction of carbon dioxide into methane, with significantly increased catalytic activity compared to undoped CdS. Platinum on the surface of CdS increases not only charge separation but also the overall surface reactivity of the photocatalyst. These changes are required for advancing the application of CdS to photocatalysis in real life because they both enhance the material's activity and stability upon solar light irradiation (Berenguer-Murcia et al., 2024) (Figure 3.5).

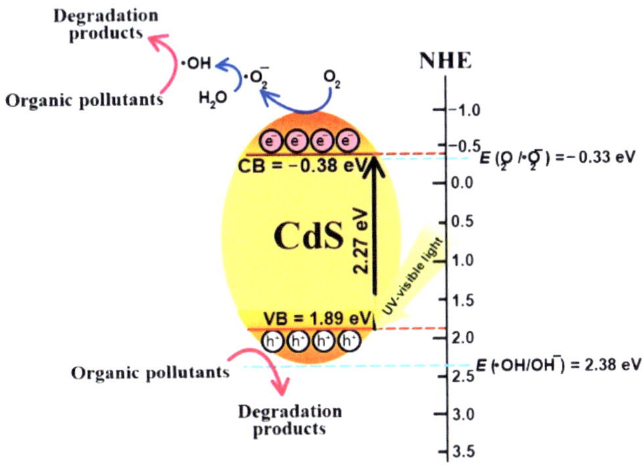

Figure 3.5. Photocatalytic mechanism scheme of CdS nanoparticles for the degradation of organic pollutants under UV-visible light irradiation.

Source: https://www.mdpi.com/1420-3049/27/22/7944.

Furthermore, the miscibility of CdS has been interesting in the context of environmental cleaning and energy transformation. Its application in photocatalytic operations is not only restricted to the degradation of simple pollutants; CdS is also responsible for solar-to-fuel conversion systems. Hydrogen or hydrocarbon production from water or carbon dioxide through CdS represents a significant step towards sustainable energy production. This exercise can be further enhanced by the development of hybrid CdS systems with other semiconductor materials, leading to higher light absorption and more efficient charge dynamics. The research efforts are focused on the development of CdS-based photocatalysts that play a dual role in energy production and environmental purification, opening up promising directions towards a cleaner and greener future. Generally speaking, the growth of CdS as a photocatalyst is indicative of its ability to address existing global challenges in environmental sustainability and energy (Klochikhin et al., 2013).

3.3. METAL-BASED PHOTOCATALYSTS

Metal-based photocatalysts are a critical class of the broader topic of photocatalysis, mainly due to their distinctive electronic structures that enable the photoabsorption and charge carrier generation with excellent efficiency. The noble metals, such as platinum, palladium, or silver, are typically introduced into these materials and enhance photocatalytic activity through mechanisms like surface plasmon resonance. The concept is that the metals facilitate electron transfer reactions, thereby raising kinetics of response in general. New advancements in photocatalyst design have evidenced that through the introduction of defects or modification of the local environment around the metal nodes, researchers can significantly enhance the photocatalytic activity of metal-organic frameworks (MOFs), which promises the realization of tailor-made materials for meeting specific photocatalytic demands, as documented in the literature (Vos et al., 2017).

In exploring the use of metal-based photocatalysts, titanium dioxide (TiO_2) has been of significant research interest due to its stability and non-toxic nature. The incorporation of metal nanoparticles into TiO_2 films significantly influences their photocatalytic activity. Research indicates that changes in film thickness can yield the best results in terms of photocatalyst activity. Specifically, 200 nm thick films performed better during water oxidation upon employing persulfate as an electron acceptor. This highlights the importance of structural optimization in enhancing photocatalytic activities, not only through compositional adjustments but also by modifying physical attributes such as film morphology, as concluded from current studies (Darr et al., 2011). The utilization of metal-based photocatalysts spans environmental cleanup to energy conversion, highlighting their versatility and importance in addressing global challenges. The synthesis of these photocatalysts typically involves careful choice of their electronic properties and reaction mechanisms. For instance, approaches to controlling the electronic structures of MOFs by introducing defects could create novel avenues for enhanced photocatalytic performances with significant advantages in specific applications. Thus, continued research on the utilization of metal-based materials is crucial for realizing photocatalytic technologies, making them central to today's research endeavors in the field. Discoveries of new synthesis routes and application-based innovations will continue to drive the development of more effective metal-based photocatalysts, establishing their position in the field of photocatalysts.

3.3.1. Platinum and Its Alloys

The role of platinum and its alloys in photocatalysis cannot be overstressed. As noble metals, platinum and its derivatives are defined by their stability and catalytic properties, which play significant roles in the speeding up of various chemical reactions under low temperatures. Such catalysts find use that goes beyond oxidizing carbon monoxide (CO) and breaking down organic pollutants. Platinum combined with other elements, especially reducible oxides, exhibits typical synergistic effects that enhance catalytic performance. For instance, the combination of platinum with SnO_2 has been shown to lead to a composite material that is highly catalytically active at temperatures as low as –27°C. At that temperature, ordinary catalysts cease to work. This ability to operate under ambient conditions efficiently is an indication of the increasing relevance of platinum-based photocatalysts for clean energy applications and environmental detoxification (Herz et al., 2025).

Figure 3.6. Platinum alloys.

Source: https://images.squarespace-cdn.com/content/v1/527ba911e4b088797b06ebdb/1648216505842-LMG8R1CQ25WAX1UCVZBM/Where+Can+I+Find+Scrap+Platinum+Buyers+Near+Me.jpg.

Under photocatalytic application, platinum performance is typically enhanced with strategic alloying and composition adjustments. Platinum interactions with metals or oxides, such as titanium dioxide, are crucial in achieving enhanced photocatalytic performance. Notably, investigations have established that the introduction of platinum nanoparticles into titanium dioxide films enhances their photocatalytic performance for water splitting and degradation of organic compounds. This improvement results from the efficient charge separation at the metal-semiconductor interface, which reduces excited charge carrier recombination rates. In addition, the customization of morphology and platinum nanoparticle organization can lead to increased light absorption and catalytic activity, positioning platinum alloys at the forefront of the evolution of next-generation photocatalyst materials (Darr et al., 2011) (Figure 3.6).

Additionally, the applications of platinum and its alloys extend beyond traditional photocatalytic reactions to include advanced processes such as CO_2 reduction and hydrogen evolution. The unique electronic character and high surface area of platinum particles give the alloys the versatility required for diverse photocatalytic uses. In recent studies, the pairing of platinum with Group VIIIA and IB metals has been explored to achieve unmatched reaction rates at low temperatures, thereby further strengthening the significance of platinum's catalytic activity. These new photocatalysts are significant in sustainable energy solutions, as they contribute to halting greenhouse gas emissions and promoting the use of renewable energy sources. As research continues to unravel the complexities of platinum alloys in photocatalysis, the potential of their innovation in a diverse range of industrial processes remains vast and holds promise (Herz et al., 2025).

3.3.2. Palladium as a Photocatalyst

Palladium has emerged as a highly active photocatalyst in the field of advanced materials because of its unique electronic and chemical properties. Being endowed with a diversified ability to catalyze numerous oxidation and reduction reactions, palladium catalysis differs significantly from conventional semiconductor photocatalysts. The transition metal exhibits excellent effectiveness in accelerating reactions during light illumination, for which it is an essential component in photocatalytic systems. For instance, the incorporation of palladium can optimize catalytic reaction efficiency, such as the degradation of organic pollutants and the production of hydrogen through water splitting. Palladium's high surface area and superior adsorption capability are the reasons behind its high photocatalytic activity, making it a worthy subject of investigation for forthcoming studies in efficient, sustainable energy solutions and environmental remediation. The interaction between palladium and semiconductors, particularly titanium dioxide (TiO_2), is crucial in achieving high photocatalytic efficiency (Figure 3.7).

Research has shown that when palladium is used as a co-catalyst, it significantly enhances the photocatalytic activity of TiO2 by facilitating charge separation and increasing the overall reactivity of the material. Through electron transfer reactions involving palladium, reactive species are formed that initiate photocatalytic reactions. In addition, pallidizing with TiO2 also facilitates the photodegradation of pollutants like rhodamine B, with a pronounced increase in degradation activity compared to utilizing TiO_2. These synergisms highlight the necessity for palladium to be incorporated in photocatalytic systems, thereby providing more effective and application-oriented means to pollution control and energy conversion schemes (Goswami et al., 2025).

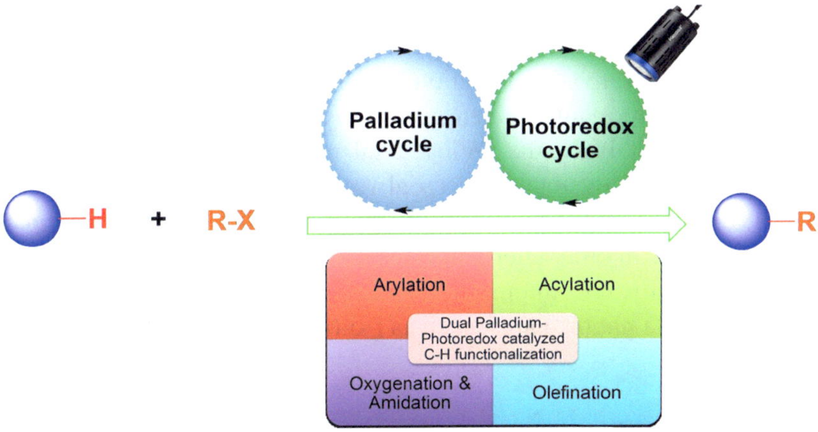

Figure 3.7. Recent progress in dual palladium–photoredox-catalyzed C–H functionalization reactions, such as C–H arylation, acylation, olefination, and heteroatom functionalization.

Source: https://chemistry-europe.onlinelibrary.wiley.com/doi/10.1002/ejoc.202200733.

Yet, the practical use of palladium as a photocatalyst is not without problems. One serious issue arises from its propensity to leach during reaction processes, thereby reducing its efficiency over time. The problem requires the establishment of strategies for immobilizing palladium on the support material, which will provide stability and reusability for extended use. Additionally, the cost of palladium is prohibitive, especially considering it is a precious metal. To address these limitations, scientists are exploring composite photocatalysts that combine palladium with other, less expensive photocatalysts, such as hematite or other clay minerals, which can enhance stability without compromising efficiency. This interaction between palladium's photonic character and integration with different materials is a prime example of the continued development of photocatalytic systems working to create sustainable solutions for the environment (Kočí et al., 2019).

3.3.3. Silver Nanoparticles (AgNPs)

The discovery of silver nanoparticles (AgNPs) in photocatalysis revolutionized various applications, particularly in antibacterial processes. AgNPs possess unique optical properties due to their surface plasmon resonance, by which they efficiently scatter and absorb light. The interaction between AgNPs and light promotes the generation of reactive oxygen species (ROS), which play a pivotal role in photocatalytic reactions. When combined with traditional photocatalysts such as TiO2 or Ru complexes, AgNPs can significantly enhance the overall photocatalytic performance. Their application not only increases the surface area of interaction but also leads to better electron transfer processes. Most importantly, when used in hybrid materials that mix AgNPs with other photocatalysts, the antibacterial effect is enhanced, indicating a potential pathway to overcome bacterial resistance in various environments. Moreover, recent work has revealed that the synergistic effects resulting from coupling AgNPs with other photocatalysts, particularly in bacterial inactivation reactions, are significant. Among these is the synthesis of a nanocomposite that pairs AgNPs with a photocatalyst such as $[Ru(bpy)_3]^{2+}$.

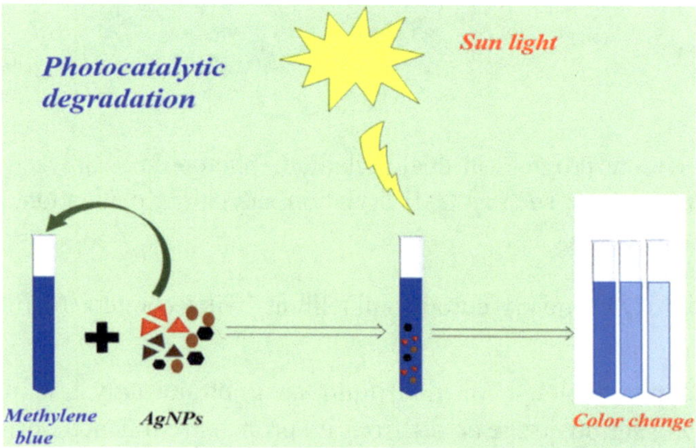

Figure 3.8. *Photocatalytic degradation by silver nanoparticles.*

Source: https://www.researchgate.net/publication/330883759/figure/fig6/AS:779411204878353@1562837520288/Photocatalytic-degradation-by-silver-nanoparticles.jpg.

Here, the AgNPs are used as light-concentrating nanoantennas to facilitate a boosted photochemical reaction upon illumination. As disclosed in recent studies, this hybrid system has also been found to be highly performative, leading to high reductions in the colony-forming units of bacteria upon exposure to visible light. The aggregation-integrated AgNPs enhance the overall photocatalytic performance, a significant breakthrough in the development of antibacterial systems. With the activation of AgNPs-induced plasmons, researchers have enhanced the bactericidal activity of conventional photocatalysts, thereby highlighting their importance in contemporary studies of photocatalysts (An et al., 2020) (Figure 3.8).

In addition, the usability of AgNPs extends to sustainable energy utilization, particularly in the reduction of carbon dioxide (CO2). Recent advancements have led to the exploitation of silver as an electrocatalyst, either as a standalone agent or blended with two-dimensional materials like graphene. Hybrid systems have demonstrated tremendous promise in converting CO2 into useful products through electrochemical reactions. These responses are effective according to the interaction of light with silver nanoparticles, which charges the system and speeds the kinetics for the reduction of CO_2. As the world focuses on addressing environmental concerns, the integration of AgNPs in photocatalytic systems to utilize CO2 is a critical area of study. It captures the common possibility of silver nanoparticles, emphasizing their role in both inhibiting microbial contamination and enhancing sustainable chemical processes in the pursuit of environmentally friendly alternatives (Erkey et al., 2020).

3.4. CARBON-BASED PHOTOCATALYSTS

Photocatalysis, a light-stimulated process, has been of great interest in recent years as the world seeks to address the growing need for sustainable energy and environmental clean-up. Among the numerous photocatalysts, carbon-based photocatalysts are rapidly becoming indispensable, thanks to their remarkable properties, which make them highly efficient and versatile. The carbon-based materials, such as carbon dots, graphene, and carbon nitride, possess the favorable properties of having high surface area, adjustable band gaps, and outstanding electron mobility. These materials can provide versatility to facilitate a large variety of photocatalytic reactions, thus broadening the scope of applications in organic synthesis, pollutant degradation, and energy conversion devices. The substantial flexibility of carbon materials to adjust their electronic properties through doping and functionalization is another plus that renders them an appropriate alternative for traditional photocatalysts, such as metal oxides. Among the key developments in carbon-based photocatalysts is the discovery of heterojunction systems, which effectively combine the capabilities of different materials to enhance photocatalytic performance. By combining carbon-based materials with semiconductors, researchers have developed hybrid photocatalysts that exhibit enhanced charge separation and improved light absorption. For example, potassium poly(heptazine imide) (K-PHI), used as a photocatalyst, has demonstrated excellent competence in forming small organic radicals upon visible light irradiation, with promising potential to drive various reactions with high selectivity and yield. It has been shown through experiments that such hybridization techniques can result in the synthesis of *γ,γ*-dichloroketones from enone substrates using K-PHI, which highlights how new photocatalytic systems can produce value-added products through efficient radical generation (Antonietti et al., 2020). These advancements place the revolutionary contribution of carbon-based photocatalysts at the forefront of sustainable chemistry. The relative performance of carbon-based photocatalysts over their classical counterparts provides evidence of their growing significance in photocatalysis. There have been various studies that have demonstrated how materials such as TiO_2, which are classically used as a model photocatalyst, although efficient, tend to require UV light to function, thereby limiting their use. Carbon-based photocatalysts can, however, be designed to absorb visible light, hence being compatible with sunlight and reducing energy input requirements. Furthermore, the flexible nature of carbon systems allows for doping with dopants, which can maximize their photocatalytic performance. This can lead to increased activity in other reactions, such as pollutant degradation and CO_2 reduction, showing the potential of carbon-based materials for further advancing photocatalysis and environmental sustainability. These findings not only contribute to material science but also provide significant potential for real application and integration (Králová et al., 2012).

3.4.1. Graphene and Its Derivatives

Graphene and its derivatives are among the most prominent materials in photocatalysis due to their excellent electronic features and large surface area. Graphene is a two-dimensional carbon allotrope with exceptional conductivity, mechanical properties, and optical properties, making it an ideal material for a wide range of applications in photocatalysis. Specifically, the ability to work with graphene through oxidation or reduction reactions generates derivatives such as graphene oxide (GO) and reduced graphene oxide (rGO), which exhibit enhanced photocatalytic efficiencies. Such modifications not only alter the electronic structure but also enable better visible light absorption, thereby enhancing charge separation and transfer during the photocatalytic reaction —a significant factor in improving photocatalyst efficiency. Moreover, incorporating graphene and derivatives into composite materials significantly enhances their photocatalytic activity. The hybrid of graphene-derived materials and traditional photocatalysts, such as titanium dioxide (TiO_2), combines the high surface area of graphene with the photocatalytic character of TiO2, thereby amplifying overall performance. These composites capitalize on the electron transfer process provided by graphene, which further acts as an electron sink to facilitate the efficient separation of charge carriers generated upon light absorption. This interaction significantly enhances pollutant degradation and the effectiveness of processes such as water splitting and CO_2 reduction. In this case, graphene offers both structural and functional advantages, making it a favorable material for the development of high-performance photocatalysts (Figure 3.9).

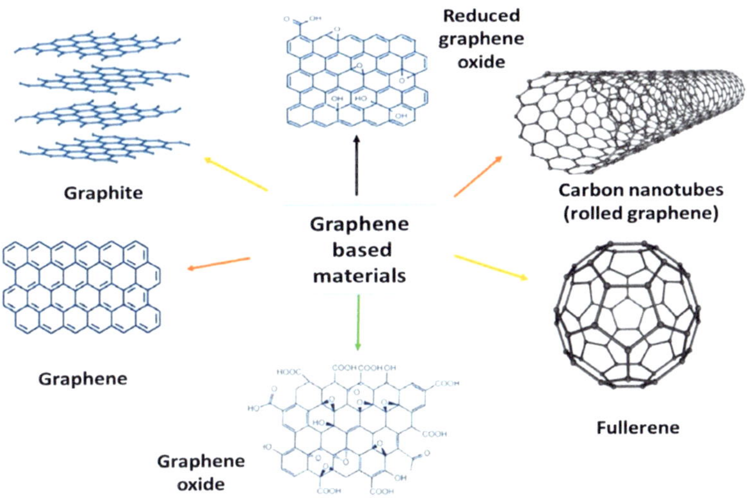

Figure 3.9. Systematic diagram of graphene and graphene-based derivatives.

Source: https://www.researchgate.net/figure/Systematic-diagram-of-graphene-and-graphene-based-derivatives_fig4_357800630.

Pioneering research, which highlights the notable work of graphene in electrochemical CO2 reduction, once again underscores its role in photocatalysis. Studies have shown that two-dimensional graphene and transition metal chalcogenides (TMCs) are promising catalysts for converting CO_2 into valuable chemicals. The conversion is particularly useful in fighting climate change and seeking sustainable sources of energy, as it minimizes the emission of greenhouse gases with the bonus of obtaining valuable products. With the enhanced electrocatalytic features of graphene, researchers aim to achieve selective reduction reactions that can be scaled to industrial dimensions. Promising results from recent investigations form the basis for more sustainable technology, categorizing graphene as not just a leader among photocatalytic materials but also one of the leading candidates in addressing global energy challenges (Jampílek et al., 2019; Erkey et al., 2020).

3.4.2. Carbon Nanotubes (CNTs)

The incorporation of carbon nanotubes (CNTs) into photocatalytic systems has proven to be a promising method for enhancing the efficiency of a wide range of semiconductor materials. CNTs possess structural characteristics of high aspect ratio and higher electrical conductivity, which can significantly benefit charge separation and transport in photocatalytic reactions. These attributes enable greater electron mobility, as well as the inhibition of electron-hole pair recombination, both of which are critical parameters for achieving efficient photocatalytic activity.

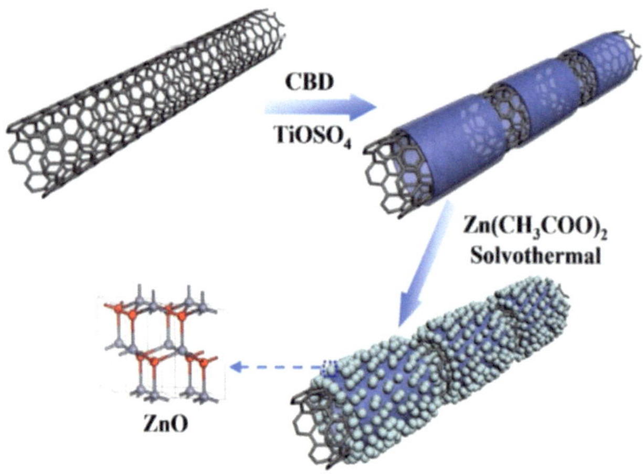

Figure 3.10. Schematic illustration of the ternary carbon nanotubes (CNTs)/TiO_2/ZnO composites.

Source: https://www.mdpi.com/2073-4344/8/4/151.

Furthermore, the compatibility of CNTs with other photocatalytic materials, such as titanium dioxide (TiO_2), ensures that synergistic effects can be achieved, enhancing the material's responsiveness to light and overall efficiency in photocatalysis. Accordingly, the addition of CNTs not only improves the photocatalytic reaction but also improves the stability and sustainability of the photocatalytic materials under practical conditions. Up-to-date research has demonstrated that CNTs can be effectively blended with traditional photocatalysts to enhance their activity during irradiation with visible light. The challenge most prevalent in photocatalysis is efficiently utilizing the solar spectrum for degradation processes. Although TiO_2 has been the subject of extensive research due to its photocatalytic potential, its own band gap restricts its activity to predominantly UV light. The incorporation of CNTs has been reported as very promising in extending light absorption into the visible spectrum, thus widening the operating range of photocatalysts. For instance, CNTs can be used as a support to enhance the dispersion of photocatalytic nanoparticles, which have the ability to facilitate electron transfer and thereby enhance photocatalytic efficiency. This integration highlights how carbon nanotubes could revolutionize photocatalytic systems by overcoming bottlenecks and expanding their use in environmental remediation (Figure 3.10).

Besides, the application of CNTs in photocatalysis is multifaceted, beyond the enhancement of light absorption and charge dynamics; it is also an essential element of pollutant degradation. Synthetic dyes are widely applied in various industries and cause severe environmental risk upon release into aquatic systems. Their photocatalytic degradation is an area of current research, with studies showing that CNT-assisted photocatalysts yield improved degradation rates compared to their traditional counterparts. By facilitating increased interaction between the dye molecules and the photocatalytic surface through high surface area and doping effects, CNTs facilitate more efficient pollutant removal. This functionality is employed to highlight the growing trend of merging carbon-based nanomaterials with semiconductors, as observed in recent advances. The combination of TiO_2 with graphene and CNTs has facilitated significant progress in synthetic dye photocatalytic degradation (D'Amato et al., 2017). Consequently, carbon nanotube functionalization in photocatalytic processes represents a significant leap forward in seeking sustainable environmental solutions.

3.4.3. Fullerenes

Fullerenes, being a category of spherical, elliptical, or tubular-shaped carbon allotropes, are newly discovered materials with great importance in photocatalysis. Their unique molecular structure, enhancing their electron-accepting ability, also facilitates charge separation, making them excellent choices for processes requiring photocatalysis. Incorporation of fullerenes into photocatalytic materials can significantly enhance the efficiency of various reactions, including those for environmental remediation. For instance, synergistic use of fullerenes with semiconductor material can lead to higher photocatalytic activity under both ultraviolet and visible light, opening up broader prospects in solar energy conversion devices. Synergy holds new promise for designing solutions for environmental problems, such as the degradation of pollutants and water purification. In investigating the potential of fullerenes as photocatalysts, researchers have highlighted their effectiveness as strong electron acceptors that facilitate the charge transfer processes required for photocatalytic activity. By facilitating the separation of photoexcited electron-hole pairs and preventing recombination, which typically impairs photocatalytic activity, fullerenes retain their photocatalytic performance. Fullerenes' versatility enables them to be applied beyond traditional photocatalysts; they can also be used as additives or modifiers to optimize the performance of existing semiconductor materials. For example, when combined with nanoporous carbon architectures, fullerenes demonstrate the ability to generate reactive oxygen species, which catalyze the photocatalytic degradation of contaminants in water solutions and enhance environmental sustainability (Fernández et al., 2016) (Figure 3.11).

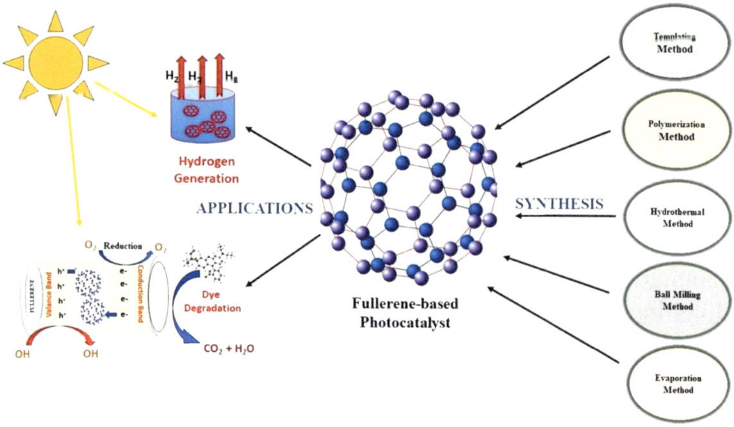

Figure 3.11. Fullerene-based materials for photocatalysis.

Source: https://external-content.duckduckgo.com/iu/?u= https% 3A%2F %2 Fchemistry-europe.onlinelibrary.wiley.com%2Fcms%2Fasset%2F6572c413-22ec-4262-8982-0c0bd a61de3 f%2Fslct202401561-toc-0001-m.jpg&f=1&nofb=1&ipt =0b7a07ef4ea2620bb64b6eb62df894bbad41616402ee6abb0b78442e12a5bdc7.

The two functionalities of boosting photocatalytic performance and facilitating the degradation of pollutants have placed fullerenes in the limelight for the creation of advanced photocatalytic systems. Besides that, recent research is unraveling new methods of integrating fullerenes with other types of carbon such as graphene to boost their applications in photocatalysis further. Fullerene association with graphene enhances the surface area for the adsorption of organic pollutants, as well as advancing charge carrier dynamics due to the high electrical conductivity of graphene. The hybrid exhibits enhanced photocatalytic activity, particularly in the degradation of synthetic dyes that pollute industrial wastewater (D'Amato et al., 2017). As work continues to find effective photocatalytic materials, the intrinsic properties of fullerenes, along with their ability to form composites with other nanomaterials, render them suitable candidates in the quest to find sustainable solutions to environmental issues. Such advances in photocatalytic technologies represent a highly significant trend toward utilizing carbonaceous materials for effective energy conversion and environmental management.

3.5. APPLICATIONS OF PHOTOCATALYSTS

The applications of photocatalysts extend far beyond normal chemical reactions, particularly in energy conversion and environmental remediation. Due to the ability to harness the sun as a source of energy to drive chemical reactions, photocatalysts can be efficiently employed to oxidize pollutants in water and air streams. For instance, titanium dioxide (TiO_2), the most extensively studied photocatalyst, has gained momentum because it can break down destructive organic contaminants to fairly less toxic compounds using visible light. New nano-photocatalytic materials have recently become more efficient, as shown by experiments, and are therefore suitable for environmental applications on an industrial scale despite the lack of efficiency in commercial photocatalysts in actual operations (Jang et al., 2015). This ongoing research explores the future potential of photocatalysts to provide sustainable solutions for global environmental issues. In addition to environmental cleaning, photocatalysts play a crucial role in converting solar energy, particularly in generating renewable hydrogen-based approaches. By decomposing water into hydrogen through sunlight exposure, photocatalysts can convert solar energy into hydrogen, a clean fuel with potential future applications in energy systems. The advances in visible-light-active photocatalysts, especially TiO_2-based composites, are noteworthy. The materials have been tailored to increase their efficiency in absorbing solar radiation, thereby boosting their activity in photoconversion reactions. The studies indicate that incorporating new composite materials enhances the electron-hole separation dynamics, a pivotal process in maximizing photocatalyst activity and catalyzing the decomposition of organic and inorganic pollutants (Afzal et al., 2014). These developments underscore the revolutionary potential of photocatalysts in addressing world energy and environmental issues. Lastly, potential applications of photocatalysts also encompass purifying indoor air and antimicrobial therapies. Their effectiveness in degrading indoor air pollutants, i.e., volatile organic compounds (VOCs) and bacteria, underscores their diversity and usefulness in urban environments. Photocatalysts can be introduced into construction materials or coatings, such that self-cleaning surfaces will be maintained through the utilization of sunlight for air purification. They not only ensure cleanliness and hygiene but also help in energy-efficient building construction. Through continuous research and development, further advancements in motile photocatalysts and novel synthesis techniques could enhance their functional properties, providing new avenues for applications in everyday life (Jang et al., 2015). In total, the broad variety of applications confirms the necessity of developing effective photocatalyst systems to meet the demands of the modern environment and energy.

3.5.1. Water Purification

In the pursuit of efficient water purification techniques, photocatalysts have been proposed as a potential solution for degrading harmful contaminants. Photocatalysis utilizes light to activate a catalyst, typically titanium dioxide (TiO_2), which catalyzes chemical reactions that result in the degradation of pollutants.

Figure 3.12. *Photocatalytic water treatment.*

Source: https://external-content.duckduckgo.com/iu/?u=https%3A%2F%2Fwww.netsolwater.com%2Fnetsol-water%2Fassets%2Fimg%2Fproduct-images%2FWhat_is_the_photocatalytic_treatment.jpg&f=1&nofb=1&ipt=c74c1b1fbc8830c366cc4386d1f77139fbcfca2cb3ed83e792c16446ffcbcad9.

The reaction is beneficial in degrading organic contaminants, including dyes and pesticides, which are highly recalcitrant to conventional treatment processes. Studies have shown that TiO2 thin films can be immobilized onto other substrates, resulting in enhanced photocatalytic activity. Using techniques such as material printing and chemical vapor deposition, researchers have managed to optimize the morphology and crystallinity of TiO_2 layers, thus accelerating the degradation of impurities in water. The advancement in fabrication techniques not only enhances efficiency but also offers an eco-friendly solution to water quality issues, significantly improving the viability of photocatalytic water purification systems (Králová et al., 2012). The efficiency of the photopurification of water by photocatalysis is more than mere structural modification of the photocatalysts. A thorough understanding of the environmental factors that influence the degradation process is crucial. These include light intensity, temperature, and the concentration of contaminants, which all make significant contributions to the photocatalytic efficiency (Figure 3.12).

For instance, investigations into the impact of varying light intensity on the performance of TiO_2 have helped provide additional understanding of the conditions for optimal work during water treatment. Additionally, new investigations have adopted new approaches that involve continuous online monitoring of pollution levels during photocatalytic reactions. Real-time assessment of decolorization processes for various pollutants is enabled by this approach, demonstrating the technology's applicability and suitability under dynamic conditions. As the research work continues, it is evident that the use of enhanced monitoring techniques would further improve the scalability and applicability of photocatalytic systems in real-world water treatment conditions (Mart et al., 2019). Emerging challenges to wastewater management require robust solutions that can address the complexity of diverse water pollutants. Photocatalytic treatment of water offers a multidisciplinary answer, one that is capable of degrading not only common contaminants but also more recalcitrant compounds, such as pharmaceuticals and industrial dyes. Recent studies indicate that the photodegradation of various azo dyes, for instance, can be effectively catalyzed through suspensions of photocatalysts, an application that attests to their capability to treat various types of wastewaters. By continuous optimization of the photocatalyst materials and processes, scientists have reported promising trends in the photocatalytic process, paving the way for commercial applications. Further research and development of double-functional photocatalyst systems, which provide both purification and disinfection capacities, could continue to revolutionize water treatment processes, offering a green, efficient, and sustainable approach to delivering clean and safe drinking water to societies worldwide (Králová et al., 2012).

3.5.2. Air Purification

Air purification has emerged as a key focus of the past decade, driven by the growing demand for improved air quality and its impact on human welfare and the environment. Among many methods employed to improve indoor air quality, photocatalytic technologies are an effective solution. Photocatalysts, particularly titanium dioxide (TiO_2), are of particular interest because they can decompose volatile organic compounds (VOCs) and other harmful substances using UV light. The photocatalytic process converts toxicants into less toxic products, thereby reducing airborne pollutants. The process not only cleans the air but also contributes to the overall enhancement of indoor environments, making them safer and healthier for users. The effectiveness of photocatalytic air cleaners depends on the nature of the selected photocatalyst and the method of its deposition. Sol-gel processing, hydrothermal synthesis, and chemical vapor deposition are the most prevalent methods for depositing TiO_2 thin films with improved morphology and surface area. Recent studies indicate that differences in the methods can significantly affect photocatalytic activity and performance. For instance, immobilization of titania onto supports like carbon or glass can enhance the reuse and stability of photocatalysts without affecting their activity in air purification processes (Králová et al., 2012). The generalized approach facilitates the design and deployment of more effective photocatalytic systems for various air quality control scenarios. Moreover, technological improvements in monitoring devices have made it possible to assess photocatalytic air purification processes more effectively. Traditional sampling methods disrupt the photocatalytic process, leading to inaccuracies in measuring the degradation of pollutants. Advances such as online monitoring of dye decolorization using light scattering technologies enable real-time analysis of photocatalytic activity without compromising the system's ability (Mart et al., 2019). Real-time feedback is crucial for optimizing photocatalytic parameters and scaling up systems. Through the improvement of these methodologies and examining diverse photocatalytic materials, researchers seek to improve the performance of air purification technologies further to address increasing demands for sustainable air quality solutions (Figure 3.13).

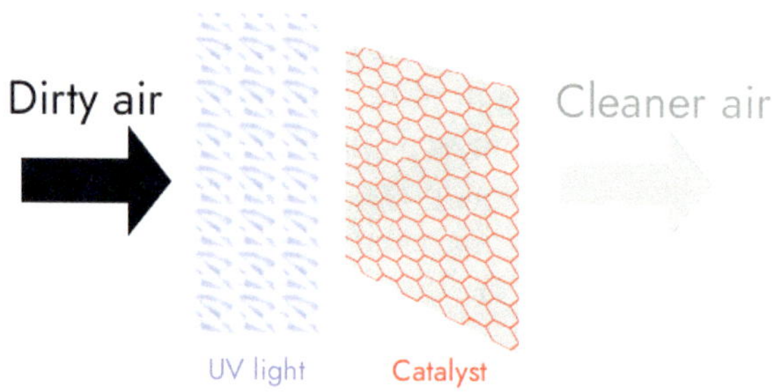

Figure 3.13. Photocatalytic air purifier .

Source: https://www.explainthatstuff.com/how-photocatalytic-air-purifiers-work.html.

3.5.3. Energy Production Through Hydrogen Generation

The potential to produce energy through hydrogen production is being increasingly recognized as a central component in global efforts to establish sustainable energy systems. Hydrogen, as a green fuel, is amenable to production through a variety of methods, and photocatalysis has emerged as one of the most promising methods. Through the application of photocatalysts that harvest sunlight, researchers are striving for the efficient decomposition of water molecules into hydrogen and oxygen. This becomes possible through the revolutionary photocatalyst design that enhances light harvesting and charge carrier mobility. Notably significant is the integration of semiconductor and metal components to enhance photocatalytic action, thereby demonstrating the potential of the technology in meeting future energy demands. The most important development in photocatalytic hydrogen evolution involves immobilizing metal oxides on semiconductor surfaces to enhance efficiency. For instance, RuO_2 deposition on one-dimensional TiO_2 nanorods has been proven effective. The composite material exhibits enhanced photocatalytic activity due to its remarkable structural properties, which facilitate visible light trapping and charge separation. The most important feature of the process is the formation of a heterointerface, which traps visible light very effectively, along with initiating photogenerated charge carrier transfer. Consequently, the method has potential for bulk production of hydrogen, addressing energy sustainability and environmental concerns, while also contributing to broader research into renewable energy (Bai et al., 2016). Additionally, research into various photocatalytic processes expands the scope of hydrogen production technology. Current research highlights the use of visible light and transition-metal-mediated photocatalysis in the production of useful energy outputs through perfluoroalkylation reactions. These answers examine various organic substrates and highlight the versatility of photocatalysts in a range of chemical reactions. By integrating diverse photochemical methods, researchers can unlock new avenues for producing hydrogen. Opportunities for improving the capacity of fluorinated reagents to influence reaction efficiency represent a thought-provoking research area. Together, these advancements underscore the importance of photocatalysis in developing effective hydrogen generation approaches, positioning this technology at the forefront of clean energy technologies (Vallejo et al., 2018).

SUMMARY

In summarizing this overview of various photocatalysts, it can be seen that material synthesis and characterization technology have enhanced the activity of the materials and usability to a great extent. Photocatalysts are a crucial tool for addressing environmental problems, particularly in terms of pollutant degradation and energy conversion. Through methods such as sol-gel and hydrothermal synthesis, scientists have synthesized new photocatalytic materials with high visible light activity. For instance, codoped titanium dioxide synthesized with platinum and chromium was reported to exhibit superior photodegradation of contaminants, such as methylene blue, demonstrating the effectiveness of composition tailoring of the photocatalyst to ensure maximum performance (Choi et al., 2010). Furthermore, immobilization of titanium dioxide on substrates through chemical vapor deposition is a promising method of improving photocatalytic ability while delivering material stability and recyclability. The comparative examination of different preparation techniques reveals differences in photocatalytic activity, highlighting the importance of preparation techniques in determining the efficiency of these materials. These advances not only optimize the photocatalytic processes but also promote further practical utilization in real environmental rehabilitation and renewable energy alternatives (Králová et al., 2012). Ultimately, the continued research and development of photocatalysts is necessary to meet the mounting demands for cleaner and more efficient technology. The union of novel materials and , as well as tap innovative approaches. With advances in research, the synthesis of scholarly research and practical applications will be crucial in unlocking the full potential of photocatalysts to address global issues. Thus, ongoing investigation of types of photocatalysts is a vital area of research that promises transformations that can revolutionize energy and environmental realities in the near future.

The quest for photocatalysts has led to a vast array of materials with unique properties and efficiencies in various applications, primarily in environmental remediation and energy conversion. Some of the prevailing photocatalysts are metal complexes such as tris(bipyridine)ruthenium(II), which has shown great promise in addressing immediate antibacterial challenges. Recent studies indicate that, when combined with plasmonic nanoantennas, the photocatalytic activity of $[Ru(bpy)_3]^{2+}$ can be significantly enhanced, resulting in a seven-log decrease in bacterial viability. This is facilitated by the new hybrid photoreactor structure, which enables the support of both the photocatalyst and the nanoparticles, thereby uncovering innovation in the area of photocatalysis for effective bactericide inactivation. Together, these sections reveal how advanced materials can be utilized to develop new antimicrobial approaches, underscoring the importance of multidisciplinary research in photocatalytic science as a key step in leveraging contemporary photocatalyst systems. The application of photochemical processes extends beyond antibacterial uses, encompassing a significant portion of the energy conversion field. Photocatalysts play a crucial role in catalyzing electrochemical reductions, particularly CO_2, to value-added chemicals, thus providing clean energy solutions. Two-dimensional (2D) materials, including graphene and transition metal chalcogenides (TMCs), have gained popularity, given that these materials display outstanding activity and selectivity in CO_2 reduction reactions. Researchers have been actively optimizing the efficiencies of these electrocatalysts, leading to significant strides in photocatalytic technology. The convergence of electrocatalysis and photocatalysis highlights the heterogeneity of catalyst structures and their diverse applications, indicating a comprehensive understanding of how different materials and morphologies can impact the forthcoming wave of developments in chemical reactions. Lastly, developments in photocatalyst research highlight an important trend towards continuous optimization of catalytic materials to address modern-day environmental and energy challenges. Sophisticated photocatalytic system designs, such as the integration of nanoparticles with traditional catalysts, enhance performance characteristics like stability and reactivity. More than merely theoretical, such systems have been applied practically, exhibiting greater catalytic activity and positioning them as central figures in both laboratory and commercial applications. Mechanistic studies of photocatalysts, such as the function of light in initiating these materials and inducing reactions, remain at the forefront. Through interdisciplinary approaches, the evolving photocatalyst ground creates new solutions to tackle the core challenges of sustainability in our society, marking a significant turning point in seeing and harnessing photocatalysts in real-world applications.

REFERENCES

1. Abdullah, N., Mohd Padzil, S. A., Mohd Salleh, R., Pillay, V., et al. (2018). Model selection and model averaging on mortality of upper gastrointestinal bleed patients. https://core.ac.uk/download/18264202.pdf (accessed on 28th September 2025).
2. Banerjee, S., Pillai, S. C., Falaras, P., O'Shea, K. E., Byrne, J. A., & Dionysiou, D. D. (2014). New Insights into the Mechanism of Visible Light Photocatalysis. *The Journal of Physical Chemistry Letters*, *5*(15), 2543–2554. https://doi.org/10.1021/jz501030xAn, X., Liu, P., Naowarojna, N., Reinhard, B., et al. (2020). Hybrid plasmonic photoreactors as visible light-mediated bactericides. https://open.bu.edu/bitstream/2144/40614/3/manuscript_1021_XA.pdf (accessed on 28th September 2025).
3. Antonietti, M., Kurpil, B., Mazzanti, S., Pieber, B., et al. (2020). Dichloromethylation of enones by carbon nitride photocatalysis.
4. Bacchini, F., Greppi, G., Mura, S., Roggero, M., et al. (2013). Advances of nanotechnology in agro-environmental studies. https://core.ac.uk/download/16749559.pdf (accessed on 28th September 2025).
5. Bahnemann, D. W., Bloh, J. Z., Kasper, C., Wagner, M., et al. (2011). Toxicological issues of nanoparticles employed in photocatalysis. https://core.ac.uk/download/52919976.pdf (accessed on 28th September 2025).
6. Bai, J., Fernandez Sanz, J., Fujita, E., Graciani, J., et al. (2016). Visible light-driven H_2 production over highly dispersed Ruthenia on rutile TiO_2 nanorods. https://core.ac.uk/download/81582504.pdf (accessed on 28th September 2025).
7. Barata Vallejo, S., Cooke, M. V., & Postigo, A. (2018). Radical fluor alkylation reactions. https://core.ac.uk/download/287880745.pdf (accessed on 28th September 2025).
8. Berenguer-Murcia, Á., Cazorla-Amorós, D., Fernández-Catalá, J., García-Santos, A., et al. (2024). Exploring Pt-impregnated CdS/TiO_2 heterostructures for CO_2 photoreduction. https://core.ac.uk/download/654361766.pdf (accessed on 28th September 2025).
9. Choi, J., Hoffmann, M. R., & Park, H. (2010). Combinatorial doping of TiO_2 with platinum (Pt), chromium (Cr), vanadium (V), and nickel (Ni) to achieve enhanced photocatalytic activity with visible light irradiation. https://core.ac.uk/download/4883950.pdf (accessed on 28th September 2025).
10. D'Amato, C. A., Giovannetti, R., Rommozzi, E., Zannotti, M., et al. (2017). Recent advances in graphene based TiO_2 nanocomposites (GTiO_2Ns) for

photocatalytic degradation of synthetic dyes. https://core.ac.uk/download/pdf/154843184.pdf (accessed on 28th September 2025).

11. Darr, J. A., Hyett, G., Kafizas, A., Kundu, S., et al. (2011). An investigation into the effect of thickness of titanium dioxide and gold-silver nanoparticle titanium dioxide composite thin-films on photocatalytic activity and photo-induced oxygen production in a sacrificial system. https://core.ac.uk/download/9040748.pdf (accessed on 28th September 2025).

12. De Vos, A., Hendrickx, K., Joos, J., Lejaeghere, K., et al. (2018). Exploring lanthanide doping in UiO-66: A combined experimental and computational study of the electronic structure. https://core.ac.uk/download/158346809.pdf (accessed on 28th September 2025).

13. De Vos, A., Hendrickx, K., Lejaeghere, K., Van Der Voort, P., et al. (2017). Missing linkers: An alternative pathway to UiO-66 electronic structure engineering. https://core.ac.uk/download/84045313.pdf (accessed on 28th September 2025).

14. Erkey, C., Hiragond, C. B., In, S. I., Kim, D., et al. (2020). Electrochemical CO_2 reduction to CO catalyzed by 2D nanostructures. https://core.ac.uk/download/287633748.pdf (accessed on 28th September 2025).

15. Faria, J., Figueiredo, J., Gomes, H., Morales-Torres, S., et al. (2014). Carbon based materials for photocatalytic solar applications in water treatment. https://core.ac.uk/download/153412669.pdf (accessed on 28th September 2025).

16. Fernández Velasco, L., Gomis Berenguer, A., Ovín Ania, M. C., Velo Gala, I., et al. (2016). Photochemistry of nanoporous carbons: Perspectives in energy conversion and environmental remediation. https://core.ac.uk/download/80864728.pdf (accessed on 28th September 2025).

17. Goswami, D. Y., Heep, A. F., Jaworske, D. A., Jones, M., et al. (2025). Enhanced TiO_2 photocatalytic processing of organic wastes for green space exploration. https://core.ac.uk/download/pdf/10573637.pdf (accessed on 28th September 2025).

18. Herz, R. K. (2025). Modeling of carbon monoxide oxidation kinetics over NASA carbon dioxide laser catalysts. https://core.ac.uk/download/pdf/42822331.pdf (accessed on 28th September 2025).

19. Jampílek, J., & Plachá, D. (2019). Graphenic materials for biomedical applications. https://core.ac.uk/download/287648849.pdf (accessed on 28th September 2025).

20. Jang, S., Kim, D. P., Parmar, J., Sanchez, C., et al. (2015). Nano-photocatalysts in microfluidics, energy conversion and environmental

applications. https://core.ac.uk/download/81578008.pdf (accessed on 28th September 2025).

21. Klochikhin, A. A., Lemke, A. A., Rempel, A. A., Rempel, A. A., Rempel, A. A., Rempel, A. A., Rempel, A. A., et al. (2013). Hybrid nanoparticles based on sulfides, oxides, and carbides. http://elar.urfu.ru/bitstream/10995/27146/1/scopus-2013-0319.pdf (accessed on 28th September 2025).
22. Kočí, K., Pavlovský, J., Prostějovský, T., Tokarský, J., et al. (2019). α-Fe_2O_3 nanoparticles/vermiculite clay material: Structural, optical and photocatalytic properties. https://core.ac.uk/download/237018357.pdf (accessed on 28th September 2025).
23. Králová, M. (2012). Self-cleaning properties of thin printed layers of titanium dioxide. https://core.ac.uk/download/30310816.pdf (accessed on 28th September 2025).
24. Romeral Martínez, J. L., Tapia Tlatelpa, T., & Trull Silvestre, J. F. (2019). In situ decolorization monitoring of textile dyes for an optimized UV-LED/TiO_2 reactor. https://core.ac.uk/download/294829789.pdf (accessed on 28th September 2025).

CHAPTER 4

MECHANISMS OF PHOTOCATALYTIC REACTIONS

Contents

INTRODUCTION

The development of photocatalytic reactions represents a significant advancement in the pursuit of sustainable environmental solutions. Because the reactions utilize the light energy to improve chemical processes, they have critical applications in energy conversion, pollutant degradation, and water purification. Photocatalysis essentially relies on the interaction of a semiconductor and light to generate reactive species that catalyze various chemical transformations.

Titanium dioxide (TiO_2) has become a premier photocatalyst due to its excellent electronic properties, chemical inertness, and high stability. Knowledge of the mechanisms behind these processes will be valuable in optimizing their efficiency and expanding their applications. Advanced techniques have now enabled researchers to investigate the multifaceted nature of TiO_2's properties, such as its crystallographic phases and surface interactions, to identify pathways that can enhance photocatalytic performance. The photocatalytic activity of TiO_2 relies heavily on its structure and electronic properties, i.e., its polymorphs: anatase and rutile. Recent research using ab initio density functional theory (DFT) has explained the relative stability of these polymorphs and established that the rutile polymorph is marginally more stable than the anatase polymorph. However, the formation energies of various dopants indicate that the alterations could have a profound effect on the TiO_2 catalytic performance. For instance, anatase is stabilized by cationic dopants, thus preventing the anatase-to-rutile phase transition that can disrupt photocatalytic activity. These findings emphasize the importance of stability in materials for the design of photocatalytic systems, aiming to maximize their performance under real-world applications (Assadi et al., 2012). Furthermore, the fabrication techniques applied to produce TiO_2 thin films are crucial in determining the photocatalytic efficiency. Sol-gel synthesis and chemical vapor deposition (CVD) are widely employed methods, offering potential control over the catalyst's thickness, morphology, and surface area. For instance, studies have found that the introduction of additives during the sol-gel process can enhance the photocatalytic activity of TiO2, as reflected in the reduced cracking and improved chemical stability of the synthesized layers (Králová et al., 2012). Furthermore, adjustment of the conditions of preparation—i.e., synthesis time and temperature—may lead to considerable differences in the photocatalytic activity of films. Overall, an integrated approach that combines material optimization with an understanding of the underlying processes is required to advance photocatalytic technology for environmental remediation and sustainable energy applications.

Definition of Photocatalysis

Photocatalysis is a process by which light energy is employed to accelerate a chemical reaction, typically through the operation of a photocatalyst. This agent catalyzes the conversion of reactants without being permanently altered itself. Photocatalysis originated from research into semiconductor materials that can be photoexcited into electron-hole pairs. These pairs are responsible for initiating redox reactions that may result in the degradation of pollutants, splitting water, or converting carbon dioxide into valuable energy sources. With environmental issues on the rise, the interest in photocatalytic processes has gained momentum, considering their capability to provide green solutions in combating pollution and renewable energy generation. Understanding the photocatalytic mechanisms of operation is critical to optimizing such reactions and enhancing their efficiencies within applications. Of defining importance in photocatalysis is the interaction between light and semiconductor material, which governs the efficiency of photocatalytic reactions. After absorbing photons by a semiconductor photocatalyst, electrons are excited from the valence band to the conduction band, forming charge carriers in terms of free electrons and positively charged holes. The separation and migration of these charge carriers are of extreme importance in driving the target redox reactions at the surface of the photocatalyst. New developments in the area, including controlled periodic illumination, have revealed ways to improve efficiency by orders of magnitude, as detailed in current studies.

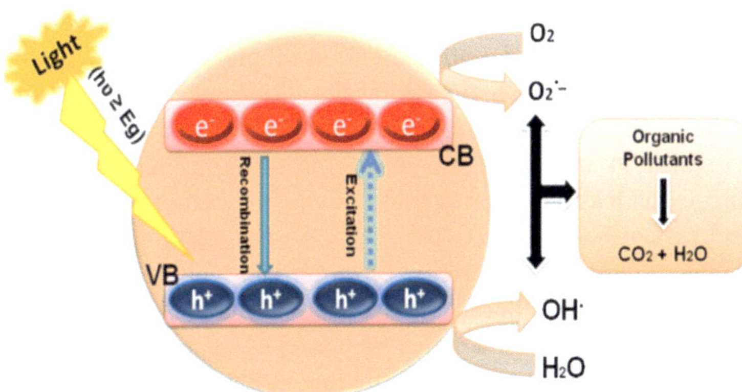

Figure 4.1. The basic mechanism of photocatalysis.

Source: https://www.researchgate.net/figure/The-Basic-mechanism-of-Photocatalysis-Reproduced-with-permission-17_fig1_336785403.

The controlled illumination has the effect of maximizing the activation of the photocatalystto enhance the overall reaction rate and product selectivity of , thereby enhancing the overall reaction rate and product selectivity of the formed products. This indicates an advanced interplay between illumination conditions and photocatalytic activity (Lawton et al., 2015). Another aspect of photocatalysis is the reduction of carbon dioxide using sunlight's energy through photoelectrocatalytic processes. It not only seeks to reverse the ill effects of the over release of carbon dioxide but also seeks to utilize the release for the formation of beneficial small molecules, basically creating a circular carbon economy. Density functional theory (DFT)- based studies have revealed the mechanistic trends in photocatalytic CO2 reduction, examining the critical role of changing interfaces between the electrolyte/catalyst and the supports/ catalyst. Understanding these complex interactions aids in the design of next-generation photocatalysts to improve performance in CO_2 reduction, facilitating sustainable energy solutions and addressing global warming challenges (Alexandrova et al., 2020). Thus, broadening our knowledge of photocatalytic processes makes way for tremendous innovation in environmental preservation and energy sustainability (Figure 4.1).

Importance of Photocatalytic Reactions in Environmental Applications

The significance of photocatalytic reactions in environmental contexts is revealed by their ability to address severe pollution issues, particularly in wastewater treatment. Photocatalytic degradation is among the most notable methods for treating industrial effluents from oil and textile processes, which generate significant levels of toxic wastewater. The challenge lies in eliminating toxic organic substances, including carcinogenic and mutagenic ones, which are also present in suspended particle form within water systems. Recent advancements, such as engineering multifunctional photocatalysts using titania (TiO2) and carbon nanotubes (CNTs), demonstrate a promising methodology for enhancing photocatalytic activity. By improving photocatalytic properties, reactor geometries, and the efficient removal of catalysts from treated water, researchers aim to enable the integration of photocatalytic treatments into high-level wastewater treatments, thereby revolutionizing pollution abatement technologies (Mason et al., 2018). Photocatalytic reaction processes, particularly with TiO_2, explain the efficacy of materials in catalyzing environmental purification processes. DFT-based studies have shown that photocatalytic efficiency is greatly determined by the stability of TiO_2 polymorphs, i.e., anatase and rutile. Anatase, which is photocatalytically more active than rutile, is significantly affected by specific dopants that stabilize its structure and enhance reactivity. Therefore, for example, fluorine doping and silicon doping were found to have a positive impact on the topography of anatase energy, preventing its transition into rutile. These results are pivotal in the design of photocatalytic materials to optimize their performance in various environmental remediation applications, demonstrating that basic research can inform applied solutions towards pollution control (Assadi et al., 2012). Lastly, photocatalytic reactions represent a crucial aspect of sustainable environmental management approaches, particularly in preventing water pollution. By harnessing light energy and activating photocatalysts, photocatalytic reactions facilitate the decomposition of toxic, long-term organic pollutants, resulting in cleaner water sources. Collaboration among materials science, chemistry, and environmental engineering is crucial to driving the development of photocatalysts and realizing their full potential. Barring the effectiveness of treatment, reusability and stability issues for the catalyst are crucial for practical use, as has been the case with the latest design concepts, such as Pickering emulsions. Not only do such advances contribute to theoretical understanding, but they also lead to material applications that support ecological renewal, marking a revolutionary stride towards less environmentally damaging industrial processes and more effective waste management systems (Mason et al., 2018).

4.1. FUNDAMENTAL PRINCIPLES OF PHOTOCATALYSIS

The foundational mechanisms behind photocatalysis are centered on the interaction between light and semiconductors, which induces chemical reactions. Photocatalytic reactions typically require the absorption of photons, which excites electrons from the valence to the conduction band, ultimately leading to the formation of electron-hole pairs. These can then participate in redox reactions in sound photocatalytic systems, leading to contaminant degradation or the production of valuable products.

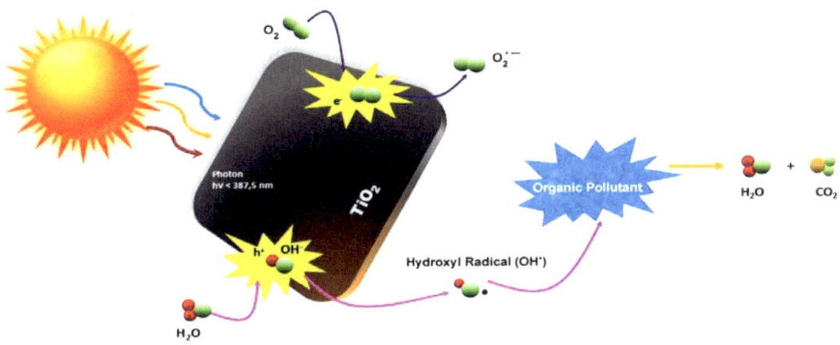

Figure 4.2. The basic principle of photocatalysis.

Source: https://external-content.duckduckgo.com/iu/?u=https%3A%2F%2Fwww.researchgate.net%2Fpublication%2F361455184%2Ffigure%2Ffig2%2FAS%3-A11431281104799737%401670262094471%2FThe-basic-principle-of-photocatalysis-illustrated-with-a-simplified-reaction-pathway.png&f=1&nofb=1&ipt=a91a546bfb0d949292e3fdd94cbcab5a003df651c406c961cea3d9f8bfaecda9.

The performance of such systems largely depends on the wavelength of light used, structural features of the photocatalyst, and reaction conditions. Advances in the area have, of late, taken into account means such as controlled periodic illumination to optimize the excitation conditions, which have been shown to be effective in enhancing the overall photocatalytic activity by prolonging the active period of such generated charge carriers (Lawton et al., 2015) (Figure 4.2).The choice of semiconductor material is primarily responsible for determining photocatalytic activity. Of all the used materials, titanium dioxide (TiO_2) is one of the most studied due to its best band gap energy and photostability. The two primary polymorphs of TiO_2 are anatase and rutile, and both possess appreciable differences in their photocatalytic activity. Ab initio DFT calculations indicate that the anatase phase is generally more reactive than the rutile phase and can generate a higher amount of reactive oxygen species, which are necessary for efficient photocatalysis. Moreover, dopants are known to have a significant impact on the stability and activity of phases. The contrasting stability of anatase in the presence of specific dopants, such as fluorine, silicon, iron, and aluminum, highlights the requirement for modifying the semiconductor's properties to enhance photocatalysis (Assadi et al., 2012). Surface properties of photocatalysts also influence the nature of specific reactions. Increased surface area and porosity can create more active sites for chemical interactions, therefore enhancing the photocatalytic efficiency. Surface modifications, such as the incorporation of co-catalysts or heterostructure formation, have been reported to improve charge separation and reduce recombination of electron-hole pairs, a bottleneck problem in most photocatalytic systems. Surface modifications can lead to improved rates of reaction and broaden the applicability of the photocatalyst to a wide range of environmental and energy applications. As research progresses, surface engineering and synthesis of photocatalysts will be essential areas of study in developing more efficient and sustainable photocatalytic processes.

4.1.1. Role of Light in Initiating Photocatalytic Reactions

The role of light in activating photocatalytic reactions is at the heart of understanding the mechanisms of photocatalytic processes. When light is absorbed by a photocatalyst, which is typically a semiconductor material, it excites valence-band electrons into the conduction band, creating electron-hole pairs. Excitation is critical because it not only explains the formation of reactive species but also determines what happens next in photocatalytic processes regarding reactivity. Its effectiveness is influenced by aspects such as the light wavelength, the band gap of the semiconductor material, and the presence of co-catalysts. Enhanced comprehension of periodic illumination, based on literature, has the promise to improve photocatalytic effectiveness through a more advanced wavelength-intensity relationship to accelerate more efficient application in environmental cleaning and energy conversion technology

(Lawton et al., 2015). Moreover, recent studies have also indicated the need to use specific wavelengths of light to facilitate maximum photocatalytic action, demonstrating that the energy of the incoming photons must be sufficient to traverse the band gap of the photocatalyst. This theory accounts for the reason why substances like titanium dioxide (TiO_2) are prevalent; their band gap is very well-suited to the energy of ultraviolet light, so that photocatalytic action will be efficient. It is also beneficial to explore the possibilities of integrating light with other technologies, such as nanoantennas, that enhance the focusing and efficiency of light in photocatalytic reactions. An example is the integration of photocatalysts with plasmonic nanoantennas, which not only enhances light absorption but also facilitates charge separation, a critical process in sustaining reactive species for extended durations (An et al., 2020). As always, the selection of materials and lighting strategy are essential to optimizing photocatalytic systems.

The response of photocatalysts to light is also at the center of more recent applications in fields such as antibacterial strategies, where light-activated materials have excellent potential. Recent findings demonstrate that certain photocatalysts, specifically tris(bipyridine)ruthenium(II), exhibit highly effective bactericidal activity that is significantly enhanced when combined with plasmonic nanoparticles. This mixture uses visible light to initiate reactions that not only accelerate bacterial inactivation but also lead to the release of antimicrobial agents into the environment. These observations support the complex yet powerful role that light plays in initiating and sustaining photocatalytic reactions, demonstrating its effectiveness in real-world applications. The encounter between light and new photocatalytic architectures thus creates new avenues to the development of next-generation materials with the potential to solve problems of relevance to contemporary environmental and health sciences (An et al., 2020).

4.1.2. Mechanism of Electron-Hole Pair Generation

The creation of electron-hole pairs is the core mechanism behind photocatalytic activity. It begins when light shines on a photocatalyst, exciting electrons from the valence band to the conduction band, thereby creating an electron-hole pair. The nature of this excitation is determined by the energy of incident photons and by the material's bandgap. For effective photocatalysis, the energy of photons must be greater than the bandgap energy to enable this transition. The longevity and stability of such electron-hole pairs are significant, as they can engage in further redox reactions required for breaking down pollutants or reducing carbon dioxide into beneficial compounds. The efficiency of such generation and subsequent reactions directly affects the performance of photocatalysts and the usability of the photocatalysts in environmental remediation and energy conversion. However, simple electron-hole pair creation is not sufficient for effective photocatalytic action; their separation and prevention of recombination are necessary for reliable catalytic activity. Recombination occurs when the electron recombines with the hole, reversing the energy obtained by excitation. To regulate it, the synthesis of photocatalytic materials involves techniques that promote the mobility of charge carriers, which can be achieved in various ways. For instance, composite materials can be engineered to contain electron-accepting intermediates, which trap electrons and increase their availability for substrate reactions. Composites such as Cu/Bi2Ti2O7/rGO have recently been demonstrated to be efficient in charge carrier trapping and have been noted to drastically improve the generation of reactive species required for pollutant degradation. Interestingly, the inhibition of recombination in e^{-}–h^{+} pairs in such materials can greatly enhance photocatalytic activity, marking a breakthrough in catalytic design (Acevedo et al., 2019) (Figure 4.3).

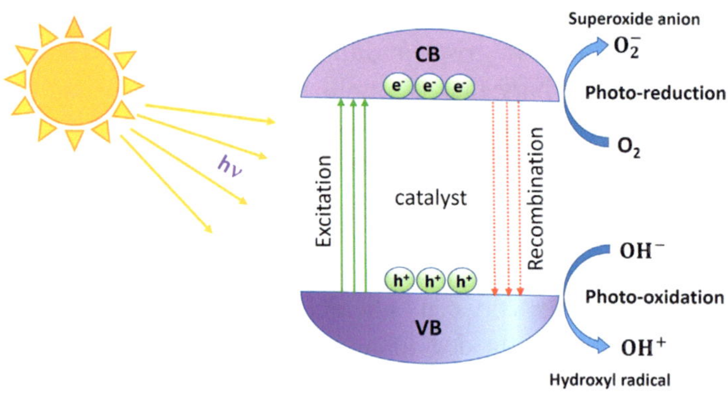

Figure 4.3. Photocatalytic reaction showing the electron-hole pair generation.

Source: https://www.researchgate.net/figure/Schematic-diagram-of-photocatalytic-reaction-showing-the-electron-hole-pair-generation_fig3_357558693.

The electron-hole pair generation mechanism is extended to intricate systems, including photoelectrocatalysis, where the incorporation of an external electric field enhances charge separation. The method has enormous potential for reducing carbon dioxide, with the possibility of using sunlight to convert CO_2 into valuable hydrocarbons. Recent photocatalytic CO_2 reduction experiments have demonstrated that the efficiency of electron-hole pair creation within defect-rich materials, such as anatase TiO2, can be highly inconsistent under a wide range of conditions. Density functional theory (DFT) calculations indicate that the electronic structure at various interfaces determines catalytic cycles and ultimately dictates product formation (Alexandrova et al., 2020). The solvent-electrolyte interaction introduces additional complexities in charge carrier dynamics, which creates opportunities for innovative approaches in the design of future photocatalysts that effectively harvest solar energy and address pressing environmental issues.

4.1.3. Energy Band Theory and Its Relevance to Photocatalysis

Knowledge of energy band theory is essential for understanding photocatalytic reaction mechanisms. Energy band theory, in essence, predicts the existence of energy levels for electrons in solids, dividing them into valence bands and conduction bands. This division provides the framework for understanding the electronic properties of photocatalysts, i.e., how light energy can be utilized by them to accelerate chemical reactions. When light stimulates a photocatalyst, electrons transfer from the valence band to the conduction band and create electron-hole pairs that are responsible for photocatalysis. The efficiency of this process primarily depends on the band gap of the material, which indicates its ability to absorb certain wavelengths of light. Thus, material band gaps with appropriate values are accountable for enhancing photocatalytic activity and are the focus of emerging technologies for energy conversion and environmental purification. Further refinement of energy band theory has significant implications for the design and optimization of photocatalysts. Materials such as metal-organic frameworks (MOFs) illustrate how structure can be optimized to fit function, particularly in terms of how electronic properties can be optimized through chemical modifications. For instance, doping UiO-66, a stable Zr-based MOF, with transition metals is capable of altering its energy levels, thereby enhancing its photocatalytic ability, as evidenced by its electronic properties. The creation of an energy scheme that combines empirical measurement with theoretical calculations helps elucidate charge transfer mechanisms, which play a vital role in activating reactants in photocatalytic processes (Vos et al., 2018). While researchers design tailored materials, awareness of the electronic world, as defined by energy band theory, will continue to be a driving force in the discovery of efficient photocatalysts to address contemporary environmental problems of relevance. Furthermore, aside from mere structure, the versatility of energy band theory also encompasses how activation steps within photocatalytic reactions, particularly carbon dioxide reduction, are comprehended. Determining the optimal electronic properties will enhance catalyst performance in converting CO_2 to chemicals and fuels. Recent first-principles-based research, relying on artificial intelligence, demonstrates this rational design concept, which determines key features in oxide-based catalysts. For example, the adsorption of molecular oxygen onto surface cations impacts the electronic structure, weakening C–O bonds and thereby enhancing CO_2 activation. This articulation of weakening of C-O bonds as a performance measure suggests a linear relationship between the electronic property described by energy band theory and photocatalytic activity (Ghiringhelli et al., 2020). It is for this reason that enhancing our knowledge in this area contributes not just to material design but also to advancing efforts in sustainable energy technologies via photocatalysis.

4.2. REACTION MECHANISMS IN PHOTOCATALYSIS

The mechanism of reaction responsible for photocatalysis is fundamental to photocatalytic material-mediated chemical changes under light irradiation. Central to these mechanisms is the photoexcitation of charge carriers—electrons and holes—upon exposure of semiconductor materials such as titanium dioxide (TiO_2) to photons. This photoactivation of light results in the splitting of electron-hole pairs, which then react with adjacent molecules, leading to oxidation and reduction reactions that drive the overall photocatalytic reaction. Such mechanisms are efficient based on factors such as the bandgap energy of the semiconductor in addition to its surface properties, which play an essential function in determining the material's photocatalytic stability and activity (Howe et al., 2015). Application of various material deposition techniques significantly influences the photocatalytic activity of TiO_2 thin films. Sol-gel processing and chemical vapor deposition (CVD) are some of these techniques, allowing for the control of film morphology, crystallinity, and surface area, which are some of the significant factors that affect photocatalytic activity. For instance, the sol-gel method allows for control over the TiO_2 sol concentration and the inclusion of additives, such as polyethylene glycol (PEG), to enhance the structure and functionality of the deposited film. On the other hand, CVD provides control over layer thickness and homogeneity.

The comparative study of photocatalytic activity from films prepared by these methods has demonstrated that optimum conditions can lead to significant enhancements in reaction rates, again highlighting the significance of synthesis methods in high-performance photocatalyst material engineering (Králová et al., 2012). In addition to synthesis methods, conditions under which photocatalysis occurs—light intensity and wavelength—are also crucial to optimizing reaction mechanisms. The spectrum of light absorption in photocatalysts determines the effective wavelengths utilized by them and thus specifies the direction of photocatalytic reactions. Intensive light sources increase the charge carrier generation rate; however, efficiency also relies on the material's ability to interact with reactive species present in the environment. Understanding the role of irradiation conditions, electron-hole kinetics, and photocatalytic surface stability is essential in designing next-generation photocatalysts. With advancing science, knowing these complex mechanisms will unleash new technologies for environmental cleanup and renewable energy conversion processes (Howe et al., 2015) (Figure 4.4).

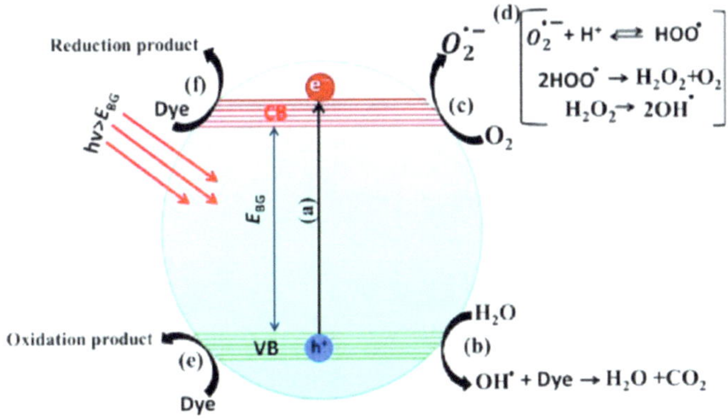

Figure 4.4. Reaction mechanism of photocatalysis.

Source: https://www.researchgate.net/figure/Reaction-mechanism-of-photocatalysis_fig25_339292325.

4.2.1. Oxidation and Reduction Processes

Central to the photocatalytic reaction mechanism are oxidation and reduction, the backbone of most catalytic technologies aimed at cleaning up the environment and generating energy. Oxidation is the process in which a species loses electrons, while reduction is the process in which electrons are gained. In photocatalysis, these half-reactions occur concurrently through photostimulated excitation of electrons in a semiconductor material. The energetic electrons subsequently reduce other chemical species, such as carbon dioxide, thereby contributing to the production of valuable commodities like methanol.

The interaction of these processes is therefore significant in maximizing photocatalytic efficiency. It has resulted in pivotal advances in materials science and catalytic methods, enabling solar energy to be utilized for chemical transformation. A notable enhancement in the efficiency of oxidation and reduction reactions is achieved in photoelectrocatalytic systems, a combination of classical photocatalysis and electrochemistry. The combined method overcomes some of the drawbacks in pure photocatalysis by increasing the rate of charge separation and reducing the electron-hole recombination rate. Particularly, recent studies have established that CO_2 can be successfully converted to small molecules in a highly efficient manner through photoelectrocatalytic hydrogenation, which harvests solar energy with very low carbon emissions (Alexandrova et al., 2020) (Figure 4.5).

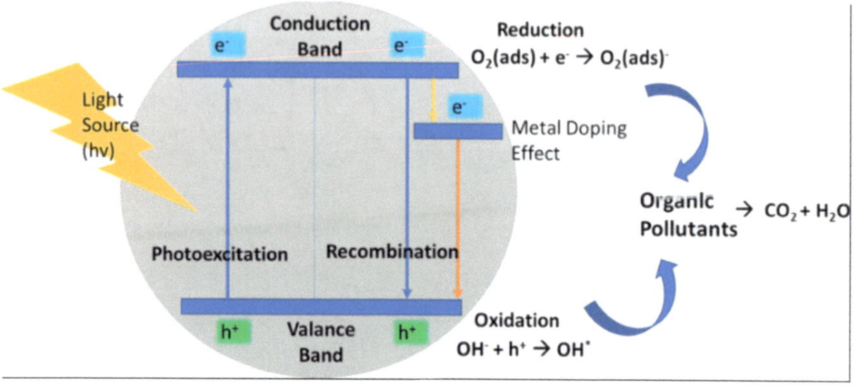

Figure 4.5. The schematic representation of photocatalytic oxidation process.

Source: https://external-content.duckduckgo.com/iu/?u=https%3A%2F%2Fwww.researchgate.net%2Fpublication%2F358110041%2Ffigure%2Ffig1%2FAS%3A1155453060952068%401652492885461%2FThe-schematic-representation-of-photocatalytic-oxidation-process.png&f=1&nofb=1&ipt=d350b462ac4d0791b863048ab3b87320183bebf353f8fed0b51262556cc22478.

By studying the electronic structure of catalytic surfaces and their interactions with solvents and electrolytes, researchers have been able to map out product-specific formation mechanisms and identify the key factors that maximize the reaction rate within these systems. Further understanding of photocatalytic oxidation and reduction has been gained from research into bimetallic catalysts, which offer a promising route to increased catalytic activity.

For instance, the investigation of palladium-copper tetrahydride anions interacting with carbon dioxide revealed efficient generation of intermediates, which play a central role in CO_2 hydrogenation (Alexandrova et al., 2020). The inclusion of the higher energy isomers within these catalytic clusters significantly influences their capability to mediate CO_2 reduction. These findings underscore the need to elucidate the molecular-level mechanism of reactions for enhancing catalytic activity and discovering novel materials for photocatalytic purposes. By focusing on the dual action of oxidation and reduction in these complex systems, scientists can devise strategies that harness solar energy and resolve environmental concerns.

4.2.2. Surface Reactions and Adsorption Phenomena

The intricacies of surface reaction and adsorption processes are crucial to understanding the mechanisms of photocatalytic reactions. Adsorption refers to the adsorption of molecules from a liquid or gaseous phase onto a solid surface, which is critical to defining the overall performance of photocatalysts. The nature of the catalyst's surface—its chemical makeup and morphology—is what controls the extent to which reactants are absorbed in photocatalytic systems. This interaction enables subsequent chemical reactions, as adsorbed species can gain the proximity necessary for effective collisions and thus accelerate reaction speeds. Adsorption mechanism research optimizes photocatalytic materials, enabling scientists to design surface properties that serve more efficiently in conversion or degradation processes. Surface reactions are also dependent on various parameters, including temperature, pressure, and the availability of different adsorbates.

The arrangement of surface sites can lead to specific reaction routes, altering the efficiency of photocatalytic activity. For instance, the effective dispersion of dopants in a semiconductor can provide active sites involved in various surface reactions. As demonstrated in more recent studies with titanium dioxide (TiO_2), bulk doping and surface species formation were induced upon the introduction of molybdenum to the TiO_2 matrix, indicating the level of interaction present. This surface chemistry background of knowledge is the key to creating advanced photocatalysts with maximum light absorption and charge segregation, which drive chemical reactions for environmental cleaning and energy conversion (Armandi et al., 2019). In practical applications, surface tailoring of photocatalysts by processes such as sol-gel synthesis and chemical vapor deposition can be extremely decisive in modifying their photocatalytic activity. These methods allow for very good control of the physical and chemical properties of the catalysts and therefore facilitate the optimization of surface reactivity and adsorption processes. For instance, during the synthesis of titanium dioxide films via sol-gel technology, modifying the sol composition, such as adding polyethylene glycol, may enhance the material's structural integrity and superhydrophilicity, which are vital for the effective degradation of contaminants. Moreover, comparative studies of different fabrication methods reveal that the photocatalytic activity can show considerable variations based on the preparation conditions, which signifies the importance of surface engineering in the design of photocatalysts (Králová et al., 2012). This correlation between surface characteristics and photocatalytic activity aligns with overall trends in materials science and presents opportunities for further studies aimed at addressing pressing environmental challenges that confront humanity.

4.2.3. Role of Reactive Oxygen Species (ROS)

In photocatalytic reactions, reactive oxygen species (ROS) play crucial roles as intermediate products that initiate a series of chemical conversions. ROS, being reactive molecules with high reactivity, including singlet oxygen, hydroxyl radicals, and superoxide anions, are mainly formed by the excitation of photocatalysts under light irradiation. Formation of ROS is needed to enhance the degradation of organic pollutants and enhance the efficiency of photocatalytic processes. The interaction of ROS with target species can lead to the disruption of chemical bonds of contaminants and therefore mineralization to less toxic, simpler molecules.

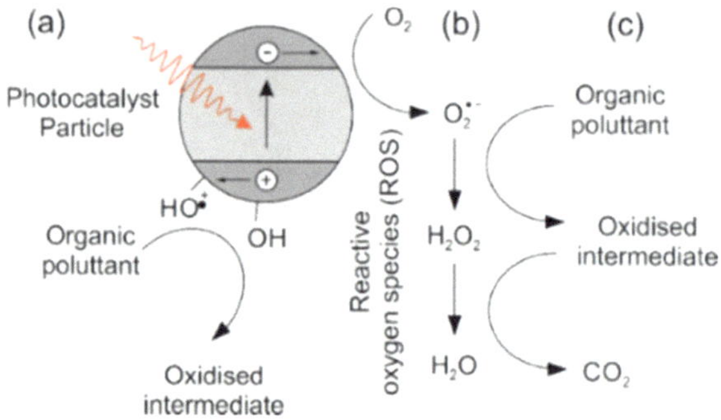

Figure 4.6. Generation of reactive oxygen species (ROS) and the process of photo-oxidation of an organic pollutant. Step (a) involves the absorption of a photon to generate an excited electron–hole pair, which can then migrate to the surface of the photocatalyst particle. In step (b), the electron can reduce atmospheric oxygen to the more labile superoxide radical. In step (c), the superoxide radical and subsequent ROS can oxidize an organic pollutant to CO_2. The holes can also form surface-bound hydroxyl radicals, which directly oxidize organic molecules.

Source: https://www.researchgate.net/figure/Generation-of-reactive-oxygen-species-ROS-and-process-of-photo-oxidation-of-an-organic_fig1_369208019.

Understanding the mechanism of ROS formation and their subsequent reactions with pollutants is crucial for effectively optimizing photocatalytic reactions and enhancing performance in environmental applications. The photocatalytic kinetics of ROS formation can be controlled by various factors, including the type of photocatalyst, NOM availability, and specific environmental conditions. For instance, certain ENMs have been shown to possess the capability of enhancing the generation of ROS and accelerating POP degradation. The interactions among light intensity, photocatalyst surface area, and the adsorption properties of the contaminant play a pivotal role in determining ROS yield. Additionally, NOM interaction with ROS would inhibit or enhance the photocatalytic removal of contaminants, as suggested by studies reporting differences in inactivation rates for microorganisms and organic compounds under sunlight (Boehm et al., 2018). These subtleties underscore the importance of comprehensive modeling in predicting the effectiveness of photocatalytic systems in real-world applications (Figure 4.6).

The photochemical conversion of certain precursors, including perfluoroalkyl acids (PFAAs), highlights the importance of ROS in photocatalytic reactions. Research indicates that ENMs have the ability to considerably increase PFAA conversion rates by primarily utilizing ROS-dependent routes (Dai et al., 2020). However, the effectiveness of ROS in total mineralization is dependent on various factors, including precursor molecule photostability, adsorption onto photocatalyst surfaces, and compatibility with other environmental factors. Understanding such mechanisms could be helpful in designing more efficient photocatalytic systems to address recalcitrant ecological contaminants. Through the utilization of ROS mechanisms, photocatalysis not only enhances pollutant degradation but also advances public health initiatives, such as water disinfection, revealing its dual function in environmental cleaning technologies.

4.3. FACTORS INFLUENCING PHOTOCATALYTIC EFFICIENCY

The efficiency of photocatalytic reactions is governed by an array of factors, one of the first of which is the identity of the photocatalyst. Photocatalysts, typically being semiconductors, vary in band gap energies, which determine how they absorb light and, consequently, how efficient they are at catalyzing photocatalytic reactions. Titanium dioxide (TiO_2), for instance, is the most widely applied photocatalyst due to its suitable bandgap and stability under UV irradiation. However, modifications to the catalyst, such as metal or non-metal doping, can enhance light absorption properties and, therefore, photocatalytic efficiency. Understanding the effect of photocatalysts' properties on their performance enables researchers to synthesize materials for specific applications, tailoring such parameters as electronic structure and surface characteristics to maximize overall reaction rates and yields in photocatalytic processes. The intensity and wavelength of irradiation light are also key contributors to the efficiency of photocatalysts. The type of the light source directly affects electron excitation in the photocatalyst, as different wavelengths correspond to other energies that can excite electrons within the bandgap. Furthermore, the lighting technology has been enhanced, particularly with the addition of controlled periodic illumination, as a means of improving the efficiency of the photocatalyst. This process enables strategic lighting control, influencing the rate of separation of photogenerated charge carriers and thereby accelerating reaction kinetics. Additionally, different light sources can alter the route and mechanism of photocatalytic reactions, emphasizing the need for selective light condition control to favor the specific photocatalyst material in use and achieve maximum efficiency during the reaction. Lastly, the ambient conditions under which the photocatalytic reaction is carried out may significantly influence total efficiency. Working temperature, pH, and presence or absence of co-existing species may curtail or promote the photocatalytic process. For instance, the pH of the reaction medium can modify the surface charge of the photocatalyst, thereby influencing adsorption properties and availability of reactants at the catalyst's surface. In addition, increasing temperature can enhance the speed of reactions by providing sufficient energy to overcome activation energies and improve the efficiency of the photocatalysis process. Additional reactive species or organic impurities introduced into the system may also lead to the production of by-products, which can compete for active sites on the photocatalyst and ultimately influence overall efficiency. Thus, it is essential to acquire an integrated comprehension of these environmental factors to effectively design and optimize photocatalytic systems for varied applications, such as environmental remediation and energy conversion (Figure 4.7).

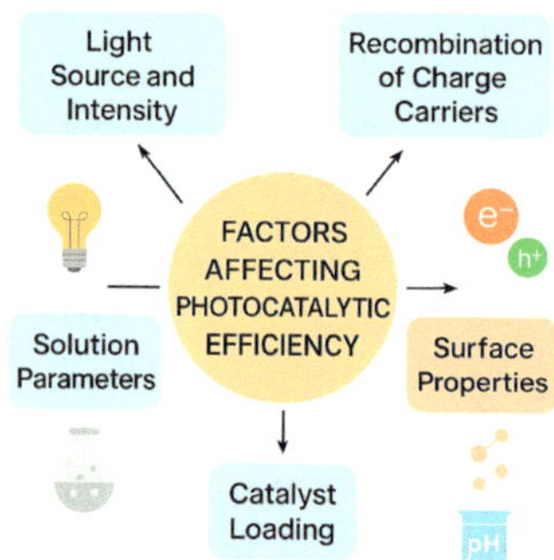

Figure 4.7. *Factors affecting photocatalytic efficiency.*

Source: https://www.researchgate.net/figure/Key-parameters-influencing-the-efficiency-of-photocatalytic-degradation-processes_fig4_391894799.

4.3.1. Light Intensity and Wavelength

The role of light intensity and wavelength in photocatalytic processes plays a crucial role in determining the efficiency of these processes. Photocatalysis is founded on the reaction between a catalyst and photons of light, where the energy from light is utilized to trigger chemical alterations. Varying light intensity has the potential to exert a tremendous effect on electron-hole pair generation rate in photocatalytic materials. More intense light results in higher photon absorption, hence more charge carrier formation. However, it is worth noting that beyond a certain point, there will be no proportionally greater improvement in photocatalytic activity due to non-linear saturation effects resulting from increases in light intensity. Therefore, it is highly essential to optimize both the light intensity and the wavelength emitted, particularly to achieve maximum photocatalytic efficiency. Additionally, the wavelength of light has a fundamental correlation with the energy levels of the photocatalyst. Different photocatalytic materials have optimal wavelengths for activity that correspond to their bandgap energy. For instance, materials that are optimized to absorb visible light can facilitate reactions that utilize sunlight more efficiently, having added value for environmental applications. Selectivity and activity of photocatalytic reactions can vary substantially based on the wavelength absorbed, which influences factors such as the types of reactive oxygen species produced. As envisioned in recent studies, photocatalyst tunability, such as carbon dots, enables adjustments that enhance their sensitivity to various wavelengths of light, thereby expanding their application in energy conversion and pollutant degradation (Boonsith et al., 2020). Research has shown that integrating light intensity and wavelength with photocatalysts can significantly enhance the reaction efficiency for degrading pollutants. For instance, in composite material experiments such as copper/bismuth titanate/reduced graphene oxide ($Cu/Bi_2Ti_2O_7/rGO$), it was observed that optimum light conditions led to improved charge carrier trapping and reduced electron–hole pair recombination, ultimately increasing the production of active radicals for photocatalytic degradation of complex drug molecules. When the composites were exposed to visible light, the photocatalytic degradation efficiency was improved, confirming the role of light properties in developing an efficient photocatalytic system. This understanding of the wavelength and intensity of light not only advances knowledge of the photocatalytic process but also presents opportunities for designing more effective materials to utilize solar energy for environmental remediation (Acevedo et al., 2019).

4.3.2. Catalyst Surface Area and Morphology

Another critical parameter for photocatalytic reactions is the morphology and surface area of the catalysts, which play an indispensable role in determining reaction rates and effectiveness. The effectiveness of photocatalytic materials depends directly on their ability to adsorb light and interact with reactants, both of which are enhanced by the higher surface area. A higher surface area entails more active sites, providing a greater opportunity for photocatalytic activity. Thus, catalyst design, including in nanostructured forms, is a significant area of research in this field. By manipulating the synthesis conditions, such as temperature and duration, scientists can prepare materials of varying morphologies that lead to improved photocatalytic activities, demonstrating the correlation between catalyst structure and function in photochemically driven chemical reactions. Secondly, advances in the synthesis route for nanostructured catalysts have resulted in materials with enhanced surface properties. For instance, titanate nanostructures synthesized via hydrothermal routes are of great value due to their high aspect ratio and tunable properties. The synthesis of titanate nanostructures (TNS) without involving conventional TiO2 precursors provides a means to form homogeneous nanotubes, significantly improving the adsorption capacity and photocatalytic activity of the materials. Experiments indicate that these TNS samples exhibit enhanced activity in the degradation of organic dyes, such as rhodamine 6G, which suggests that their specific morphology is directly responsible for the enhanced photocatalytic efficiencies (Franco et al., 2012). Therefore, being aware of the synthesis and resulting morphology of catalysts is essential in designing photocatalytically efficient systems. Apart from surface area and morphology, the synergy of semiconductor materials with different catalytic properties indicates a good potential for enhancing photocatalytic activity. The synthesis of nanocomposite materials, such as the synergism of titanate nanofibers with nanocrystalline ZnS and Bi2S3 semiconductors, is a promising approach to promote photocatalytic reactions. These composites utilize the peculiarity of different material properties, thus achieving maximum adsorption and photocatalytic activities. For instance, the stratified morphology of the ZnS/TNF and Bi_2S_3-ZnS/TNF nanocomposites facilitates organic dye interaction with structures like methylene blue, thereby ultimately increasing a much higher overall removal efficiency through a sequential adsorption-photocatalytic degradation mechanism (Silvestre et al., 2015). Such integration of morphology is not only beneficial in elucidating the catalytic role but also assists in formulating innovative solutions to environmental remediation through efficient photocatalytic action.

4.3.3. Environmental Conditions (pH, Temperature)

Environmental parameters, particularly temperature and pH, play significant roles in governing photocatalytic reactions. The pH may affect the surface charge of photocatalytic materials, hence affecting the adsorption of reactants. For instance, in photocatalytic processes where titanium dioxide (TiO_2) is used, various pH levels alter the charge characteristics of the oxide surface, fostering or inhibiting the interaction with target pollutants. As depicted in studies of the sol-gel method for the synthesis of TiO_2 films, pH changes can directly influence the photocatalytic activity of the films deposited. The research highlighted that optimum pH ranges produced enhanced photocatalytic activities of degradation of toxic compounds, such as 2,6-dichlorindophenol, necessitating fine-tuning optimization of pH at the time of synthesis and utilization of photocatalysts for ensuring maximum efficiency in purifying the environment (Králová et al., 2012). Temperature is yet another important parameter that affects photocatalytic processes due to its direct impact on the kinetic energy of reactants and the rates of light-excited reactions. Increased temperatures can enhance charge carrier mobility and improve the efficiency of electron-hole pair formation in photocatalysts. For example, in hydrothermal synthesis reactions, an increase in temperature during the production of TiO_2 nanoparticles has been shown to impart increased photocatalytic activity due to enhanced crystallinity and reduced defects in the material's structure. Furthermore, control studies have established that temperature fluctuations significantly affect the degradation rate of various organic contaminants, necessitating a careful assessment of photoreactor design and operational conditions for optimal performance (Choi et al., 2010).

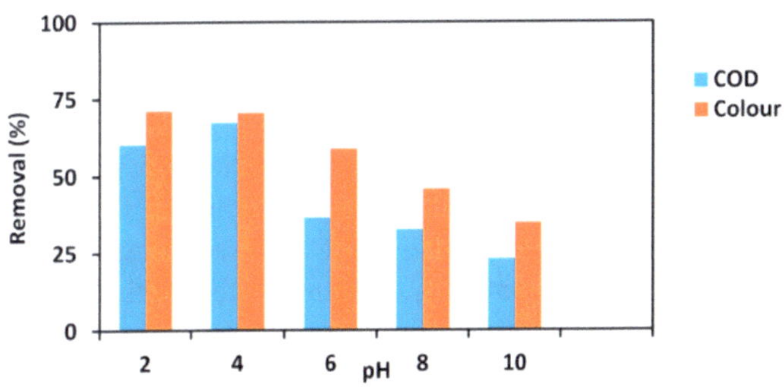

Figure 4.8. Effect of pH on photocatalytic degradation (process conditions: TiO_2 dosage – 500 mg/L, reaction time – 3 h, UV light irradiation – 64 W).

Source: https://www.researchgate.net/figure/Effect-of-pH-on-photocatalytic-degradation-process-conditions-TiO2-dosage-500-mg-L_fig2_342321887.

Lastly, the interplay between pH and temperature conditions complicates the process of optimizing photocatalytic reactions even further. Synergistic effects between these two factors tend to augment or diminish photocatalytic activity in most instances. For example, while acidic conditions can increase reactivity in some photocatalysts, extreme temperatures can exacerbate dissolution or degradation of the catalyst itself under adverse conditions. Thus, understanding how pH and temperature intersect to regulate charge dynamics in photocatalytic systems is essential for advancing environmental technology application improvements. Not only does this data help illuminate the development of future catalysts, but it also guides functional uses in real-world situations, where conditions are less than optimal (Králová et al., 2012) (Figure 4.8).

4.4. CHALLENGES AND LIMITATIONS

Photocatalytic processes are often promoted for their potential to transcend environmental pollutants, yet several issues prevent them from being used on a large scale. One of the primary limitations is presented by the efficacy of photocatalyst materials under varying sources of illumination. Most photocatalysts, such as titanium dioxide (TiO_2), work most effectively under ultraviolet (UV) light, which limits their utility since UV light accounts for only a small percentage of the solar spectrum. Additionally, improvements in TiO2 modification—through doping with compounds such as nitrogen—have been encouraging in increasing activity under visible light. Inconsistencies in performance, however, such as those observed between the varying performances of nitrogen-doped titanium dioxide and the initial TiO2, are reflective of the issue of optimizing these materials to deliver uniform results across different environments (Steiner et al., 2017). There is another critical issue with photocatalytic processes: their tendency to deactivate. Overextended operating time can result in the photochemical degradation of the catalysts themselves. Production of reactive oxygen species by photocatalysis can ultimately lead to surface degradation of the catalysts. Compounding this, the presence of by-product degradation of targeted contaminants routinely interferes with photocatalytic activity, creating a feedback loop that reduces effectiveness. While modifications, for instance, the incorporation of metals like copper, have been imagined to enhance catalytic activity, there is evidence to show that certain additives do not offer anticipated gains (Steiner et al., 2017). Therefore, the identification of long-lived photocatalysts that retain their activity when used repeatedly over extended periods remains a key stumbling block to practical limitations. In addition to material performance issues, the nature of the photocatalytic reaction itself creates operational challenges with system design and scale-up. Photocatalytic systems often require precise control of environmental conditions, such as pH and temperature, for optimal catalytic activity. The problem of maintaining these parameters in actual application settings becomes limiting to large-scale use. Working on top of this challenge is how strongly photocatalysis's efficacy relies on the chemical composition of pollutants. Variance in amino acid concentration, for example, may impact the rate of inactivation upon UV irradiation and photocatalysis treatment and thus influence overall system performance. In addition, considerations in optimizing reactor designs reveal a fine balance between light distribution and catalyst exposure, further contributing to practical challenges. These are advanced considerations that underscore the pressing need for further research to define the materials and operational practices involved in photocatalytic applications more precisely (Figure 4.9).

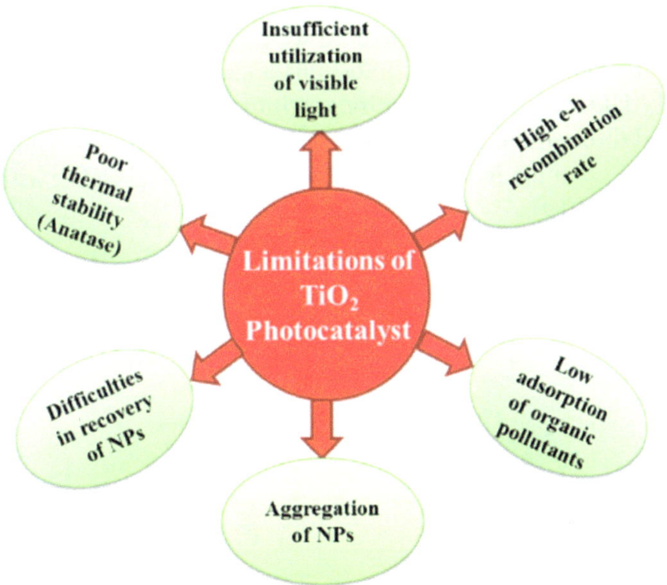

Figure 4.9. A schematic diagram showing the limitations of TiO_2 photocatalysts.

Source: https://www.researchgate.net/figure/A-schematic-diagram-showing-the-limitations-of-TiO2-photocatalysts_fig2_336070065.

4.4.1. Stability and Reusability of Photocatalysts

Photocatalyst stability is a crucial parameter that establishes its overall practicality and efficiency in photocatalytic reactions. Photocatalysts must endure operational stresses over extended periods without severe degradation to maintain their active surface area and catalytic activity. One promising approach is the incorporation of stable materials, such as carbon nanotubes or magnetic nanoparticles, into traditional photocatalytic systems. For instance, hybrid nanomaterials like TiO_2-CNTs and magnetic Fe_3O_4-TiO_2-CNTs are more stable, with the reaction sites being open while ensuring the promise of ease of recovery after the reaction. Such innovations are important not only in achieving prolonged photocatalytic activity but also in reducing the loss of precious materials, which is a significant economic concern in the development of photocatalysts (Mason et al., 2018). Reusability is another critical feature in the development of effective photocatalysts because the opportunity for material regeneration with minimal loss of performance is required for large-scale applications. The development of photocatalysts that are easy to separate and reuse without sacrificing chemical activity provides a significant advantage. Techniques such as utilizing magnetic properties allow photocatalysts to be easily recovered from reaction mixtures, facilitating green operations in wastewater treatment. In hybrid photocatalyst studies, for example, the ternary $Cu/Bi_2Ti_2O_7/rGO$ composites, significant improvements in reusability have been realized through the effective trapping of charge carriers. This minimizes electron-hole recombination, thereby maximizing the catalyst life with recycling (Acevedo et al., 2019).

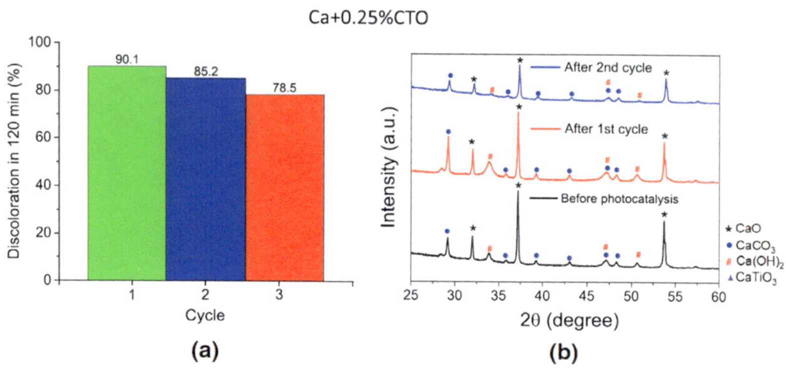

Figure 4.10. Stability and reusability of the heterojunction. (a) Photocatalytic efficiency of the reused Ca + 0.25% CTO heterojunction for discoloration of RhB dye over three cycles. (b) XRD analysis of the Ca + 0.25% CTO sample after reuse.

Source: https://www.researchgate.net/figure/Stability-and-Reusability-of-the-heterojunction-a-Photocatalytic-efficiency-of-the-reused_fig11_344806098.

These materials enhance the development of photocatalytic technology for environmental applications. Investigating photocatalyst stability and reusability is also directly associated with the mechanism of photocatalytic processes, providing opportunities for more efficient designs. Understanding these mechanisms enables the fine-tuning of photocatalytic materials to achieve maximum performance under real-world conditions. For example, photocatalysts with efficient charge carrier trapping abilities can enhance pollutant degradation rates while conferring stability over multiple cycles. As researchers continue to discover varied composite materials and reaction conditions, advances in the design of photocatalysts can help achieve more effective treatments for toxic organic pollutants in industrial wastewater. By focusing not only on the stability but also on the recyclability of photocatalysts, the community of photocatalysts can work toward achieving sustainable solutions that serve the demands of environmental remediation (Mason et al., 2018; Acevedo et al., 2019) (Figure 4.10).

4.4.2. Scalability of Photocatalytic Processes

The scalability of photocatalytic processes remains a crucial question in the overall field of photocatalytic reactions, as it directly affects the feasibility of applying these technologies on an industrial scale. Scalability is achieved in photocatalytic systems through not only efficacy but also through efficiency in extensive processes. The recent advances in heterogeneous photocatalysis, particularly with TiO_2, reflect a movement towards attaining scalability. The ability to use TiO_2 in selective $C(sp^3)$–F bond formation demonstrates its long-term turnover frequency and stability, which are essential characteristics for scaling up from photocatalytic reactions. These features fulfill the conditions of sustainability by enabling the use of mild fluorine donors catalytically, a novel route to synthesizing fluorinated products (Hammond et al., 2018). These developments suggest that the use of optimum catalysts enables scalability, promoting the wider adoption of photocatalytic technology. The scalability of photocatalytic processes is another motivation provided by the development of high-efficiency materials through design. For instance, mesoporous silicon nanostructures have been found to exhibit outstanding photocatalytic properties as well as stability, particularly in the degradation of organic pollutants like methyl orange. The special properties of these materials, including high surface area and unique pore topology, play a crucial role in promoting charge migration and mass transfer during photocatalytic reactions. The structural order of these nanomaterials, as indicated in recent research, significantly enhances mass diffusion and the accessibility of active sites, thereby improving overall photocatalytic efficiency (Ahire et al., 2016). These developments not only benefit the scalability of photocatalytic processes but also open up opportunities for new applications in the conversion and storage of solar energy. Ultimately, solving challenges in the scalability of photocatalytic processes requires a multifaceted approach involving not only designing efficient catalysts but also optimizing reaction conditions and system design considerations. Integration into continuous flow systems is an exciting opportunity for scaling up such chemical reactions. Such systems can obtain enhanced throughput and efficiency while maintaining reaction conditions necessary for extensive periods of time. Ongoing research must remain focused on optimizing light-harvesting devices and reactor designs to attain maximum deployment of photocatalytic systems in reality. By targeting these constituents, scientists are able to that the transition of photocatalytic technologies from the lab to commercially driven processes is successful, thus contributing towards the creation of sustainable synthesis procedures as well as environmental cleaning technologies.

4.4.3. Economic Considerations and Cost-Effectiveness

The integration of photocatalytic reactions into various industrial processes is also of significant economic consequence that impacts their cost-effectiveness in general. Photocatalysis has been cited as a method for enhancing efficiency in chemical synthesis and environmental remediation, particularly when compared to conventional processes. By using sunlight, these processes can reduce energy costs and minimize greenhouse gas emissions significantly, making it an appealing idea for investment in terms of sustainability. However, the initial capital expenditure of establishing photocatalytic systems can be costly, particularly when paying for premium materials and reactor designs. Hence, an economic analysis must be undertaken to determine the break-even and long-term gains from reduced energy and raw material consumption expenses. These factors need to be considered by industries that intend to transition to photocatalytic technologies and must be integrated into broader sustainability plans. Additionally, the application of nanotechnology in photocatalytic reactions demonstrates its cost-saving nature in enhancing agricultural productivity while mitigating environmental degradation. The application of nanomaterials involves a twofold benefit—improved efficiency of pesticides and fertilizers, with the possibility of renovating degraded lands. Such innovations are relevant since they facilitate more sustainable agriculture that is necessary in facing the rising food demands of a continuously expanding global population. Bacchini et al. (2013) make a point that while some nanomaterials pose potential toxicological risks, their appropriate application can lead to overall environmental quality and agricultural yield enhancement. The economic implications of adopting these technologies can lead to a spectacular reduction in the cost of environmental cleanup and enhanced agricultural productivity, hence making the investment in nanotechnology a worthwhile proposition in both cost-effective and ecological assessments. In addition, the economic climate of producing transportation fuels from nonconventional sources will be favorably impacted through the development of enhanced photocatalytic technologies. The development of efficient catalysts and new processing technologies, as stated by Ball et al. (2025), ensures that waste materials are converted into precious energy resources. This approach, in addition to resolving waste disposal problems, positions the industry for competitive success in the new renewable fuel market. The cost feasibility of using such nonconventional feedstocks economically depends on achieving trace impurity levels and maximizing the life of catalysts, which directly translates into cost reduction in production. As such industries pursue these breakthroughs, it is now becoming inevitable to balance the cost of the early phase with the anticipated long-term economic benefits, paving the way for a greener energy future anchored in photocatalysis.

SUMMARY

In describing the photocatalytic reaction mechanisms, it is evident that the processes have a promising potential in alleviating some of the most severe environmental issues, particularly concerning the reduction of carbon dioxide. Advancements in the field of photocatalytic technology, particularly through processes such as photoelectrocatalytic hydrogenation, offer strategic avenues for harnessing solar energy to convert CO_2 into valuable products. Such conversions not only result in a reduction of atmospheric CO2 levels but also enable a sustainable energy storage cycle. As highlighted by recent studies, the electronic and mechanistic intricacies involved in such reactions underscore the dual role of photocatalytic systems in facilitating both efficiency and selectivity in product formation, underscoring the significance of such advancements in practical applications (Alexandrova et al., 2020). Furthermore, the exploration of photocatalytic material structural properties, such as titanium dioxide, highlights the interface chemistry at the catalyst surface. Understanding the electronic structures and dynamics of various catalytic interfaces can also significantly improve the efficiency of photocatalytic systems. The variations observed in electrolyte-catalyst and solvent-catalyst interactions provide insight into maximizing reaction channels. This difference is critical, as photocatalyst performance and stability can be enhanced by adjusting surface properties to promote favorable product formation and minimize energy loss. The findings related to defect engineering in semiconductors—most significantly those of methanol production as a primary product—point to the advanced approaches scholars are using to enhance these systems (Alexandrova et al., 2020). Furthermore, recent findings on the catalytic process for bimetallic hydride clusters further indicate conflicting paths for CO2 hydrogenation, emphasizing the need for new methodologies in materials science. The ability to efficiently produce intermediates like $PdCuCO_2H_4$ demonstrates the capacity to employ intricate structures in efficiently driving catalytic activity. The exploration of higher energy isomers as catalysts raises significant questions in the design and development of CO_2 reduction reaction catalysts. These results contribute to the broader field of photocatalytic research and highlight the ongoing necessity for a thorough exploration of reaction mechanisms. Lastly, the culmination of these works supports the vision that strategically designed photocatalytic systems can potentially be game changers in energy conversion and environmental remediation processes—particularly within the framework of sustainable development.

The photocatalytic reaction mechanisms themselves are key to comprehending how such reactions can be very efficient in producing desired effects in environmental applications, particularly in pollutant degradation. Fundamental is the explanation of the role of hydroxyl radicals generated during the photocatalysis process. The radicals are nonspecific and have been found to attack amino acids more frequently than nucleotides, demonstrating that the mechanisms of viral inactivation are determined by the capsid structure as a protein. These results suggest that understanding the interactions between these reactive species and target molecules is crucial for optimizing photocatalytic processes. Another point of importance is the function of specific amino acids, specifically tyrosine, in sensitizing viruses to photocatalytic destruction. This information will be used to direct the development of target-specific photocatalytic therapies for viral diseases, emphasizing molecular composition to optimize efficacy and efficiency (Abbaszadegan et al., 2015). Another critical application of photocatalytic reactions is their employment in the reduction of carbon dioxide (CO_2), a concern in light of global warming. Photoelectrocatalytic reduction of CO_2, a process that combines the principles of photocatalysis with electrochemical reactions, is a new method of utilizing solar energy to convert CO_2. It has been established that the electronic structure of the catalyst at different interfaces is crucial to determining the efficiency and selectivity of product formation in the reaction. The application of density functional theory (DFT) has further elucidated these mechanisms, enabling a comprehensive understanding of why methanol is the predominant product under certain conditions. This highlights the intricacies of catalytic cycles at both solvent/catalyst and electrolyte/catalyst interfaces, underscoring the multifaceted nature of photocatalytic processes (Alexandrova et al., 2020). In its review of photocatalytic reaction processes, it is evident that the chemical composition of reactants as well as the structural properties of catalysts have a significant influence on the outcome. The discussion of the degradation of viral proteins by photocatalysis and CO_2 reduction methods highlights the flexibility of these processes in addressing actual environmental issues. As more is learned about the fundamental principles underlying these mechanisms, there are greater opportunities for innovation in the development of improved photocatalysts and treatment systems. These advances not only enable pollution abatement efforts but also enhance methods for the sustainable utilization of resources, making them more efficient, thereby further entrenching the imperative function of photocatalytic reactions in today's scientific research (Abbaszadegan et al., 2015; Alexandrova et al., 2020).

REFERENCES

1. Mayer, B. K., Yang, Y., Gerrity, D. W., & Abbaszadegan, M. (2015). The Impact of Capsid Proteins on Virus Removal and Inactivation during Water Treatment Processes. *Microbiology Insights*, *8s2*. https://doi.org/10.4137/mbi.s31441Acevedo, R., Anbarasan, P. M., Gkanas, E., Priyadharsan, A., et al. (2019). High efficient catalytic degradation of tetracycline and ibuprofen using visible light driven novel Cu/$Bi_2Ti_2O_7$/rGO nanocomposite: Kinetics, intermediates and mechanism. https://core.ac.uk/download/228156750.pdf (accessed on 28th September 2025).
2. Ahire, B., Canham, L., Canham, T., Chandrasekaran, N., Chen, W., Coxon, P., et al. (2016). Hydrogen-terminated mesoporous silicon monoliths with huge surface area as alternative Si-based visible light-active photocatalysts. https://core.ac.uk/download/77027128.pdf (accessed on 28th September 2025).
3. Alexandrova, A. N., Blankenhorn, M., Bowen, K. H., Liu, Y., et al. (2020). CO_2 hydrogenation to format and formic acid by bimetallic palladium-copper hydride clusters. https://core.ac.uk/download/323082880.pdf (accessed on 28th September 2025).
4. Alexandrova, A. N., Cronin, S. B., Gong, X. Q., Li, X., et al. (2020). (Photo)electrocatalytic CO_2 reduction at the defective anatase TiO_2 (101) surface. https://core.ac.uk/download/323311157.pdf (accessed on 28th September 2025).
5. An, X., Liu, P., Naowarojna, N., Reinhard, B., et al. (2020). Hybrid plasmonic photoreactors as visible light-mediated bactericides. https://open.bu.edu/bitstream/2144/40614/3/manuscript_1021_XA.pdf (accessed on 28th September 2025).
6. Armandi, M., Bonelli, B., Ditaranto, N., Esposito, S., et al. (2019). Application of reverse micelle sol-gel synthesis for bulk doping and heteroatoms surface enrichment in Mo-doped TiO_2 nanoparticles. https://core.ac.uk/download/pdf/196286710.pdf (accessed on 28th September 2025).
7. Assadi, M. H. N., Hanaor, D. A. H., Li, S., Sorrell, C. C., et al. (2012). Ab initio study of phase stability in doped TiO_2. http://arxiv.org/abs/1210.7555 (accessed on 28th September 2025).
8. Bacchini, F., Greppi, G., Mura, S., Roggero, M., et al. (2013). Advances of nanotechnology in agro-environmental studies. https://core.ac.uk/download/16749559.pdf (accessed on 28th September 2025).
9. Ball, D. W., Chianelli, R. R., Hepp, A. F., Kulis, M., et al. (2025). Novel catalysts and processing technologies for production of aerospace

fuels from non-petroleum raw materials. https://core.ac.uk/download/pdf/42722367.pdf (accessed on 28th September 2025).

10. Ban-Weiss, G., Chen, S., Destaillats, H., Everman, W., et al. (2019). De-pollution efficacy of photocatalytic roofing granules. https://core.ac.uk/download/294625302.pdf (accessed on 28th September 2025).
11. Bellardita, M., Corsino, S. F., Loddo, V., Palmisano, L., Parrino, F., Torregrossa, M., & Viviani, G. (2019). Sequential biological and photocatalysis based treatments for shipboard slop purification: A pilot plant investigation. https://core.ac.uk/download/224993434.pdf (accessed on 28th September 2025).
12. Boehm, A., Davies-Colley, R., Dodd, M., Kohn, T., et al. (2018). Sunlight-mediated inactivation of health-relevant microorganisms in water: A review of mechanisms and modeling approaches. https://core.ac.uk/download/323080860.pdf (accessed on 28th September 2025).
13. Boonsith, S., Kim, D. C., Sakdaronnarong, C., Sangjan, S., et al. (2020). Recent developments in synthesis and photocatalytic applications of carbon dots. https://core.ac.uk/download/323383933.pdf (accessed on 28th September 2025).
14. Choi, W., Hoffmann, M. R., Kim, K., Park, H., et al. (2010). Photo reductive dissolution of iron oxides trapped in ice and its environmental implications. https://core.ac.uk/download/4884997.pdf (accessed on 28th September 2025).
15. Dai, Y., Guo, X., Hoffmann, M. R., Wang, Y., et al. (2020). Photochemical transformation of perfluoroalkyl acid precursors in water using engineered nanomaterials. https://core.ac.uk/download/323085206.pdf (accessed on 28th September 2025).
16. De Vos, A., Hendrickx, K., Joos, J., Lejaeghere, K., et al. (2018). Exploring lanthanide doping in UiO-66: A combined experimental and computational study of the electronic structure. https://core.ac.uk/download/158346809.pdf (accessed on 28th September 2025).
17. Dubnová, L., Dvoranová, D., Edelmannová, M., Kočí, K., et al. (2020). Successful immobilization of lanthanides doped TiO_2 on inert foam for repeatable hydrogen generation from aqueous ammonia. https://core.ac.uk/download/322825201.pdf (accessed on 28th September 2025).
18. Franco, A., Papa, A. L., Silvestre, A. J., Erdem, B., Zielińska, B., Viana, B. C., Huang, C., et al. (2012). Synthesis of titanate nanostructures using amorphous precursor material and their adsorption/photocatalytic properties. http://arxiv.org/abs/1201.3644 (accessed on 28th September 2025).

19. Ghiringhelli, L. M., Illas, F., Levchenko, S. V., Mazheika, A., et al. (2020). Ab initio data-analytics study of carbon-dioxide activation on semiconductor oxide surfaces. https://core.ac.uk/download/491525678.pdf (accessed on 28th September 2025).

20. Hammond, C., & Tarantino, G. (2018). Catalytic formation of $C(sp^3)$-F bonds via heterogeneous photocatalysis. https://core.ac.uk/download/287579966.pdf (accessed on 28th September 2025).

21. Howe, R. F., & Rhydderch, S. (2015). *In-situ* EPR studies of reaction pathways in titania photocatalyst-promoted alkylation of alkenes. https://core.ac.uk/download/29151174.pdf (accessed on 28th September 2025).

22. Králová, M. (2012). Self-cleaning properties of thin printed layers of titanium dioxide. https://core.ac.uk/download/30310816.pdf (accessed on 28th September 2025).

23. Lawton, L. A., Prabhu, R., Robertson, P. K. J., Tokode, O., et al. (2015). Controlled periodic illumination in semiconductor photocatalysis. https://core.ac.uk/download/33591761.pdf (accessed on 28th September 2025).

24. Mason, E. (2018). Synthesis and photocatalytic performance of TiO_2-CNT and magnetized Fe_3O_4-TiO_2-CNT multifunctional hybrids: A Pickering emulsion platform for organic degradation. https://core.ac.uk/download/211346833.pdf (accessed on 28th September 2025).

25. Silvestre, A. J., Albuquerque, M., Augugliaro, V., Barrocas, B., Bem, J., Cabrita, M., Chen, Y., et al. (2015). Synthesis of titanate nanofibers co-sensitized with ZnS and Bi_2S_3 nanocrystallites and their application on pollutants removal. http://arxiv.org/abs/1503.08693 (accessed on 28th September 2025).

26. Steiner, M. G. (2017). Photocatalytic decomposition of phenol under visible and UV light utilizing titanium dioxide-based catalysts. https://core.ac.uk/download/84124263.pdf (accessed on 28th September 2025).

CHAPTER 5

PHOTOCATALYTIC REACTOR DESIGN AND LIGHT SOURCES

Contents

INTRODUCTION

Photocatalytic processes have drawn significant interest due to their prospects for applying photocatalytic devices in environmental purification, energy transformation, and organic synthesis. At the core of these processes is the photocatalytic reactor, which is the means by which photonic sources of light interact with semiconductor materials. The construction of these reactors is crucial as it identifies the efficiency and effectiveness of the photocatalytic reactions within them. The application of effective light sources and reactor design for geometrical optimization can enhance mass transfer and light penetration, thereby enhancing reaction rates. Recent advancements in reactor designs, particularly in photochemical microreactors, demonstrate the ability to optimize geometries for improved gas-liquid mass transfer and photon absorption during catalysis. Operation and simplification of such devices reflect an impressive leap towards demystifying photocatalytic processes and achieving greater efficiency (Günther et al., 2016).

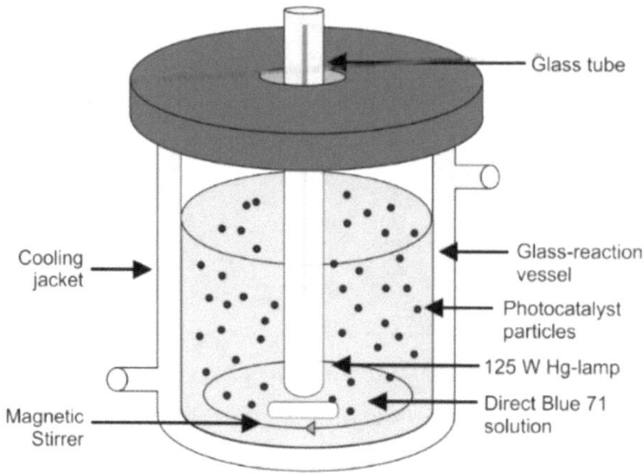

Figure 5.1. Photocatalytic reactor design.

Source: https://www.researchgate.net/figure/Fig-1-Photocatalytic-Reactor-Design_fig1_330982923.

Photocatalytic reactions are aided by light sources, as photons give the energy necessary to initiate the catalytic process. The effective use of both artificial and natural light sources can significantly impact the efficiency of photocatalytic reactions. As semiconductor photocatalysis has been enhanced, controlled periodic illumination is a method to optimize the energy transfer in these reactions. Besides increasing light absorption by the photocatalyst, this method can further result in increased reaction rates and yields of products. These techniques enable more precise control over the photocatalytic process, utilizing specific wavelengths of light that are best suited for the semiconductor materials in use. Empirical evidence supporting this paradigm shift demonstrates its potential implications for multiple applications in photocatalysis (Lawton et al., 2015). With continued work on photocatalytic reactor design, it is increasingly important to consider the interaction between reactor design and light source optimization. Continued research into new materials and reactor designs is crucial for achieving sustainable and efficient photocatalytic processes.

Of interest are advances in microflow reactors as examples of how ease of operation and modularity can significantly extend the repertoire of photoredox reactions. This access encourages ongoing research and experimentation among nonspecialist users, thereby widening the scope of photocatalytic technology. The value of integrating light source development with the best reactor design confirms the potential for innovation in the area. Throughout the conversation, understanding the intricate relationship between design choice and reaction efficiencies will become integral to future development in photocatalytic technology (Günther et al., 2016) (Figure 5.1).

Importance of Photocatalytic Reactors

Photocatalytic reactors hold significant relevance in addressing global challenges associated with water treatment and environmental remediation. Lack of clean drinking water affects millions of individuals worldwide, and hence, effective treatment methods are of utmost urgency. Photocatalysis using semiconductor materials, such as titanium dioxide, has been demonstrated to be an effective tool for degrading toxic microorganisms in water, including *E. coli* and *Cryptosporidium*. By harnessing solar energy to drive these reactions, photocatalytic reactors not only reduce the time required for traditional disinfection methods but also enhance the overall efficiency of the process. Such seminal development is not merely imperative, as highlighted by the need to deliver safe drinking water to both developed and developing nations, offering a crucial benchmark for subsequent developments in photocatalytic reactor design aimed at enhancing efficiency and sustainability (Byrne et al., 2014). Photocatalytic reactor design extends beyond mere performance; it is accompanied by optimization of multiple parameters to deliver peak photon flux and catalyst activity. Central among the critical factors are the irradiated volume-to-total volume ratio, catalyst loading, and flow rate, all of which significantly contribute to the efficiency of solar-based photocatalysis. Enhanced photon capture can be achieved with immobilized catalysts compared to conventional slurries because the former minimizes mass transfer limitation and maximizes catalyst stability. Furthermore, breakthroughs in material synthesis have led to innovative reactor designs that effectively harness solar energy. With advancements in research, the application of such designs is becoming increasingly evident, particularly in their ability to effectively reduce disinfection times, rendering photocatalytic reactors inevitable tools for future water treatment technologies (Byrne et al., 2014). In addition to water cleaning, photocatalytic reactors demonstrate impressive potential for application in environmental processes, such as CO_2 reduction. Recent work has demonstrated the efficacy of two-dimensional materials, such as graphene and transition metal chalcogenides, in the photocatalysis of CO_2 conversion to chemical feedstocks of merit. This is a double-win scenario, whereby photocatalytic reactors not only address the pressing needs for water disinfection but also contribute to mitigating the sources of climate change and ensuring resource sustainability. The continuous development of photocatalytic processes, including the formation of active and efficient catalysts, underscores the importance of innovation in reactor design and operating conditions. Coupled with these developments, photocatalytic reactors have the potential to be the cornerstone of an integrated approach to address global environmental concerns and stimulate the development of renewable energy use.

Overview of Light Sources Used in Photocatalysis

Photocatalysis, a process of advanced oxidation, largely relies on specific light sources to activate photocatalysts, enhancing the degradation of contaminants and organic content. Ultraviolet (UV) light sources, such as mercury vapor lamps, have been the traditional option historically since they can effectively excite traditional photocatalysts like titanium dioxide (TiO_2). Mercury exposure-related environmental and health concerns, however, necessitate shifting to alternative light sources. Recent advancements have introduced the study of visible-light-responsive (VLR) photocatalysts, which can utilize natural sunlight or LEDs, thereby increasing the efficiency and safety of photocatalytic systems and reducing the total operating cost of conventional UV sources. The integration of these newer technologies is central to the design of more environmentally friendly photocatalytic reactors that will be employable in different environments, from rural locations to urban wastewater treatment facilities. The incorporation of visible light sources within photocatalysis offers unique benefits, particularly in optimizing reactor designs. For instance, current studies indicate that VLR-TiO_2 photocatalysts can be excited using ambient solar light or extremely efficient blue LEDs, which offer a vast potential for reducing the energy demand of photocatalytic reactions. By adjusting the band gap energies of catalysts, researchers, such as those at the Kennedy Space Center, demonstrated more efficient photocatalytic activity in situations where traditional UV light sources would be impractical. The enhanced photocatalytic efficiency of VLR materials enables the construction of reactors that can operate under a wider range of conditions, making them suitable for both Earth and space applications. The possibility of harnessing sunlight or artificial light sources to their maximum potential indicates a key breakthrough for greener photocatalytic technologies. Moreover, the universality of applying light sources to photocatalysis has driven research on hybrid systems that combine diverse light technologies for enhanced pollutant breakdown. Combined photocatalytic systems, including either electrohydraulic discharge or sonophotocatalysis with conventional light sources, best represent this pioneering venture. These combinations have proved effective in treating wastewater containing resistant organic pollutants. Through the incorporation of multi-component methods into reactor design, scientists have been able to overcome the drawbacks of individual treatment processes, resulting in improved efficiencies and decreased operational costs. Consequently, as photocatalytic reactor engineering continues to evolve, the careful selection of light sources remains critical for achieving significant reductions in major environmental pollutants and optimizing photocatalysis for effective wastewater treatment and pollution control (Abou-Ghazala et al., 2014; Bauer et al., 2025).

5.1. TYPES OF PHOTOCATALYTIC REACTORS

Photocatalytic reactor design is essential in improving the efficiency of photocatalytic processes for environmental and chemical applications. The most commonly known is external illumination reactors, which utilize an external light source to catalyze the process. The reactors typically assume varied geometric arrangements, such as annular or circular geometries, to maximize light distribution and provide uniform irradiation of the photocatalyst. The choice of materials used in the reactor's construction, particularly glass, is essential for effective light penetration and protection against potential chemical corrosion. As cited in recent research, efficient design considerations can have a heavy impact on the overall efficiency of the photocatalytic reaction towards organic pollutant breakdown. The findings highlight the importance of reactor design for specific applications, thereby optimizing the impact of semiconductor photocatalysis in real-world scenarios (Lee et al., 2004). Thin-film reactors, with the photocatalyst coated as a thin film upon a substrate, constitute another significant category of photocatalytic reactors. The configuration provides a larger area for light-pollutant interaction and adsorption, which is a critical factor for achieving higher reaction rates. Thin-film reactors, in the majority of cases, can be designed in a manner that increases light intensity exposure with less shadowing effect. In addition, thin-film reactors have been found effective in treating various contaminants, such as azo dyes found in wastewater from the textile industry. The advancement of monitoring technologies over the last few years has made it possible to conduct real-time assessments of pollutant degradation in these reactors, providing information on process kinetics and reaction mechanisms. This new approach represents a significant step towards more efficient and optimized reactor designs that can be tailored to specific applications and pollutant types (Mart et al., 2019) (Figure 5.2).

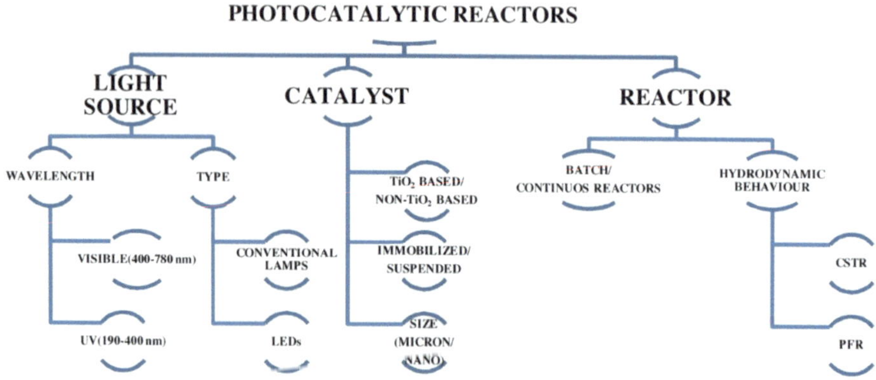

Figure 5.2. The design of photocatalytic reactors is pivotal in enhancing the efficiency of photocatalytic processes.

Source: https://www.sciencedirect.com/science/article/abs/pii/S0263876219305647.

Designing solar photocatalytic reactors is a new milestone in the field of photocatalysis, where chemical reactions are initiated by renewable energy sources. Solar photocatalytic reactors can be designed as either non-concentrating or concentrating systems, each offering advantages based on application and geographical location. Non-concentrating reactors tend to use diffuse sunlight and are thus suitable for decentralized applications, while concentrating solar photocatalytic reactors collect and focus sunlight to achieve elevated reaction temperatures and rates. Such designs could quite significantly enhance the efficiency of semiconductor photocatalysis, particularly in sun-abundant areas. Although researchers are still working to create new designs and operating conditions, the future for solar photocatalytic reactors is promising, as they are expected to transform wastewater treatment operations, making pollution control sustainable and efficient (Lee et al., 2004).

5.1.1. Batch Reactors

Batch reactors play a key role in photocatalytic reactor design and are an effective platform for a broad array of chemical transformations, particularly in environmental remediation and organic synthesis. Batch reactors also contrast with continuous flow systems in that reaction conditions like light intensity, temperature, and concentration of reactants can be rigorously controlled, which is highly essential for optimal photocatalytic efficiency. These reactors are useful for small-scale operations and research environments where flexibility and adaptability are most desirable. Secondly, real-time observation of reaction development provides better insights into the kinetics involved, thereby making it possible to produce more efficient photocatalytic processes. Operation with sophisticated monitoring techniques in batch reactors significantly improves the assessment of photocatalytic activities (Figure 5.3).

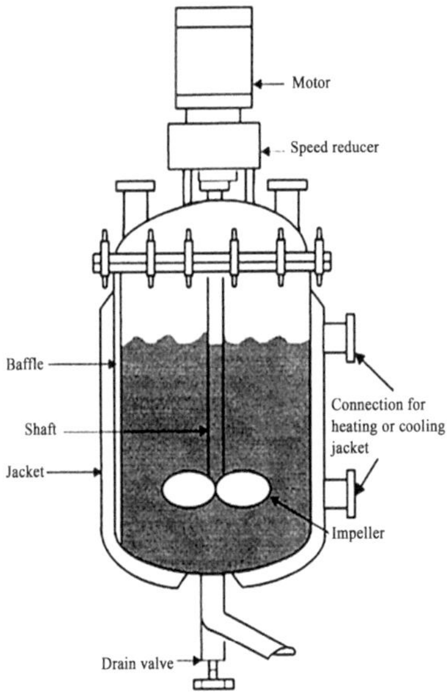

Figure 5.3. Schematic diagram of batch reactor.

Source: https://www.researchgate.net/profile/Enas-Ahmed-5/publication/ 259 169 382/figure/fig9/AS:267863705518104@1440875092597/Schematic-diagram-of-batch-reactor.png.

For instance, *in situ* online monitoring of decolorization in dye degradation research shows how batch reactors are maximized. By examining the spectrum measured *in situ* by light scattering from the photocatalyst in suspension, researchers can accurately measure absorbance and decolorization. This method is not only effective in reducing disturbances caused by sample withdrawal but also serves as a representative example of a robust approach for studying reaction dynamics in complex batch reaction systems. Such breakthroughs positively reflect the significant role that batch systems are playing in the advancement of photocatalytic technology, particularly in environmental design (Mart et al., 2019). In addition, batch reactors are used as an essential platform to study new radical generation processes in photocatalysis, which are significant for expanding the scope of synthesis pathways. The ground-breaking methodology of generating dichloromethyl radicals within a potassium poly(heptazine imide) photocatalyst through irradiation by visible light illustrates the potential of batch systems to facilitate new chemical transformations. By enabling the coupling of γ,γ-dichloroketones from various enones, such reactors thus prove to be efficient in delivering valuable organic matter. The capability of the quasi-homogeneous flow-photoreactor to markedly enhance productivity is additional proof of batch reactors' versatility to integrate into more complex reaction systems and, as a result, strengthen their role in pharmaceutical as well as environmental chemistry (Antonietti et al., 2020).

5.1.2. Continuous Flow Reactors

Flow reactors are of paramount significance in photocatalytic reaction development, particularly due to their capability to enhance the interaction of the reactants and achieve maximum exposure to light. Through ordinary batch reactors, discontinuous feeding of the reactant may cause inefficient mixing and long reaction duration.

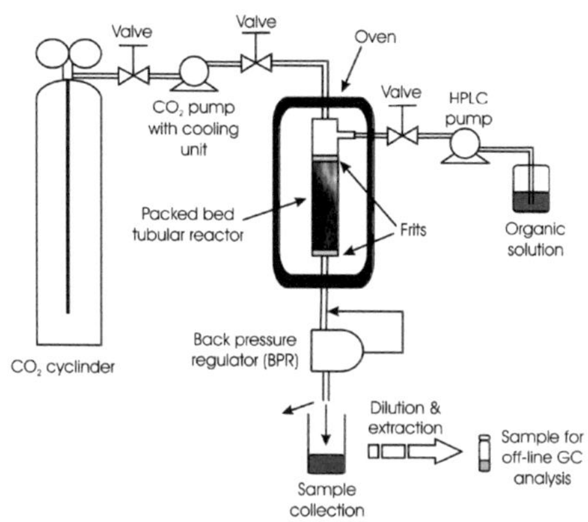

Figure 5.4. Schematic diagram of the continuous flow system.

Source: https://www.researchgate.net/figure/Schematic-diagram-of-the-continuous-flow-system-The-reactor-shown-is-the-39-mm-id_fig2_259456911.

Continuous flow systems, however, ensure a constant and homogeneous supply of the reactants, which is vital in achieving higher reaction efficiency. Such a configuration minimizes dead volumes and facilitates better mass transfer, particularly where rapid mixing is a prerequisite in gas-liquid reactions. All these advances significantly reduce processing time, minimizing reaction times from hours in conventional batch systems to minutes in flow reactors. This underscores the importance of integrating continuous flow technology into photocatalytic reactor design for improved efficiency (Günther et al., 2016) (Figure 5.4).

Continuous flow reactor design can be maximized to enhance semiconductor photocatalysis performance significantly. Included among the key considerations are the physical design of the reactor, i.e., the type of flow patterns within the reactor, which could directly influence light availability for photocatalytic reactions. Annular and tubular geometries can optimize the path of light, maximizing the number of photons absorbed by the photocatalyst. As surface illumination is a highly significant variable in reaction rate determination, accurately integrating light sources—whether LED-based or solar-powered—can also enhance the efficacy of photocatalytic reactions within continuous flow reactors. Such design principles are the very foundation for designing efficient photocatalytic continuous flow reactors to address intricate problems in environmental cleaning (Lee et al., 2004). Further, the application of continuous flow reactors in photocatalytic reactions offers significant advantages in scalability and operational durability. The modularity of the apparatus facilitates easy integration into current installations, providing a platform for realizing improved photocatalytic technologies in viable applications. Continuous operation allows for consistent reaction conditions, which tend to present lower energy consumption compared to batch operation. By utilizing the features of continuous flow configurations, researchers and engineers can help develop more efficient municipal waste management and water treatment systems. As the technology in photocatalytic reactors advances, the role of continuous flow reactors will be increasingly prominent in paving the way for future solutions to environmental problems (Lee et al., 2004; Günther et al., 2016).

5.1.3. Fixed-Bed Reactors

Fixed-bed reactors have emerged as a frontrunner in photocatalytic reactor design to achieve the maximum degradation of organic pollutants. Fixed-bed reactors use a fixed bed of photocatalyst material, usually semiconductor particles like titanium dioxide (TiO_2), which are irradiated with light sources to facilitate photocatalytic reaction. The complexities of the fixed-bed reactor design relate to factors such as illumination distribution, fluid flow, and the interaction between pollutants and photocatalysts. Penetration of light into the reactor is essential; thus, the configuration often employs a transparent housing material for maximum illumination. Ideal reactor design couples the reaction kinetics and mass transfer properties such that the pollutants are effectively contacted with the photocatalyst, enhancing reaction efficiency and rate (Lee et al., 2004). Scalability and operation lifetime of the systems also pose challenges in the design of fixed-bed reactors. The packing structure of the photocatalyst within the reactor significantly influences mass transfer efficiency; a close-packed structure may generate low flow velocities, resulting in potential channeling effects or dead zones. Further, the design of the reactors should ensure easy maintenance and regeneration of the photocatalyst, as catalyst deactivation can have a dramatic influence on performance. With better reactor design—e.g., application of pulsed or intermittent irradiation and introduction of fiber optics for better delivery of light—researchers are already discovering new ways to break through these limitations. For instance, the establishment of fiber-optic cable reactors shows the potential for more effective immobilization of photocatalysts and greater availability of surface area (Hoffmann et al., 1997) (Figure 5.5).

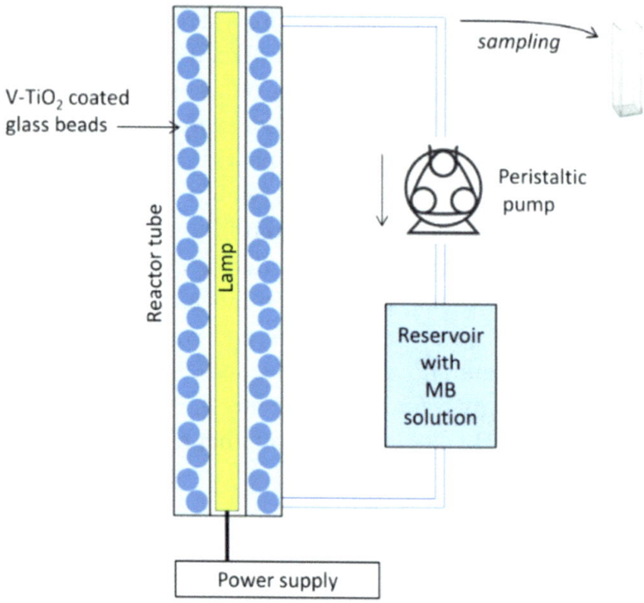

Figure 5.5. Schema of fixed-bed photocatalytic reactor.

Source: https://www.researchgate.net/figure/Schema-of-fixed-bed-photocatalytic-reactor_fig9_356617177.

The use of solar light sources in fixed-bed reactors presents new challenges and opportunities. The application of natural sunlight for photocatalytic purposes necessitates designs that effectively harvest solar radiation, commonly using concentrating collectors. Such systems may include tracking systems to achieve optimal capture of light throughout the day. Increased quantum efficiencies achieved with newer designs illustrate the practical benefits of integrating solar energy with fixed-bed reactor design. The singular advantage of using natural light is that it reduces operational costs and facilitates environmental sustainability in remediation processes. Designing products that can accommodate variable light levels and operating conditions will be key in the promotion of fixed-bed reactor use in photocatalytic pollutant degradation (Lee et al., 2004).

5.2. DESIGN CONSIDERATIONS FOR PHOTOCATALYTIC REACTORS

The design of a photocatalytic reactor is based on several fundamental concerns that directly influence the efficiency and effectiveness of the photocatalytic process. An important one is the reactor's configuration, which plays a major role in determining the photoreactor's hydrodynamics and mass transfer rates. For instance, various configurations like batch, continuous flow, and annular reactors all possess different sets of merits and demerits. Particularly, annular reactors have been observed to achieve maximum contact between the photocatalyst and impurities, thereby increasing degradation rates, as demonstrated in experiments involving photocatalytic phenol degradation in aqueous suspensions under various conditions (Neto et al., 2009). Such configurations must also be aligned with the intended application to ensure the reactor operates under the desired conditions for achieving the maximum rate of photocatalytic reactions. A second important design feature in photocatalytic reactor design is the selection of appropriate light sources. The early photocatalytic systems employed mainly ultraviolet (UV) light for exciting the catalysts, such as titanium dioxide, which does not allow high flexibility in operation due to the fact that it possesses only a limited activation spectrum. Recent innovations focused on visible-light-responsive catalysts, which will enable the use of more wavelengths of light, i.e., solar or LED light sources (Bauer et al., 2025). This transition not only enhances the overall economic viability and efficiency of photocatalytic processes but also mitigates problems of safety and environmental issues created by mercury UV lamps. The utilization of visible light sources in reactor systems significantly expands the types of possible applications while simultaneously improving energy efficiency, hence, more sustainable processes. Finally, the choice of photocatalyst material is a critical design parameter that influences reaction kinetics, efficiency, and reactor performance. The limitation of UV light stimulation disadvantages traditional photocatalysts like TiO2; therefore, new materials capable of working with optimal visible light are being explored (Bauer et al., 2025). The preparation of catalysts such as Ag-doped TiO_2, with reduced band gaps, proves the importance of material science in the synthesis of photocatalysts. Additionally, factors such as immobilization techniques for catalysts, which enable the photocatalyst to retain its activity throughout the reaction process, are essential for multiple runs (Neto et al., 2009). Thus, photocatalytic reactor design must adopt a holistic approach, considering light source compatibility, reactor design, and catalyst properties to offer maximum performance in all applications.

5.2.1. Material Selection for Reactor Construction

Material selection for the fabrication of reactors is a critical factor that significantly dictates the effectiveness and longevity of photocatalytic reactors. The materials, besides being chemically stable to exposure to high-energy light, must also possess enhanced thermal resistance.

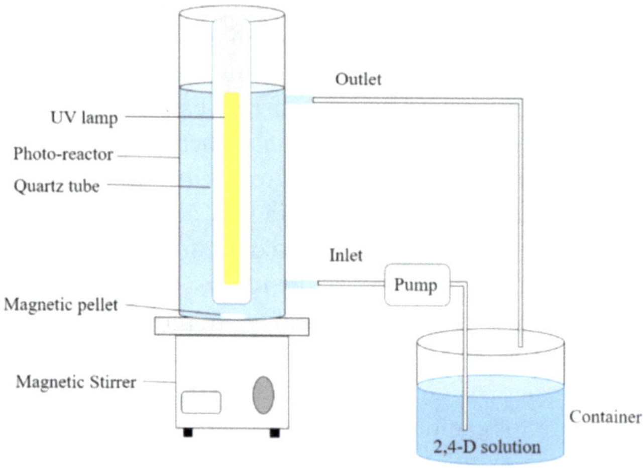

Figure 5.6. Selection of materials for reactor construction is a critical factor that significantly impacts the efficiency and longevity of photocatalytic reactors.

Source: https://www.researchgate.net/figure/Schematic-representation-of-a-continuous-photocatalytic-reactor_fig1_349665039.

This is particularly vital as the risk of high temperatures and reactive conditions accompanies photocatalysis. For instance, titanium dioxide is usually used as a material owing to its photonic and catalytic properties; however, its use in reactor designs implies that its resistance when exposed to light and heat has to be a consideration. Compatibility of the materials with the reactants and potential byproducts requires a deeper understanding to prevent degradation or contamination, which could cause reactor performance to deteriorate over time. In addition to chemical stability, the mechanical characteristics of selected materials must take priority. Photocatalytic reactors must withstand operational strains like pressure fluctuations and thermal expansion. Strict analysis of the material in terms of elasticity, tensile strength, and corrosion resistance is necessary (Figure 5.6).

For example, inclusion of materials with high resistance to oxidative conditions can enhance the overall durability of the reactor system. It is equally crucial to consider the risk of scaling and fouling, which can occur due to deposition of reactants or by-products on reactor surfaces. Utilization of durable and appropriate materials can lead to lower maintenance costs and extended periods of operation, thereby improving the economic viability of photocatalytic technology overall. Material choice decision-making in photocatalytic reactors is also very crucial regarding sustainability. As concerns over the environment gain greater prominence, concern over the use of eco-friendly and recyclable materials has grown. This is particularly relevant for reactor design lifecycle assessment as the environmental effects of construction materials may have a profound impact on the overall sustainability of the technology. Integration of research findings of controlled ecological life support systems (CELSS) emphasizes the need for interdisciplinary thinking in optimizing photocatalytic reactor designs (Carden et al., 2025). In addition, workshop guidelines on ecology and nutrition for CELSS development emphasize the integrated understanding of sustainability principles that can be applied in material choice (Carden et al., 2025). By emphasizing the sustainable choice of materials, engineers can contribute to the greater goal of reactor development that not only functions efficiently but is also green.

5.2.2. Reactor Geometry and Configuration

From the reactor design for a photocatalysis perspective, reactor configuration and geometry are critical elements in determining the system's efficiency and overall performance. Reactor geometry defines the distribution of light, the flow pattern of the reactant, and the exposure of the catalyst to the irradiation. For instance, a good reactor can allow light to interact effectively with the photocatalyst, which translates to improved contaminant degradation. The geometry must permit homogeneous distribution of light such that the entire area of the photocatalyst is activated uniformly. Specifically, reactors with various geometric shapes, such as internal baffles or structured surfaces, can enhance mixing and light capture, resulting in enhanced photocatalytic performance in pollutant degradation processes. Moreover, reactor design selection can contribute significantly to operating parameters, such as residence time and scalability. Flow reactors, for instance, have the advantages of constant product quality and the ability to treat larger amounts of reactant. Batch reactors may be more suitable where extensive treatment of specific pollutants is required. Design must consider not just the type of photocatalyst used but also its application, such as in water or air treatment. As depicted through the use of spinning disk reactors, which excel in nanoparticle synthesis, specific geometries of the reactor can maximize mixing and reduce reaction times significantly. This highlights the significance of thorough screening of the reactor geometries that are unique to the target photocatalytic process (Chianese et al., 2015) (Figure 5.7).

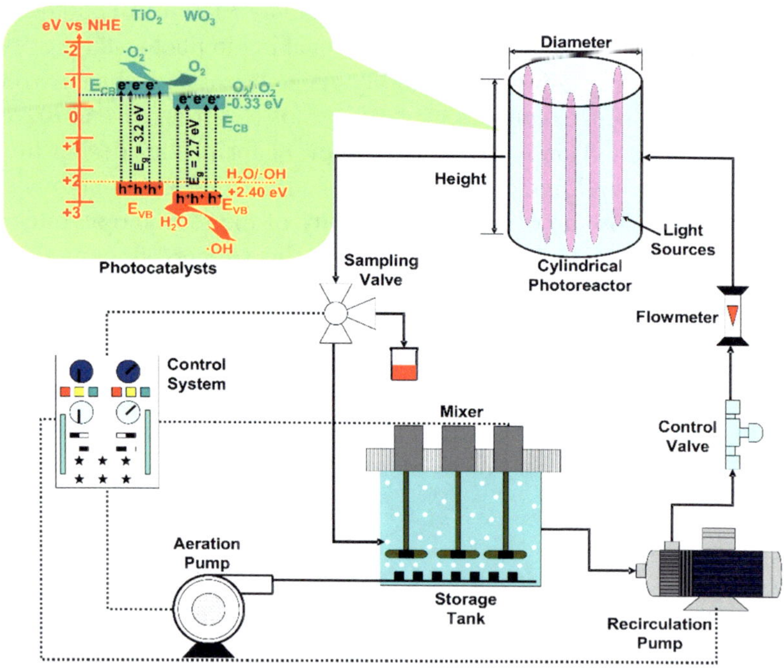

Figure 5.7. Components of cylindrical photoreactors (TiO_2/WO_3 heterostructure photocatalyst).

Source: https://www.mdpi.com/2073-4344/11/5/556.

Further, advancements in measurement techniques, such as *in situ* monitoring, have facilitated the ability to comprehend the reactor conditions during photocatalysis better. The integration of technologies that permit real-time monitoring of decolorization rate and absorbance is quite helpful when optimizing the performance of a reactor. For example, using light scattering to perform photocatalyst dynamics measurements *in situ* inside the reactor eliminates the need for sample extraction, providing more precise control over process conditions. By measuring the reactor's geometry and flow conditions, combined with advanced monitoring techniques, scientists can enhance the effectiveness of photocatalytic systems. This multilateral approach acknowledges the importance of balancing reactor design intricacies with performance, thus creating a way for improved photocatalytic uses (Mart et al., 2019).

5.2.3. Integration of Photocatalysts Within the Reactor

Introduction of photocatalysts into a photocatalytic reactor is key to increasing reaction efficiency and efficient light absorption. Photocatalysts, often materials such as titanium dioxide, are catalytic agents that initiate a chemical reaction upon exposure to light. Their use must be well planned to offer maximum surface area and interaction with light, which can be achieved through various mechanisms such as immobilization onto supports or embedding into composite materials. Recent advancements in the synthesis of nanocomposite films exemplify this approach, where halloysite clay and nanocrystalline TiO2 are mixed to form a stable photocatalytic surface capable of efficiently facilitating photochemical reactions. Such advancements not only improve the stability and dispersibility of the photocatalysts in the reactor but also ensure an even distribution of light, therefore enhancing the overall catalytic activity for chemical processes like the degradation of dyes and the removal of pollutants. To further increase photocatalyst effectiveness, recent research highlights the use of composite films that encapsulate photocatalytic material within a structured medium. Scientists have succeeded in synthesizing films using sol-gel methods, composed of TiO2 nanoparticles homogeneously dispersed on halloysite, which yield high light absorption and catalytic activity. The integration has a reduction in particle agglomeration common in standard photocatalytic systems, which decreases effectiveness. The integrated films have been explained in structural properties, supporting low-level cracking and robust surface properties that are suitable for photocatalytic processes. Such films have demonstrated high-performance azo dye degradation, supporting their application as effective photocatalysts in reactors. The ability to optimize the loading of halloysite and incorporate metallic additives further attests to the versatility of such integration, with potential enhanced photocatalytic efficiencies for wastewater treatment (Kaplani et al., 2014). Additionally, advances in photoreactor design are also crucial to enable the practical application of integrated photocatalysts. Photoreactor optimization is based on knowledge about the dynamics of radiation transport in the reactor and the in-built properties of the used photocatalysts. Innovative methods, such as numerical modeling, can model light distribution and calculate optimal reactor shapes that maximize photocatalytic interaction. Furthermore, more recent work has enlightened the potential of additive manufacturing processes to create tailored reactor prototypes, particularly to intended photocatalytic reactions. Such designs enhance reactor working flexibility by making them customizable based on the involved chemical reactions. Lastly, the integration of engineered photocatalysts with optimized reactor geometries opens up ways to enhance solar-powered chemical syntheses as well as pollution clean-up techniques. This strategy of global integration ensures the attainment of sustainability goals by means of sophisticated photocatalytic technologies (Kant et al., 2023).

5.3. LIGHT SOURCES FOR PHOTOCATALYSIS

Optimal selection of appropriate light sources is an essential element of design and optimization of photocatalytic reactors. Traditional light sources like ultraviolet (UV) lamps have been conventionally favored since they are capable of providing a suitable spectrum for the excitation requirements of common photocatalysts like titanium dioxide (TiO_2). However, constraints on the application of UV radiation—such as high energy requirements and safety issues—make alternative technologies highly appealing.

Figure 5.8. Photocatalysis setup with UV lamps.

Source: https://external-content.duckduckgo.com/iu/?u=https%3A%2F%2Fwww.researchgate.net%2Fprofile%2FSabyasachi-Mishra%2Fpublication%2F228619833%2Ffigure%2Ffig1%2FAS%3A301824646107141%401448972011702%2FPhotocatalysis-setup-with-UV-lamps.png&f=1&nofb=1&ipt=efcedd64e144556e813b344a84cd0251464e1c763fd4350bf501a7a0793b8807.

Advances in light-emitting diode (LED) technology over recent years provide an attractive option. LEDs can be made to emit specific wavelengths, so photocatalytic reactions of particular interest are possible without the wasteful energy loss inherent in broad-spectrum sources. Furthermore, their strength and power savings position LEDs as viable light sources, potentially reducing power expenses in photocatalytic systems and providing a greener edge to photocatalysis. Of equal significance is controlled periodic illumination, which has gained empirical support in enhancing photocatalytic effectiveness. Developed early in the 1990s, the technique involves modulation of light irradiation to optimize semiconductor photocatalysis. By controlling the duration and intensity of light exposure, scientists have greatly improved the photocatalytic reaction efficiency. This strategic approach is based on utilizing the inherent characteristic of semiconductor materials, such as TiO_2, which are photo-sensitive. The outcomes of various studies have illustrated that controlled light increases reaction rates and broadens the horizon of photocatalysis applications across multiple sectors, including water purification and environmental remediation. Hence, integrating the technique into reactor design can potentially facilitate more effective and economically viable photocatalytic processes (Lawton et al., 2015) (Figure 5.8).

Secondly, using solar energy in photocatalytic reactions is a very advanced approach to clean water purification techniques. As the demand for clean drinking water globally continues to increase, technologies that utilize solar power remain highly relevant, particularly in areas where access to traditional power supplies is not readily available. Solar-irradiation-assisted photocatalysis can be employed as a viable method for annihilating the pathogens from infected water, addressing the two-fold challenge of water scarcity and health risks. Research highlights the effectiveness of combining solar power with high light-absorbing and photon flux photocatalyst reactor designs, which optimizes the purification time to nearly zero. In addition, advancements in catalyst technology, including immobilized photocatalysts, have shown promise in enhancing the overall performance of solar-powered systems. Thus, the future of photocatalytic water treatment lies in the sacred union of efficient light sources and innovative reactor designs unique to sustainable water quality management practices (Byrne et al., 2014).

5.3.1. Natural Sunlight as a Light Source

The application of natural sunlight as a light source in photocatalytic reactor design is increasingly viewed as having the potential to solve energy sustainability and mitigate climate change. Being a renewable and inexhaustible source of energy, sunlight provides a reliable source of excitation necessary for photocatalytic reactions that facilitate the process of converting carbon dioxide (CO_2) into useful chemical fuels. Present research indicates that combining natural sunlight with photocatalytic reactions not only enhances CO2 reduction efficiency but also enables the implementation of these technologies on a broader scale for other environmental applications. This application of sunlight aligns with efforts to discover sustainable energy alternatives based on existing natural resources, offering a revolutionary alternative to conventional chemical production routes. Besides, the application of photocatalytic technology with solar light has continued to develop, particularly in the field of photoelectrocatalytic (PEC) systems. The systems generate chemical products under outdoor solar illumination, showing their potential use beyond laboratory setups (Figure 5.9).

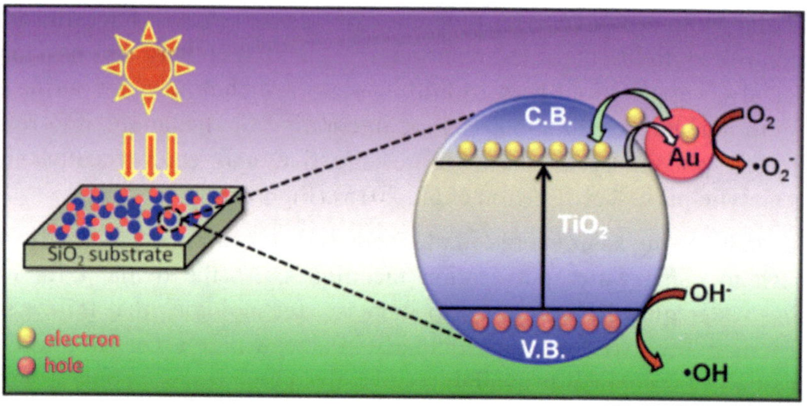

Figure 5.9. Schematic representation of the photocatalytic degradation process of MB dye under sunlight with the Au–TiO_2 nanohybrid.

Source: https://www.researchgate.net/figure/Schematic-representation-of-photo-catalytic-degradation-process-of-MB-dye-under-sun-light_fig11_335223843.

The effectiveness of PEC systems is primarily attributed to improvements in photocatalyst design, such as doping and nanostructured material synthesis, which enhance light absorption and charge transfer. Such integration facilitates enhanced CO_2 conversion rates, making PEC a significant potential for sustainable chemical manufacturing. Utilizing sunlight in these processes not only supports cleaner energy solutions but also provides the opportunity for new reactor designs based on practical efficiencies. Furthermore, skilled investments in solar-driven chemical processes are evident in funding trends from organizations like the European Commission, which have seen increasing investments in research and development in photocatalytic and photoelectrocatalytic technology. Since 2019, disbursements of more than 75 million euros reflect a societal shift towards embracing solar energy solutions for synthesizing chemicals. Evaluating technology readiness level projects allows the researchers to identify turning point opportunities and challenges associated with scaling the systems for use in commerce. The transition to utilizing natural sunlight in photocatalysis not only encapsulates a promising strategy for combating greenhouse gas emissions but also fosters economic growth through making solar-powered chemistry the champion of sustainable industrial processes. This trend highlights the vital role of natural sunlight as a substitute light source in inspiring photocatalytic reactor design and technological progress.

5.3.2. Artificial Light Sources: LEDs and UV Lamps

Photocatalytic reactor design relies heavily on the use of artificial light sources, in the form of LEDs and UV lamps. The light sources play a central role in activating photocatalysts, which are required for pollutant degradation in several applications, including wastewater treatment. Traditional UV lamps, such as mercury vapor fluorescent sources, have been popular because they can emit light at wavelengths that effectively excite catalysts like titanium dioxide.

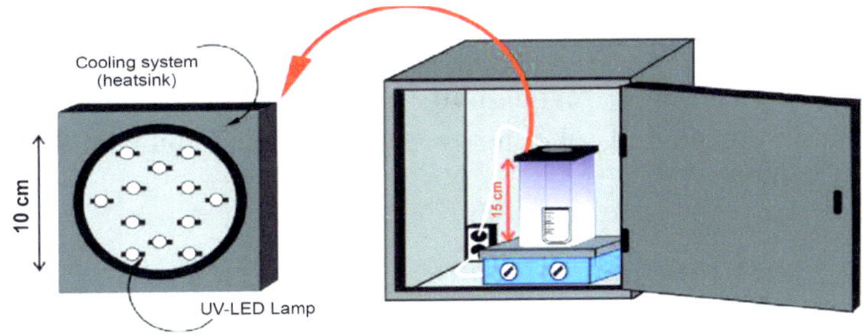

Figure 5.10. Design of photocatalytic reactor with UV-LED lamp system.

Source: https://external-content.duckduckgo.com/iu/?u=https%3A%2F%2Fwww.researchgate.net%2Fpublication%2F358288529%2Ffigure%2Ffig2%2FAS%3A1-1431281145664805%401681454369722%2FDesign-of-photocatalytic-reactor-with-UV-LED-lamp-system.png&f=1&nofb=1&ipt=b64c3732d93c3dcbb7dd18b55f4e1cd850dd8d5b2c31895bbdeded315a1c2c99.

However, their widespread use is tempered by security concerns, particularly in environments like space exploration where mercury exposure is a significant risk, prompting scientists to look for alternatives such as LEDs, which are characterized by efficiency and security (Bauer et al., 2025). The creation of visible-light-responsive (VLR) titanium dioxide has massively expanded the number of artificial light sources available for photocatalyst use. This innovation enables the use of highly efficient LEDs, reducing reliance on UV sources and decreasing electricity use—a significant aspect of reactor design. The shift to LEDs not only enhances safety by eliminating harmful chemicals but also aligns with the call for green methods in photocatalytic systems. LEDs can be designed to emit wavelengths that are optimal for matching the absorption characteristics of optimized photocatalysts, yielding the highest photocatalytic performance for volatile organic compounds and other processes for degrading contaminants. This light source technology revolution is a paradigm that can significantly enhance the applicability and availability of advanced oxidation processes in various environments (Tlatelpa et al., 2019) (Figure 5.10).

Moreover, the geometry of photocatalytic reactors using diverse light sources should be considered during design, including reactor geometry and working conditions. The geometric configuration of LEDs or UV lamps should ensure homogenous irradiance in the photocatalytic surface to ensure effective reaction. Successful modeling of light distribution has been proposed to enhance reactors by ensuring homogeneously activated photocatalysts. The use of novel techniques, such as producing LED arrays from models of constant irradiance, exemplifies the innovative approaches scientists take to optimize photocatalytic reactors. Hence, with the development of next-generation lighting technology, they have the potential to revolutionize photocatalytic processes, facilitate environmental remediation, and foster energy-efficient industrial processes (Bauer et al., 2025).

5.3.3. Comparison of Light Source Efficiencies

The efficiency of the light source is a crucial parameter in photocatalytic reactor operation and design, particularly when comparing different semiconductor materials. Light sources have other spectral distributions, which significantly affect the photocatalytic reaction. For instance, ultraviolet lamps, which are typical in laboratory use, provide intense light that can efficiently excite electrons in semiconducting materials such as titanium dioxide. However, this technology is power-intensive and may not prove to be an economic option for bulk applications. New-generation light-emitting diodes (LEDs) are, however, gaining attention because they use less energy, last longer, and can emit a specific wavelength depending on the absorption characteristics of photocatalysts. In optimizing reactor design and enhancing photocatalyst performance, particularly for pollutant degradation in water treatment systems (Lee et al., 2004), recognizing differences among light sources is critical. The comparison of advanced versus conventional light sources in this scenario indicates advances in photocatalytic reactor technologies. Specifically, researchers have been examining the integration of solar power into reactor designs, which is a green alternative to traditional UV light sources. Concentrating sunlight-based photoreactors can improve photocatalytic reaction efficiency while reducing the need for electricity. Periodic illumination methods under control represent yet another instance of how light source applications are being modified to enhance photocatalytic efficiency. These methods allow more strategic photon delivery, possibly improving the reaction kinetics with increasing time. Research into solar-powered photocatalytic systems unearths a significant breakthrough in developing environmentally friendly solutions to address key water quality problems and achieve more efficient and cost-effective purification techniques (Lawton et al., 2015). In conclusion, relative efficiency comparisons of light sources in photocatalytic reactors reveal certain advantages and disadvantages for each type. Conventional UV lamps possess high catalytic activity, but their power input and economic costs may deter widespread adoption. Advancements in LED technology and solar energy harnessing present promising avenues for enhancing both the environmental acceptability and efficacy of photocatalytic processes. Continued search for efficient light sources is not only essential for improving the performance of these reactors but also for aligning them with the world's goals for saving energy and reducing environmental impacts. With progress in research and innovations, the application of optimized light sources will ultimately be a central issue in the future of photocatalytic reactor design, offering improvements in water treatment and pollutant degradation systems (Lee et al., 2004).

5.4. OPTIMIZATION OF LIGHT UTILIZATION

The utilization efficiency of photocatalyst light is necessary for optimal efficiency in photocatalytic reactions. Photocatalytic reactors use the energy from light to drive chemical reactions primarily for wastewater treatment and hydrogen production. The design of the reactor largely determines the light absorption properties of the photocatalyst and the overall photon trapping efficiency. Good optimization involves adopting a multi-dimensional strategy, including the customization of the photocatalyst's surface area and morphology to enhance light interaction.

In addition, the use of light-scattering layers in the reactor can distribute light evenly, hence improving the efficiency of photocatalytic reactions. Therefore, optimizing light usage not only accelerates reaction rates but also contributes to making photocatalytic systems more sustainable overall. This underlines the need for new design approaches that balance aims such as energy efficiency and environmental protection. Besides reactor design, the selection of light sources is just as crucial to optimizing light usage in photocatalytic reactions. Natural sunlight is optimal; however, artificial sources like LEDs are also reasonable options due to their tunability in wavelengths and lower energy requirements. Engineering the spectral light output of sources to align with the absorption features of a photocatalyst in question can significantly enhance the kinetics of the reaction (Figure 5.11).

Figure 5.11. Schematic illustration of photocatalytic reactor setup.

Source: https://www.researchgate.net/figure/Schematic-illustration-of-photocatalytic-reactor-setup_fig16_339941138.

For instance, using LEDs with emission wavelengths closely aligned to the photocatalyst's absorption wavelength can provide the highest energy utilization, thus reducing losses. Second, utilizing periodic illumination, which simulates natural processes, can further enhance the performance of photocatalytic reactors. Through optimal usage of different light sources and modulation techniques in combination, researchers can design effective systems that optimize the energy conversion efficiency in photocatalytic operations. Finally, integration of novel materials and technologies into photocatalytic reactors can lead to paradigm-shifting gains in light utilization. Nanostructured photocatalysts, for example, those synthesized via metamaterials, can manipulate light at the nanoscale, leading to enhanced absorption rates and consequent catalytic activity. New synthesis strategies for catalysts, as indicated by recent investigations, are crucial for ensuring long-term operational stability at high activity levels. Utilizing renewable energy resources, in combination with improved process designs, enables the production of sustainable energy solutions. Predictive models that consider optimal configurations and light source-photocatalyst interactions, as proposed by recent research, can guide the development of custom-designed reactors that are capable of efficiently utilizing light. Through embracing these new techniques, the utilization efficiency of light in photocatalytic systems can radically transform the domains of renewable energy and environmental cleaning, ushering in cleaner technological utopias (Berrill et al., 2013; Ball et al., 2025).

5.4.1. Light Intensity and Distribution

The performance of photocatalytic reactors is incredibly reliant on light distribution and intensity, which are critical factors in the optimization of the photocatalytic reaction. The intensity of the light has a direct impact on the rate of the photocatalytic reaction, where a higher photon flux will usually result in higher production of reactive species critical in pollutant degradation. Besides, the distribution of light within the reactor is essential to provide an even illumination of the photocatalytic surface. Ineffective light distribution can result in hot spots or light-starved areas that limit reaction efficiency. Therefore, reactor design permitting even light penetration is essential. All variables, such as choice of light source, reactor shape, and light intensity, and properties of the photocatalytic material, must be considered for enhancing light intensity and achieving uniform distribution inside the system (Lee et al., 2004). Second, the choice of light source plays a vital role in determining both intensity and spectral nature of the light used in the photocatalytic process. Familiar light sources are ultraviolet (UV), visible light, and sunlight, each with its own strengths and weaknesses. For instance, UV lamps can provide high intensity over a broader range of wavelengths well-suited to maintaining photocatalytic reactions, but they may be less energy efficient in wide-area applications. Alternatively, using sunlight as a light source offers a renewable option, but it has the limitations of fluctuating intensity and seasonality, which must be addressed in reactor design to achieve maximum efficiency. The light source will react differently with the photocatalyst; therefore, understanding the photonic properties of the materials involved is essential to optimize reaction rates and product yield, ultimately leading to better reactor design practices (Mart et al., 2019). In parallel with the selection of light sources, advancements in photocatalyst materials have considerably enhanced the capability for efficient light absorption and effective degradation of contaminants. Nanostructured photocatalysts such as metal-doped titanium dioxide exhibit improved photocatalytic activity due to enhanced light-harvesting ability. The surface area, morphology, and electronic properties of these materials play a critical role in establishing their activity under varied light conditions. Thereafter, making the light source consistent with the inherent absorption properties of the photocatalyst can significantly increase the process efficiency. Finally, the interaction between light intensity, distribution, and photocatalytic material properties should be investigated in a controlled manner in reactor designs to realize maximum photocatalytic performance for various applications, such as water treatment and air purification (Lee et al., 2004).

5.4.2. Wavelength Selection for Maximum Absorption

The photocatalytic reaction fundamentally depends on the photocatalyst's good light absorption, which in turn affects its efficiency in pollutant degradation and overall reactor performance. Understanding the principles of wavelength selection is crucial because different photocatalysts have different absorption characteristics in different light spectra. For instance, typical titanium dioxide photocatalysts are limited to ultraviolet light due to their broad band gap of 3.2 eV. They hence must be modified to be used with visible light for improved efficiency. Recent studies have focused on developing visible-light-responsive (VLR) photocatalysts, such as Ag-doped TiO_2, which have a reduced band gap of 2.72 eV, allowing for excellent absorption in the visible region. This strategic choice of wavelength is optimized not only to maximize the operating range of the photocatalyst but also to align with the limits of accessible ambient solar radiation, thereby improving sustainability in photocatalytic operations (Bauer et al., 2025). Further, integration of effective light sources in photocatalytic reactor design is key to optimizing wavelength choice for maximum absorption. While traditional mercury-vapor lamps are efficient at exciting ordinary TiO_2, LEDs—specifically LEDs of particular wavelengths that fall on the photoabsorption maxima of the photocatalyst—are more innovative solutions (Figure 5.12).

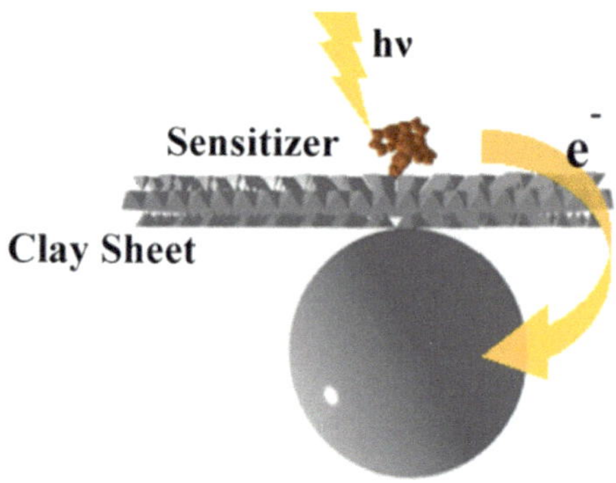

Figure 5.12. Visible-light-responsive photocatalytic flow reactor.

Source: https://pubs.acs.org/doi/10.1021/acsami.5b03128.

For instance, high-efficiency blue LEDs can be used to excite high-performance VLR photocatalysts such as Ag-doped TiO_2, enhancing photocatalytic performance in visible light. In addition to reducing energy wastage by selecting a mode of operation, this selective mode can adequately address safety concerns related to mercury-based light sources, particularly in confined areas such as spacecraft (Bauer et al., 2025). Thus, selecting a suitable light source is crucial to enhance the operational efficiency of photocatalytic reactors. Furthermore, the overall system design must consider the interaction between the photocatalyst and the light source to achieve optimal absorptive efficiency. The location and orientation of the photocatalyst within the reactor, as well as the sources of light, must allow for uniform distribution of light to ensure maximum activation. Past studies have indicated that integration of biological and physicochemical processes for waste-gas treatment can improve the performance of photocatalytic systems, particularly during transient shock-load conditions encountered in industrial operations. Through integrated strategies like gas-phase conditioning pre-treatment before biological treatment, the system keeps pollutant loads within manageable levels in the bioreactor, thereby optimizing the strengths of both processes. This total comprehension of the photocatalytic processes underlines the importance of wavelength selection and reactor engineering in the development of effective environmental remediation solutions.

5.4.3. Strategies for Enhancing Light Penetration

In the interest of improving the penetration depth of light in photocatalytic reactors, numerous strategies can be utilized to achieve maximum performance of photocatalytic processes. One of the key strategies is the manipulation of reactor design, wherein geometrical structure and material dictate light distribution and intensity. For instance, integrating reflective surfaces and transparent materials can maximize the exposure of photocatalysts to incident light, resulting in increased interaction during the photocatalyst reaction. The spatial distribution of light sources relative to the reaction medium is crucial for optimal light penetration, especially with higher concentrations of absorbing species that might otherwise hinder efficient light transmission. Yet another approach to enhancing light penetration involves using photocatalysts capable of absorbing a broader range of light, including visible wavelengths. Advanced materials, such as metal oxide-functionalized titanium dioxide (TiO_2), have been proven promising by enhancing the range of light absorption. For example, loading of a trace amount of RuO_2 onto TiO_2 nanorods not only extends the visible-light response of the photocatalysts but also enhances the charge separation efficiency of the resultant heterointerface. These advancements can lead to improved photocatalytic activity, as demonstrated by studies on these new materials, which highlight the critical correlation between photocatalyst material properties and light absorption behavior (Bai et al., 2016). Finally, optimizing reactor operating conditions will also play a significant role in light penetration. pH, reactant concentration, and the inclusion of additives have a significant influence on determining the overall efficiency of the photocatalytic process. Current studies suggest that adjusting these parameters can minimize the light absorption of reactants and allow for better light penetration. For instance, control of pH could affect the photocatalysts' condition and interaction with light, and thus photocatalytic activity. Further, additives capable of increasing the mobility of charge carriers or modifying the photocatalyst's surface could facilitate increased utilization of light and consequently increase the efficiency of the overall reaction (Busse et al., 2015). These approaches form a comprehensive system for optimizing light penetration in photocatalytic reactor design, ultimately resulting in improved degradation processes.

5.5. APPLICATIONS OF PHOTOCATALYTIC REACTORS

Applications of photocatalytic reactors are as diverse as they are broad, for they are used in numerous industries and pollution abatement processes. Perhaps the most prominent application is wastewater purification, particularly the removal of toxic dyes from textile effluents. Scientists have used heterogeneous photocatalysis to demonstrate that it is feasible to break down toxic azo pigments, which are a serious environmental hazard if left untreated. Recent advancements in photocatalytic reactor design allow for *in situ* observation of decolorization reactions through the utilization of online approaches. The new technologies not only enhance the efficiency of dye removal but also minimize the risk of disturbances in reactor volume or catalyst concentration during treatment, thereby maximizing the overall process efficiency (Mart et al., 2019). The application of real-time monitoring technology is a significant milestone towards optimizing photocatalytic wastewater treatment systems. In addition to wastewater treatment, photocatalytic reactors are also being used to optimize chemical compound synthesis through gas-liquid photoredox catalysis. Microreactor design enables high-performance visible-light-mediated continuous-flow reactions with much shorter reaction times compared to batch procedures. These microreactors are inexpensive to fabricate from commercially sourced components and are modular in nature. They also provide enhanced gas-liquid mass transfer, which is crucial for effective photocatalytic reactions. Applications include, for example, continuous-flow synthesis of valuable compounds such as disulfides and trifluoromethylated heterocycles. Therefore, the ease of operation and efficiency of such microreactors demonstrate how photocatalytic reactor innovation can lead to leaps in chemical synthesis efficiency (Günther et al., 2016). These innovations show the increasing importance of photocatalytic systems for solving both synthetic and environmental issues. The technological potential of photocatalytic reactors also extends to air cleaning, as TiO_2 photocatalysts can be employed to annihilate airborne toxins like VOCs and pathogens. The application of photocatalytic systems for improving indoor air quality is a promising option in controlling health risks emanating from air pollutants. With the reactors constructed to optimize light absorption and catalytic surface area, these systems enhance the rates of toxic substance degradation. The use of photocatalytic surfaces in building materials or air purification can generate a continuous self-cleaning process, resulting in cleaner indoor environments. With ongoing research expanding on the efficacy and durability of photocatalyst materials, their application to air purification is bound to gain more momentum. This development is accompanied by a sustainable strategy for indoor pollution management, which aligns with increasing global health awareness (Günther et al., 2016; Mart et al., 2019).

5.5.1. Environmental Remediation

The need for effective environmental remediation techniques arises from the accelerated levels of contamination and ecosystem degradation due to industrialization and urbanization.

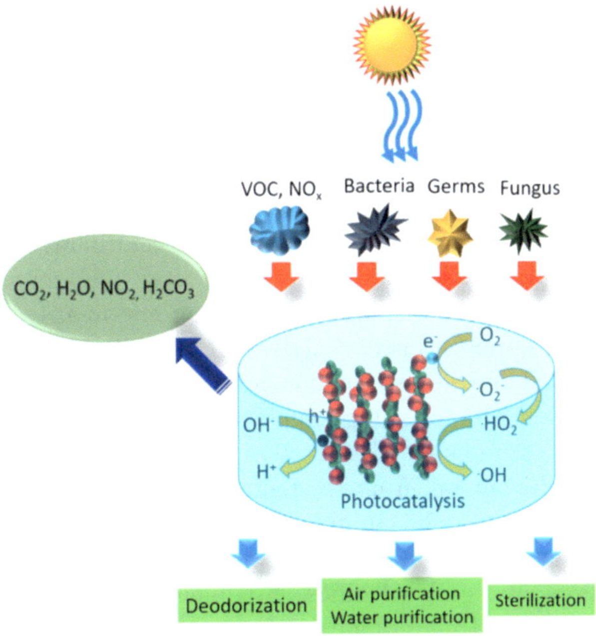

Figure 5.13. Schematic representation of environmental remediation by photocatalysis.

Source: https://www.researchgate.net/figure/Schematic-for-the-representation-of-environmental-remediation-by-photocatalysis_fig1_317280372.

Among the new solutions to counter such challenges, photocatalytic reactions have emerged as the key area of research. Powered by solar light, photocatalysis offers a promising means for the decomposition of chemicals in water and air. The application of semiconductor materials, predominantly titanium dioxide (TiO2), enables chemical reactions that can break down toxic substances into less harmful chemicals. Existing studies focus on optimizing photocatalytic materials, indicating that better design could significantly enhance their remediation efficiency, making technology a viable alternative to conventional techniques that are costly and environmentally unacceptable (Jang et al., 2015). Photocatalytic reactor design is crucial to maximize the use of photocatalysts for cleaning up the environment. Various reactor designs, such as tubular and packed bed reactors, have been experimented with to maximize the contacting surface area between light, photocatalysts, and pollutants. Light illumination is the key factor since it has an immediate role in stimulating the photocatalyst and, by default, the rate of degradation. In addition, the integration of new materials, such as nanocomposite films, has significant potential. For instance, utilizing halloysite clay blended with TiO_2 enhances the structural properties of photocatalysts, and this leads to increased efficiency in the annihilation of complex pollutants such as dyes in industrial wastewater (Kaplani et al., 2014) (Figure 5.13).

This new technology not only helps in effective environmental cleaning but also depicts the ongoing development in the design of photocatalytic reactors. Light sources used in photocatalytic reactors are critical in influencing degradation rates of pollutants, with visible light and ultraviolet light being the most common ones. The incorporation of new light sources, including LEDs and solar concentrators, would contribute enormously to the energy efficiency of these systems. Photocatalysts that can utilize a broader spectrum of light are required to maximize solar energy capture and enhance the overall degradation reaction. Moreover, the complexity inherent in using new materials and light sources positions the photocatalytic reactor not just as a remediation tool, but as a multicomponent device requiring the stringent handling of all design factors. With research continuing to investigate reactor design synergy, photocatalytic materials, and light sources, environmental remediation can develop at extremely rapid velocities, providing real solutions to the most pressing ecological challenges of the day (Jang et al., 2015).

5.5.2. Energy production through hydrogen generation

The search for renewable resources has fast-tracked interest in hydrogen production, especially using novel methods like photocatalysis. In the process, semiconductor materials in a reactor system harness sunlight for the decomposition of organic compounds or water to yield hydrogen efficiently. Current advancements reveal that titanium dioxide (TiO_2) is a very active photocatalyst when immobilized on carriers such as Al_2O_3/SiO_2 foam, facilitating sustained operation and ensuring easy separation after reaction. In ultraviolet light tests, immobilized TiO_2, particularly lanthanide-doped species, demonstrates superior hydrogen production activities, which confirms that catalyst design plays a key role in maximizing energy yield and reaction kinetics (Dubnová et al., 2020). Thus, developing photocatalytic reactors with effective illumination sources and durable materials is an important path to creating a stable hydrogen energy supply. The design features of reactors for photocatalytic hydrogen generation are particularly significant. The choice of illumination source, either natural or artificial light, and reactor configuration are primary considerations. For instance, annular or circular photoreactor geometries may enhance light penetration and maximize surface area contacts among reactants and photocatalyst. Such configurations facilitate effective reaction kinetics and maximize hydrogen production. Furthermore, semiconductor photocatalytic kinetics adhere to Langmuir-Hinshelwood principles, which emphasize the limitation of molecule adsorption at the catalyst's surface as a function of the reaction process (Lee et al., 2004) (Figure 5.14).

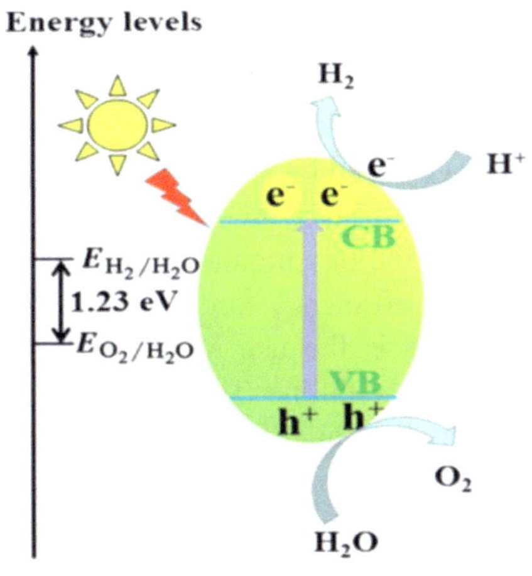

Figure 5.14. Photocatalytic hydrogen generation using a suitable semiconductor.

Source: https://www.mdpi.com/2079-6412/14/3/366.

It is crucial to understand these kinetics, as they determine the optimum operating conditions and upgrade requirements for optimal energy generation through hydrogen production. Furthermore, the viability of hydrogen production relies on the selective selection of photocatalysts and reactor materials that can withstand extended contact with light and corrosive conditions. With the use of novel semiconductor materials and system-scale reactor designs, highly increased production rates and energy efficiency are possible. Some doped phases of TiO_2, according to the research, increase the conduction band conductivity and also improve effective hydrogen production upon light irradiation. Advanced photocatalytic reactors not only aid in achieving increased yields but also emphasize the importance of scalability and cost-effectiveness in hydrogen production technologies. As the world keeps moving towards greener sources of energy, enhanced photocatalytic reactor technology is a principal system in mass hydrogen production and hence fueling a hydrogen economy.

5.5.3. Industrial Applications in Chemical Synthesis

The role of photocatalytic reactors in industrial chemical synthesis processes has generated tremendous interest due to their ability to enhance reaction efficiencies and reduce environmental impacts. The systems operate by means of light activation of photocatalysts to facilitate a wide variety of chemical conversions, e.g., carbon dioxide conversion into valuable chemicals. In particular, the high-temperature CO_2 electrolysis and CO_2 hydrogenation processes demonstrate the capability of the photocatalytic reactor to address the twin challenges of fossil fuel reserve depletion and greenhouse gas emissions. Experiments have indicated that the energy efficiency of CO_2 electrolysis for methanol synthesis is up to 41%, nearly double that of typical CO_2 hydrogenation reactions. This efficiency is especially notable when powered by renewable energy sources, representing a promising prospect for industrial synthesis applications (Al-Kalbani et al., 2015). Moreover, the integration of nanomaterials with photocatalytic systems significantly enhances their efficiency and performance for most chemical processes. A classic example is the use of nanocomposite films consisting of halloysite clay and nanocrystalline TiO_2, synthesized through a sol-gel technique followed by thermal treatment to impart the best structural properties. The composite films possessed better photocatalytic activity, demonstrating their ability to degrade organic pollutants such as Basic Blue 41 dye from aqueous media. The incorporation of silver nanoparticles further improves the photocatalytic efficiency of such films, illustrating nanotechnology's power in optimizing chemical synthesis methods in industrial processes (Kaplani et al., 2014). Apart from the efficiency improvement in chemical synthesis, photocatalytic reactors also contribute to greening industrial processes by enabling cleaner production pathways. Optimizing reactor designs to maximize light sources and enhance the interaction between light and photocatalysts is critical for achieving maximal reaction rates. With the potential of photocatalysis being utilized, industries can move away from traditional fossil-fuel-based processes to environmentally less destructive ones. This new strategy not only addresses the pressing issue of climate change but also opens up new avenues for sustainable chemical synthesis, establishing photocatalytic reactors as a foundation stone for future industrial chemical processes.

SUMMARY

In summarizing this research on photocatalytic reactor design and light sources, it is essential to notice the groundbreaking achievements in optimizing these systems for environmental applications. The development of a solar-powered fiber-optic reactor represents a significant improvement, as sunlight is effectively harnessed to drive photocatalytic reactions. The enhanced performance of this system, as observed in empirical observation, reflects that solar illumination offers faster response times than man-made UV sources. Such enhancements, hence, reflect the potential of solar-powered systems for *in situ* pollution degradation, particularly in rural settings. This new method can dramatically enhance the sustainability and efficiency of photocatalytic processes, highlighting the importance of further research in this area (Hoffmann et al., 1997). Another parameter to consider is the role of controlled regular illumination as a method for improving photocatalytic activity. By regulating the timing of light incidence, researchers have proven that this method can considerably enhance the activity of semiconductor photocatalysts. This concept, initially conceived from research on photosynthesis, has since been effectively translated to photocatalysis, suggesting that the timing of light availability can positively affect reaction kinetics. Literature in its support shows a potential route for improving the efficiency of photocatalytic reactors, as set by results from milestone studies. Therefore, the synergy between reactor design and controlled light is an improvement that needs to be practiced and investigated more (Lawton et al., 2015). Briefly, the intersection of new reactor design and purposeful light delivery is essential for making photocatalytic technologies happen. The emphasis on the effectiveness of solar energy in driving reactions not only promotes an environmentally benign means of environmental cleanup but also highlights the versatility and flexibility of photocatalytic systems. Moreover, the implementation of methods such as controlled periodic illumination can be maximized to enhance reaction efficiencies, offering new routes to next-generation photocatalytic reactor design. To further address major environmental problems, future development and research into reactor design using photocatalysis will be key to making such technology viable and efficient for implementation in many real-world applications. The study of such approaches is central to unlocking the general potential of photocatalysis for addressing global pollution and energy challenges.

Photocatalytic reactor design is fundamental to photocatalysis performance, particularly with respect to pollutants such as pathogenic microorganisms in water. Current debates have centered on how to maximize parameters such as photon flux and reactor shape to achieve optimal performance in such systems. By effectively increasing the surface area of photocatalysts, reactor designs can facilitate maximum light penetration and optimal contact between the photocatalyst and the pollutants. The use of immobilized catalysts rather than powdered slurries can also lead to increased mass transfer rates and system stability. Optimum reactor design is therefore a critical factor that can significantly contribute to the overall efficiency of photocatalytic applications, particularly in facilitating the degradation of complex organic molecules and enhancing disinfection processes. In judging the role of light sources towards photocatalytic reaction, the role played by light wavelength and intensity in photocatalytic efficiency must be taken into consideration. Different photocatalysts, such as titanium dioxide and zinc oxide, have varying absorbance characteristics that dictate their performance under given light conditions. For instance, although titanium dioxide is effective under ultraviolet light but inefficient under visible light, its application in given environments is limited. The application of visible-light-absorbing materials or nanocomposite systems holds the potential to eliminate these limitations and broaden the application range of photocatalytic reactors for the effective utilization of solar light. Moreover, optimizing light distribution in photocatalytic reactors can enhance photocatalytic efficiency, thereby improving total disinfection and degradation rates, which are critical for water treatment and other environmental processes (Abbaszadegan et al., 2015). The promise of photocatalytic reactor technology is premised on overcoming challenges of commercialization and scale-up of the systems. While early studies indicate promising results, there is still much to be done to derive effective means of operation at large scales. Foremost among these are standardizing testing protocols, scaling reactor design to various environments, and overcoming photocatalytic material limitations, such as its visible light absorption properties. In addition, materials synthesis research and the identification of novel catalysts will be key factors in enhancing the efficiency of these systems. Additional breakthroughs in interdisciplinary research coupling engineering theory with scientific investigation will enable the development of functional photocatalytic systems that address contemporary water treatment challenges, significantly contributing to global clean drinking water requirements (Abbaszadegan et al., 2015).

REFERENCES

1. Mayer, B. K., Yang, Y., Gerrity, D. W., & Abbaszadegan, M. (2015). The Impact of Capsid Proteins on Virus Removal and Inactivation during Water Treatment Processes. *Microbiology Insights, 8s2*. https://doi.org/10.4137/mbi.s31441Abou-Ghazala, A., Fujishima, A., Kaur, A., Mahvi, A. H., Kutepov, A. M., Stasinakis, A. S., Chefetz, B., et al. (2014). A review of combined advanced oxidation technologies for the removal of organic pollutants from water. https://core.ac.uk/download/145190182.pdf (accessed on 28th September 2025).
2. Al-Kalbani, H., García, S., Wang, H., Xuan, W., et al. (2015). Comparative energetic assessment of methanol production from CO_2: Chemical versus electrochemical process. https://core.ac.uk/download/152439005.pdf (accessed on 28th September 2025).
3. Antonietti, M., Kurpil, B., Mazzanti, S., Pieber, B., et al. (2020). Dichloromethylation of enones by carbon nitride photocatalysis.
4. Aquino Neto, F. R., Azevedo, E. B., Dezotti, M., Tôrres, A. R., et al. (2009). TiO_2-photocatalyzed degradation of phenol in saline media in an annular reactor: Hydrodynamics, lumped kinetics, intermediates, and acute toxicity. https://core.ac.uk/download/pdf/37459700.pdf (accessed on 28th September 2025).
5. Bai, J., Fernandez Sanz, J., Fujita, E., Graciani, J., et al. (2016). Visible light-driven H_2 production over highly dispersed ruthenia on rutile TiO_2 nanorods. https://core.ac.uk/download/81582504.pdf (accessed on 28th September 2025).
6. Ball, D. W., Chianelli, R. R., Hepp, A. F., Kulis, M. J., et al. (2025). Novel catalysts and processing technologies for production of aerospace fuels from non-petroleum raw materials. https://core.ac.uk/download/pdf/42722367.pdf (accessed on 28th September 2025).
7. Bauer, B. M., Coutts, J. L., Devor, R. W., Hintze, P. E., et al. (2025). Visible-light-responsive photocatalysis: Ag-doped TiO_2 catalyst development and reactor design testing. https://core.ac.uk/download/pdf/42693355.pdf (accessed on 28th September 2025).
8. Berrill, T., Bundschuh, J., McCreath, R., Mushtaq, S., et al. (2013). A rural agricultural-sustainable energy community model and its application to Felton Valley, Australia. https://core.ac.uk/download/16670895.pdf (accessed on 28th September 2025).
9. Busse, N., Czermak, P., Herrenbauer, M., Lekelefac, C. A., et al. (2015). Photocatalytic based degradation processes of lignin derivatives. https://core.ac.uk/download/33372036.pdf (accessed on 28th September 2025).

10. Byrne, A. J., Dionysiou, D. D., Dunlop, P. S. M., Ibanez, P. F., et al. (2014). Solar photocatalysis for water disinfection: Materials and reactor design. https://core.ac.uk/download/pdf/60775675.pdf (accessed on 28th September 2025).

11. Carden, J. L., & Mason, R. M. (2025). Controlled Ecological Life Support System: Research and Development Guidelines. https://core.ac.uk/download/pdf/42856444.pdf (accessed on 28th September 2025).

12. Carden, J. L., & Mason, R. M. (2025). Guiding the Development of a Controlled Ecological Life Support System. https://core.ac.uk/download/pdf/42867040.pdf (accessed on 28th September 2025).

13. Chianese, A., De Caprariis, B., Stoller, M., Verdone, N., et al. (2015). CFD model of a spinning disk reactor for nanoparticle production. https://core.ac.uk/download/54526774.pdf (accessed on 28th September 2025).

14. Dubnová, L., Dvoranová, D., Edelmannová, M., Kočí, K., et al. (2020). Successful immobilization of lanthanides doped TiO_2 on inert foam for repeatable hydrogen generation from aqueous ammonia. https://core.ac.uk/download/322825201.pdf (accessed on 28th September 2025).

15. Erkey, C., Hiragond, C. B., In, S. I., Kim, H., et al. (2020). Electrochemical CO_2 reduction to CO catalyzed by 2D nanostructures. https://core.ac.uk/download/287633748.pdf (accessed on 28th September 2025).

16. Farràs, P., Galushchinskiy, A., González-Gómez, R., McCarthy, D. L., et al. (2022). Progress in development of photocatalytic processes for synthesis of fuels and organic compounds under outdoor solar light. https://core.ac.uk/download/516205676.pdf (accessed on 28th September 2025).

17. Hoffmann, M. R., & Peill, N. J. (1997). Solar-powered photocatalytic fiber-optic cable reactor for waste stream remediation. https://core.ac.uk/download/33122294.pdf (accessed on 28th September 2025).

18. Jang, S., Kim, D. P., Parmar, J., Sanchez, C., et al. (2015). Nano-photocatalysts in microfluidics, energy conversion and environmental applications. https://core.ac.uk/download/81578008.pdf (accessed on 28th September 2025).

19. Kalamaras, E., Maroto-Valer, M., Shao, M., Wang, W., et al. (2018). Solar carbon fuel via photo electrochemistry. https://core.ac.uk/download/288360687.pdf (accessed on 28th September 2025).

20. Kant, P. P. (2023). Optimizing Photocatalysts and Photoreactors for Solar Fuel Synthesis. https://core.ac.uk/download/596775364.pdf (accessed on 28th September 2025).

21. Kaplani, E., Panagiotaras, D., Papoulis, D., Rapsomanikis, S., et al. (2014). Nanocrystalline TiO_2 and halloysite clay mineral composite films prepared by sol-gel method: Synergistic effect and the case of silver modification to the photocatalytic degradation of basic blue-41 azo dye in water. https://core.ac.uk/download/41990680.pdf (accessed on 28th September 2025).

22. Kennes, C., Rene, E. R., & Veiga, M. C. (2012). Combined biological and physicochemical waste-gas cleaning techniques. https://core.ac.uk/download/61913340.pdf (accessed on 28th September 2025).

23. Lawton, L. A., Prabhu, R., Robertson, P. K. J., Tokode, O., et al. (2015). Controlled periodic illumination in semiconductor photocatalysis. https://core.ac.uk/download/33591761.pdf (accessed on 28th September 2025).

24. Lee, S. K., & Mills, A. (2004). Detoxification of water by semiconductor photocatalysis. https://core.ac.uk/download/9013846.pdf (accessed on 28th September 2025).

25. Romeral Martínez, J. L., Tapia Tlatelpa, T., & Trull Silvestre, J. F. (2019). In situ decolorization monitoring of textile dyes for an optimized UV-LED/TiO_2 reactor. https://core.ac.uk/download/294829789.pdf (accessed on 28th September 2025).

26. Tapia Tlatelpa, T. (2019). Optoelectronic Optimization of Photocatalytic Processes for Wastewater Treatment. https://core.ac.uk/download/237156747.pdf (accessed on 28th September 2025).

CHAPTER 6

PHOTOCATALYSIS IN C–C BOND FORMATION

Contents

INTRODUCTION

Photocatalysis has emerged as a groundbreaking technique in organic synthesis, particularly in the formation of carbon-carbon (C–C) bonds, where standard practices generally suffer from limitations in both efficiency and selectivity. At its core, photocatalysis is a process that uses light to activate a catalyst, which subsequently initiates chemical reactions under mild conditions. This approach has gained popularity because it can be driven by renewable energy sources, unlike conventional thermal approaches that tend to require high pressure and temperature. The advent of visible-light photocatalysis has significantly broadened the scope of substrates to the point where it can efficaciously transform a wide range of organic molecules into complex forms, showing its potential as an efficacious instrument in synthetic chemistry. One very significant use of photocatalysis in C–C bond synthesis is observed in the arena of perfluoroalkylation reactions. Such reactions produce perfluoroalkyl radicals, which can be added to or used to replace other organic substrates. Recent research focuses on the significance of visible-light-promoted processes, specifically the employment of transition-metal complexes and organic dyes (Vallejo et al., 2018).

This diversity in methodology not only makes it easy to create new C–C bonds but also permits access to functionalized compounds that are otherwise challenging to synthesize. Continuous discovery in photocatalytic systems has led to higher selectivity and reactivity, further establishing photocatalysis as a driving force behind the evolution of modern organic synthesis. Moreover, photocatalysis drives greener fluorination strategy design by addressing significant problems in fluorinated compound synthesis. Traditional methods rely on heavy metals and stoichiometric reagents that are environmentally hazardous and harmful to health (Hammond et al., 2018). By using heterogeneous photocatalysts like TiO_2 in combination with mild fluorinating reagents, researchers have demonstrated competent decarboxylative fluorination of carboxylic acids. This approach not only provides highly efficient fluoroalkylated products but also exhibits excellent turnover frequencies and stability. With advancements in photocatalytic approaches being made by the scientific community, such advancements are set to revolutionize the future of chemical synthesis, particularly within areas calling for regioselectivity and sustainability (Figure 6.1).

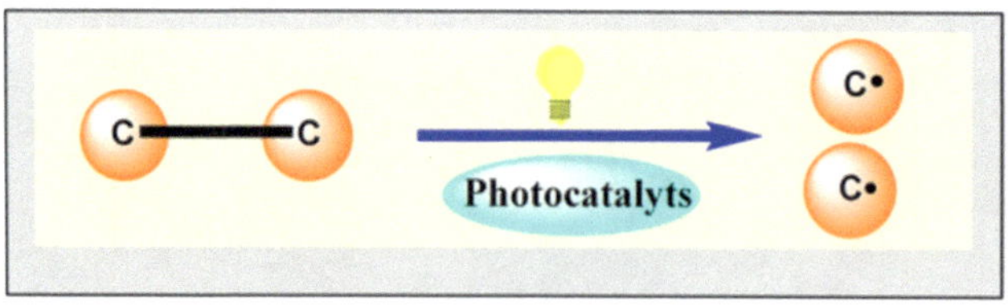

Figure 6.1. *Photocatalytic carbon–carbon bond activation has emerged as a promising tool for organic synthesis, especially in the preparation of bioactive molecules.*

Source: https://chemistry-europe.onlinelibrary.wiley.com/doi/10.1002/slct.2022 02679.

Definition and Principles of Photocatalysis

Photocatalysis is a mechanism by which a chemical reaction is facilitated by illumination by light, most often in the ultraviolet or visible region, to activate a catalyst. The root of photocatalysis lies in electron-hole pair generation in a semiconductor material upon illumination by light energy. Photocatalysis plays a foundational role in most applications, including organic transformations, environmental cleanup, and energy conversion. The semiconductor, typically titanium dioxide (TiO_2), possesses certain features that allow it to act as a catalyst for reactions when it is excited, overcoming the energy barrier of specific chemical reactions. Specifically, the materials band gap is crucial in photocatalysis efficiency since it controls its light absorbability and capability to generate reactive species required for the creation of bonds within organic molecules. The photocatalytic mechanisms can be encapsulated in a few primary processes like absorption, charge separation, and subsequent chemical reactions. When a photon is absorbed, an electron from the valence band is excited to the conduction band, resulting in the creation of an electron-hole pair. Proper separation of charge carriers is necessary to sustain high photocatalytic activity and avoid recombination. These influences, such as surface area, morphology of photocatalysts, and even dopant and co-catalyst nature, can effectively make such a separation larger and enhance reactivity. Advanced calculation methods have also been employed to model these processes, allowing scientists to design and control the electronic properties of photocatalysts and thereby maximize their activity in reactions like C–C bond formation. By studying these interactions at a nanoscale level, scientists aim to design more effective photocatalytic systems that will work in visible light irradiation (Chiodo et al., 2010; Valentin et al., 2017). Besides, the application of photocatalysis in C–C bond formation is just one example of its versatility and significance in organic synthesis. By energizing electrons, photocatalysts enable the production of radical species that can engage in bond-forming chemistry, thereby providing a green avenue for the assembly of complex organic molecules. The process is resistant to the pitfalls of traditional approaches because it allows the production of valuable products under less stringent conditions. Through the union of photocatalysis and C–C coupling reactions, the potential of environmentally clean processes with renewable energy sources as feedstocks is achieved. With ongoing research and development of photocatalytic materials, increasing knowledge of the principles of photocatalysis will surely open new areas of innovation in synthetic chemistry, accelerating its use in both academia and industry.

Historical Context and Development of Photocatalytic Methods

The creation of photocatalytic methods finds its roots in early research on semiconductor materials, particularly titanium dioxide (TiO_2), which was discovered for its unique ability to absorb light energy for chemistry. Found for the first time in the 1970s, TiO_2s photocatalytic properties became better recognized as studies hastened in the late 20th century. The integration of photocatalysis in organic synthesis, specifically C–C bond-forming reactions, was a paradigm shift in chemical practice. Researchers only then appreciated the value of these methods for constructing compounds under lenient conditions, offering a green solution compared to traditional chemical reactions that typically rely on harsh reagents and high energy inputs. This development not only addresses the increasing environmental problems associated with chemical manufacture but also enables new applications in renewable energy and waste management.

Recent improvements have been achieved in the knowledge of photocatalytic mechanisms, including charge separation and transfer processes. These advancements, coupled with the development of new photocatalysts by synthesis, have facilitated more effective and selective efficiencies and selectivities in many chemical reactions. The addition of co-catalysts, as well as the manipulation of light wavelengths, has also improved the optimization of these materials to enable the tuning of reaction pathways with precision. Of note, research has now expanded to include the study of emerging light sources, such as LEDs, that hold promise for more energy-saving photocatalytic reactions. This strategic focus on reaction condition and catalyst optimization has been crucial in bridging photocatalytic processes from theory to practical use in synthetic organic chemistry, where C–C bond formation is central. A recent collection of work illustrates the expanding scope of photocatalytic applications beyond traditional organic synthesis.

For instance, the photochemical conversion of lignin derivatives demonstrates the efficiency and potential of photocatalytic techniques in generating valuable chemicals, such as phenolic products and aromatic hydrocarbons, which are crucial in the manufacture of many materials. The potential of lignin as a renewable material has gained ground, with photocatalysis being a key technology for the transformation of lignocellulosic biomass into valuable chemicals. As recently highlighted in literature, the incorporation of photocatalysis and biocatalysis processes underscores the synergistic potential of these technologies in the sustainable production of chemicals. The secret to achieving the desired product yield without causing maximum environmental disruption highlights the promising future of the photocatalytic process, echoing the long journey this new method has undertaken as a revolutionary force in modern chemistry (Busse et al., 2015; Gherardi et al., 2018).

Importance of C–C Bond Formation in Organic Synthesis

Carbon-carbon (C–C) bond formation is at the heart of organic synthesis because it represents a key step in the construction of a vast array of organic compounds, including medicines, agrochemicals, and materials. The ability to make C–C bonds with high efficiency is the basis for synthetic chemistry's method to construct complex molecular architecture from simpler precursors. These changes are not solely academic; they have far-reaching implications in the discovery of new drugs and in improving the efficiency and selectivity of known chemical processes. The development of photocatalysis technologies has sparked an engineering revolution in C–C bond assembly, enabling reactions to be performed under mild conditions without resorting to severe reagents or extensive post-reaction purification. These tactics are crucial in the quest for sustainability in organic synthesis, offering highly efficient and environmentally friendly approaches. Recent studies have highlighted new methods to generate small organic radicals, which play a major role in facilitating C–C bond formations through photocatalysis.

For instance, chloroform can be easily converted to dichloromethyl radicals with the assistance of a heterogeneous potassium poly(heptazine imide) (K-PHI) photocatalyst under visible light irradiation, which illustrates the universality of photo-driven reactions in organic synthesis. This process not only allows for the effective production of γ,γ-dichloroketones but also allows for the subsequent synthesis of high-value organic compounds such as substituted furans and pyrroles. As radical intermediates are transient, utilizing them through photocatalytic approaches widens the synthetic chemist's arsenal, leading to new pathways that can influence the design and construction of complex molecules (Antonietti et al., 2020). Therefore, photocatalysis is a tremendous leap in precision and flexibility of C–C bond formation. The impact of photocatalysis on C–C bond formation extends beyond merely enhancing reaction efficiency; it also facilitates the exploration of new chemical space. By developing techniques that exploit radical chemistry, researchers can access previously unexplored structural and functional diversity, leading to new materials and drugs. These advancements show the evolving scenario of organic synthesis, whereby former boundaries are being remolded by new machinery. The funding and support of initiatives such as the Engineering and Physical Sciences Research Council (EPSRC) also reflect the collaborative process required in identifying and advancing the field (Howe et al., 2015). Such strategic examination not only enhances practical application but also facilitates appreciation of more fundamental or lower-level chemical interactions, further underlining the critical role of C–C bond construction in future organic synthesis.

6.1. TYPES OF C–C BOND FORMATION REACTIONS

The formation of carbon-carbon (C–C) bonds is a fundamental reaction in organic chemistry, with various synthetic approaches being applied to make it feasible. Among the known classes of reactions is the coupling of organic radicals that are generated through various photocatalytic processes. Visible-light-facilitated approaches, for instance, render feasible the activation of substrates and the subsequent formation of C–C bonds via radical intermediates. These approaches facilitate the fusion of a wide range of organic functionalities, allowing for a broad range of products with particular properties. Furthermore, recent years have seen advances in catalysis move the spotlight to visible-light transition-metal-mediated photocatalysis. This approach creates accessible pathways to intricate molecular scaffolds through radical intermediates, which play pivotal roles in the assembly of C–C bonds in other organic molecules (Vallejo et al., 2018). The other major category of C–C bond forming involves the utilization of organometallic reagents, particularly cross-couplings such as the Suzuki and Heck reactions. Organometallic substrates react with electrophiles with excellent selectivity and yield the generation of new carbon centers. These coupling reactions typically involve the activation of C–X bonds (X = halogen or other leaving groups) and are facilitated by photocatalytic processes, which can enhance yield and expand the substrate scope.

The utilization of visible light in these reactions allows more mild reaction conditions and improved functional group tolerance, which represent benefits over traditional thermal approaches. Current studies demonstrate how photocatalysis can effectively enable the formation of C–C bonds in challenging pairs of substrates and therefore expand the chemical space that is available to synthetic chemists (Antonietti et al., 2020). Apart from the reactions mentioned above, radical addition onto alkenes is another critical methodology for constructing C–C bonds through photocatalysis. This methodology often exploits the formation of radicals from minor organic precursors for effective alkylation of alkenes. For example, dichloromethyl radicals generated using a novel photocatalytic strategy have been employed in C1-elongation of enone skeletons for the synthesis of valuable target molecules like γ,γ-dichloroketones. These intermediates can be subjected to additional chemical transformations for the production of complex organic molecules like substituted furans and pyrroles. These approaches indicate the remarkable contribution of photocatalysis to enhancing the yield and selectivity of radical reactions, creating scopes for new synthetic routes in organic chemistry (Antonietti et al., 2020). By integrating radical processes of various types, photocatalysis has emerged as a revolutionary tool for the synthesis of C–C bonds, thus enriching the field of organic synthesis.

6.1.1. Cross-Coupling Reactions and Their Significance

Cross-coupling reactions have revolutionized synthetic organic chemistry by offering highly efficient methods for carbon-carbon (C–C) bond construction. Cross-coupling reactions are the coupling of two dissimilar carbon centers, facilitated by catalysts, to form more complex molecular structures. This capability is most critical in pharmaceuticals and materials science, where constructing complex organic molecules is generally a precursor to function. Use of transition metals as catalysts in such reactions has made it possible to have greater control over reaction conditions, selectivity towards the substrate, and yield.

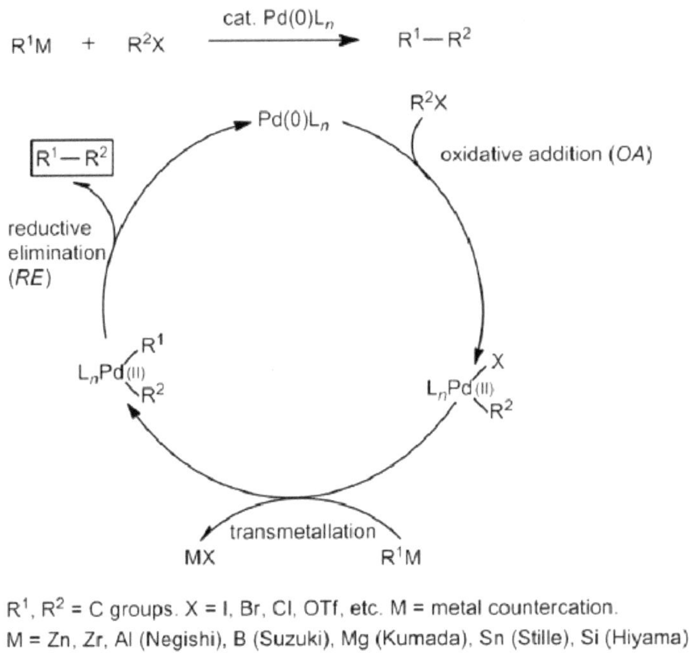

Figure 6.2. Mechanism of Pd-catalyzed C–C cross-coupling reactions.

Source: https://www.researchgate.net/figure/Scheme-1-Mechanism-for-Pd-catalyzed-C-C-cross-coupling-reactions_fig1_263737657.

The consequences of successful C–C bond formation are extensive, significantly reducing the time and materials required in complex synthetic routes, ultimately expanding the limits of contemporary organic synthesis into new fields of chemical innovation and sustainability. The introduction of photocatalysis in cross-coupling reactions has been a milestone in the field, harnessing visible light to trigger processes that were previously required under harsh conditions or with stoichiometric agents. Photocatalysis not only benefits the efficiency of C–C bond formation but also aligns with the growing interest in green chemistry and sustainable environments. Solar energy has increasingly been used as a clean source of powering such reactions, thereby obviating waste and minimizing the use of toxic reagents. Organic chemists can perform cross-coupling reactions under milder conditions through the use of photoredox catalysts such as ruthenium or iridium complexes, thereby enhancing the safety and environmental impact of synthesis approaches. As observed, the adoption of sustainable techniques aligns with the chemical industry's general objective of adopting greener approaches in synthetic organic chemistry (Kudoahor et al., 2024) (Figure 6.2).

Furthermore, the use of visible light to drive cross-coupling reactions has led to new strategies for activating recalcitrant substrates, such as unstrained aliphatic rings, which traditionally have high C–C dissociation energies. This new era of photocatalysis as a light-activated reaction catalyst highlights its creativity and durability in the face of conventional restraints. For instance, recent developments have disclosed new methods of functionalizing complex scaffolds, e.g., quinazolinones, that are in high demand due to their biological activities (Kudoahor et al., 2024). In addition, the potential of these reactions is amplified by expanding the substrate scope, which results in excellent yields and regioselectivities. This attests to the criticality of cross-coupling reactions in the future advancement of synthetic methodology and their role in drug discovery and materials science.

6.1.2. Radical-Based C–C Bond Formation Methods

The discovery of radical-based C–C bond-forming methods is amongst the most significant advances in organic synthesis, and particularly in photocatalysis. The technique takes advantage of the generation of carbon-centered radicals, which are highly versatile intermediate species to employ in the construction of complex molecular skeletons. Current research has emphasized a broad variety of novel methods for synthesizing perfluoroalkyl radicals by using visible-light-induced photocatalysis, amongst other processes. Such radicals can react by substitution and addition with every organic substrate, allowing the essential C–C bond formation required for synthesizing functionalized molecules. Illustrating this, reactions involving carbon-carbon multiple bonds, isocyanides, and many organic compounds have gained interest. As suggested by Barata Vallejo et al., application of more fluorinated alkylating reagents illustrates the expanding role of such procedures in product fine-tuning within modern-day synthetic chemistry (Vallejo et al., 2018). The radical-based techniques for the construction of C–C bonds have also illustrated the potential of photocatalysis to produce small organic radicals, important and essential intermediate within many synthetic applications. These radicals enable functional group installation and electronic property tuning, thereby enhancing the overall pharmacokinetic profiles of target molecules. One notable development in this area comprises the photoinduced generation of dichloromethyl radicals from chloroform by a heterogeneous potassium poly (heptazine imide) (K-PHI) photocatalyst upon irradiation with visible light. This methodology efficiently elongates the enone backbone by C1-extension to yield γ,γ-dichloroketones with striking efficiency for a wide range of substrate types. The employment of quasi-homogeneous flow-photoreactors also increases productivity, with a remarkable improvement over traditional batch processing (Antonietti et al., 2020) (Figure 6.3).

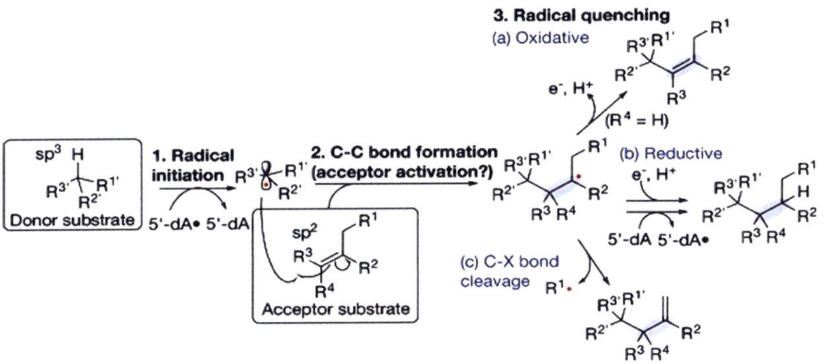

Figure 6.3. Radical-based C–C bond formation methods.

Source: https://ars.els-cdn.com/content/image/3-s2.0-B9780124095472147456-f14745-02-9780081026908.jpg.

These advancements help to highlight the promise of radical-based methods in photocatalytic C–C bond formation to transform the practice of organic synthesis. The integration of C–C bond formation via radicals into the broader framework of photocatalysis not only enhances the efficiency of the chemical reaction but also expands the scope of accessible transformations. The diverse functionalities available through radical intermediates provide new options for synthetic chemists, enabling the synthesis of a wide range of valuable compounds. By capitalizing on the characteristic reactivity patterns of such radicals, researchers can create strategies that circumvent traditional limitations in organic synthesis. Further, continuing investigation into new photocatalysts and reaction conditions will undoubtedly lead to the discovery of even more effective methods for radical generation and resultant manipulation, paving the way for an ever more sustainable and versatile approach to C–C bond formation. Continued funding to advance this field is a testament to the value of radical-based approaches to redefining the routes to novel and complex organic molecules.

6.1.3. Enolate Chemistry in Photocatalytic C–C Bond Formation

The theory of enolate chemistry is a premise for the understanding of photocatalytic C–C bond construction. Enolates, formed by the deprotonation of carbonyl compounds, are nucleophilic intermediates with the potential to play an essential role in complexity at the molecular level when utilized in photochemical processes. The unique reactivity of enolates allows them to be subjected to multiple transformations, and most of the time, the transformations lead to the creation of varied and complex carbon skeletons. In photocatalysis, the input of light energy can stabilize enolate species, allowing them to participate in reactions and ultimately form ketones or aldehydes. Synergy between light-driven processes and enolate chemistry not only expands the scope of synthetic approaches but also opens new perspectives for efficient C–C bond build-up, which is central to organic synthesis. The development of photocatalytic systems has sufficiently supported enolate chemistry to enable new C–C bond-forming methodologies. Through the utilization of photoredox catalysts, researchers have succeeded in harnessing visible light to produce energized states which are responsible for the formation of enolate reactive intermediates. This supports enhancing the reactivity of enolates, making it possible to perform selective coupling reactions under mild conditions. Moreover, recent research has demonstrated the various applications of enolate-derived species in carbon-carbon bond-forming reactions, highlighting the success of photocatalysis in creating molecular diversity. Through the investigation of light-enolate intermediate interactions, new synthetic methods have been developed that surpass traditional thermal processes. These methods demonstrate the potential of these procedures to transform synthetic organic chemistry, highlighting the evolving nature of modern science (Figure 6.4).

Along with this, enolate chemistry in photoreaction facilitates the creation of complex scaffolds that are not otherwise feasible. Preparing functionalized heterocycles through enolate coupling with other electrophiles, facilitated by photochemical activation, is one of the key directions. Sulfonyl radical formation, as detailed by Angnes et al. (2017), demonstrates how light can initiate the formation of radicals and then further act upon enolate species to create useful structural templates. The application of photochemical protocols not only enhances the efficiency of C–C bond assembly but also renders creative solutions to synthetic challenges. Overall, the synergy of photocatalysis and enolate reactivity represents a profound revolution in synthetic capability. It initiates a new era where sophisticated organic molecules can be systematically assembled with precision and economy, highlighting the revolutionary power of light in organic synthesis.

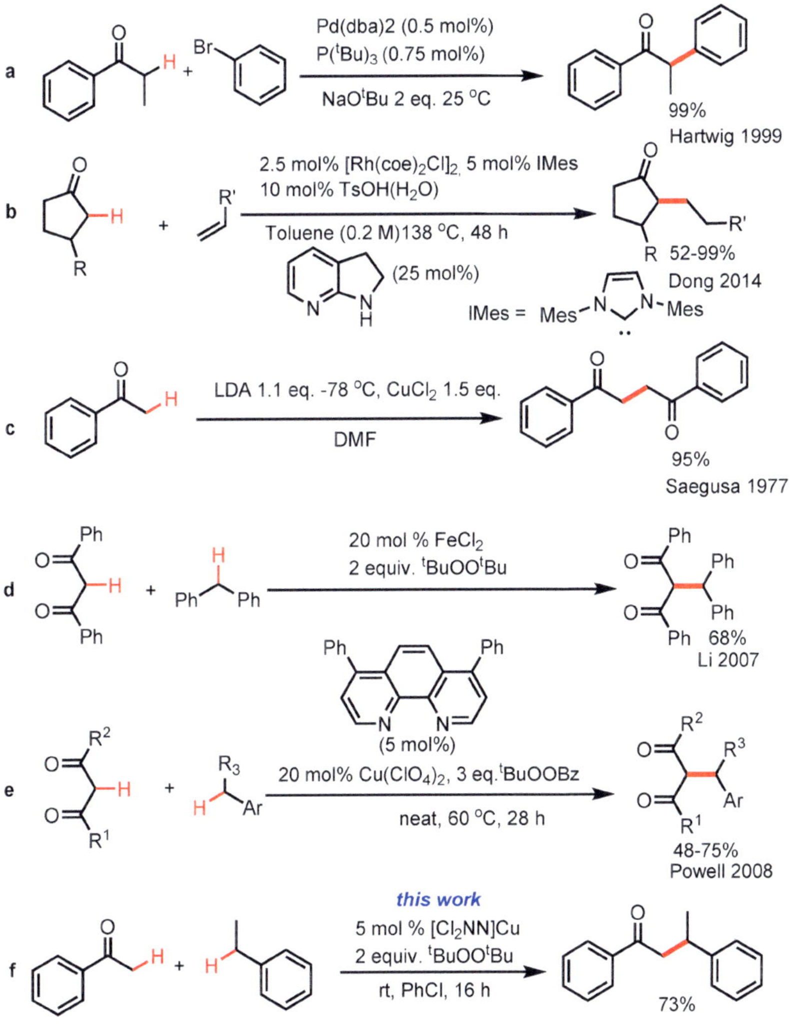

Figure 6.4. *Approaches to C−C bond formation via enolates.*

Source: https://www.researchgate.net/figure/Approaches-to-C-C-bond-formation-via-enolates_fig1_385390569.

6.2. APPLICATIONS OF PHOTOCATALYSIS IN ORGANIC SYNTHESIS

In recent years, photocatalysis has emerged as a groundbreaking approach to organic synthesis, particularly for carbon-carbon (C–C) bond formation. With the ability to harness light energy to trigger and propagate chemical reactions, new approaches can be invented that are generally greener and more selective than traditional synthetic procedures. Photocatalytic processes can generate reactive intermediates, e.g., low-molecular-weight organic radicals, which are key intermediates for constructing complex molecular architectures. A notable case is decchloromethylation of chloroform using a potassium poly(heptazine imide) (K-PHI) photocatalyst. Such a process allows C1-extension of enone backbones and yields important γ,γ-dichloroketones. They are competent intermediates for bifunctional building blocks in the synthesis of substituted furans and pyrroles, exemplifying the efficiency and flexibility brought about by photocatalysis (Antonietti et al., 2020). Also, the introduction of metal-organic frameworks (MOFs) such as UiO-66 into the area of photocatalytic reactions has doubled the number of feasible synthetic methods available to chemists. The structural attributes of UiO-66 can be well controlled by introducing defects, which improve the electronic qualities of the materials. In particular, missing linker defects have been shown to improve the efficiency of the ligand-to-metal charge transfer process and thereby increase the overall photocatalytic activity. This new approach demonstrates that the electronic structure of photocatalysts can be adjusted not only to enhance their activity but also to explore new synthetic pathways. Accordingly, this approach tends towards the utilization of photocatalysts designed for enhanced utilization in specific reactions of carbon-carbon bond formation, and it pinpoints the importance of precise design in the advancement of photocatalytic methods (Vos et al., 2017). The ongoing exploration of photocatalysis in organic synthesis also holds enormous promise for the design of novel reactions that enhance the efficiency, selectivity, and sustainability of chemical transformations. With visible light and judiciously chosen photocatalysts, researchers cannot only enhance yield and reduce reaction times but also access complex molecular architectures that were otherwise hard to synthesize. The coupling of small organic radicals generated through photocatalytic processes to other substrates is an emerging field in synthetic organic chemistry that can be expanded to produce a wide range of functionalized compounds. With the development of photocatalytic techniques, their application in the construction of C–C bonds will likely lead to quick progress in many fields like pharma, materials science, and green chemistry, which bespeaks the spectacular function of photocatalysis in contemporary organic synthesis.

6.2.1. Synthesis of Pharmaceuticals and Agrochemicals

In the case of pharmaceutical and agrochemical synthesis, the synthesis of complex organic molecules has been of great significance. Development of photocatalysis, particularly in the construction of C–C bonds, has revolutionized synthetic methodologies, allowing for more efficient and environmentally friendly pathways. Through this approach, reactions can be performed under milder conditions, which not only conserve energy but also reduce the formation of poisonous waste products. As researchers advocate for greener solutions, visible light photocatalysis is another viable option compatible with green chemistry principles. Utilization of light energy for organic reactions is also a very feasible option in novel drug and agrochemical design, which ultimately enhances their physical properties as well as efficacies in applications. This approach not only accelerates the drug discovery process but also provides avenues for the development of green agrochemicals for sustainable agriculture. Another significant aspect of this new synthesis is the incorporation of difluoromethyl groups via late stage difluoromethylation reactions. This strategy has demonstrated how CF2X group introduction can profoundly alter the lipophilicity, metabolic stability, and solubility of organic compounds, which are crucial parameters in agrochemical formulation and pharmaceutical chemistry. To this purpose, visible-light-photocatalyzed difluoromethylation has been found to demonstrate efficient application on both unsaturated carbon-containing substrates and (hetero)aromatic compounds. As science advances, more work is needed to determine the range of substrates available for these reactions, aiming to expand the toolkit chemists have to design drugs with improved properties (Figure 6.5).

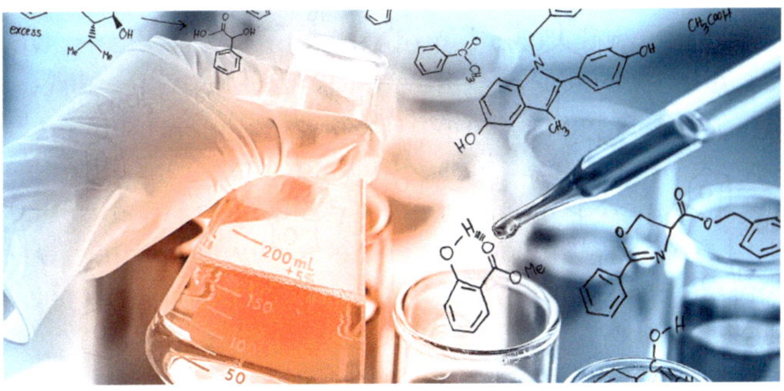

Figure 6.5. Synthesis of pharmaceuticals and agrochemicals.

Source: https://master-chimie.univ-grenoble-alpes.fr/medias/photo/master-2-organic-synthesis-for-pharmaceutical-and-agronomical-industries_1494863642244-jpg?ID_FICHE=116993.

The integration of such advanced reactions within drug design is a significant advancement in drug and agrochemical synthesis, promoting increased efficiency and minimized environmental impact. Moreover, the application of biocatalysts, such as flavin-dependent nitroreductases, adds a layer of complexity to agrochemical and pharmaceutical synthesis. These enzymes possess unique catalytic properties which allow them to selectively reduce nitro compounds and *α,β*-unsaturated carbonyl compounds with high chemo-, regio-, and enantioselectivity.

The combination of photocatalytic systems with nitroreductases allows efficient conversion of diverse substrates, ultimately yielding valuable compounds like anthranils and quinolines. The integration of these biocatalytic strategies into the photocatalytic system not only enhances the scope of synthesis processes but also offers an eco-friendly alternative for productively achieving complex target molecules. With advancements in the field of photobiocatalysis, it holds the promise to transform conventional synthetic pathways to produce medicines and crop protection agents more precisely and sustainably, paving the way for a new generation of medicinal chemistry and agricultural technologies.

6.2.2. Development of New Materials Through C–C Bond Formation

The creation of new materials through C–C bond forming has been a significant focus of interest in synthetic chemistry, particularly in light of the evolving scenario of photocatalysis. Photocatalysis facilitates the photoreduction of simple precursors into complex products using light energy, making it an environmentally friendly method of organic synthesis. This route is particularly fascinating as it allows the creation of small organic radicals, which are key intermediates in most synthesis pathways. Additionally, the discovery of new photocatalysts, such as potassium poly(heptazine imide), improves efficiency and effectiveness in such reactions, allowing for the production of valuable compounds with tailored functionalities (Antonietti et al., 2020). Through innovative strategies, researchers are continuously generating novel materials capable of leveraging advancements for beneficial purposes in pharmaceuticals and material science.

The incorporation of C–C bond assembly into increasingly complex systems has led to progress in achieving the selective functionalization of organic molecules. Photocatalysis not only enables the formation of such bonds but also facilitates the functionalization of the resulting materials, thus becoming a bountiful ally in synthetic organic chemistry. The ability to utilize visible light to activate photocatalysts opens a door for novel ways of introducing functional groups, thus enhancing the diversity of the products that can be attained. For instance, there has been recent evidence of the ability to utilize dichloromethyl radicals derived from chloroform under visible light to achieve the production of γ,γ-dichloroketones from enones. This kind of transformation is characteristic of how precise control of radical species can produce high yields of target substrates, improving the overall utility of photocatalytic processes in material synthesis (Antonietti et al., 2020).

Besides, the expanding scope of fluorinated compounds in various applications highlights the need for photocatalysis-assisted C–C bond formation. Traditional methods of fluorine introduction are hampered by less than perfect selectivity or reliance upon heavy metal toxins. To challenge these limitations, new photocatalytic methods have been conceived, most notably those involving TiO_2 as a catalyst. This approach allows for the selective formation of $C(sp^3)$–F bonds by decarboxylative fluorination with high turnover frequencies and efficiency. The ability to carry out these transformations under mild conditions not only makes synthesizing fluorinated compounds greener but also is a quantum leap towards greener chemical reactions. The continuous evolution of photocatalytic techniques for the construction of C–C bonds highlights the contributory significance of this branch in developing new materials according to the needs of modern society (Figure 6.6).

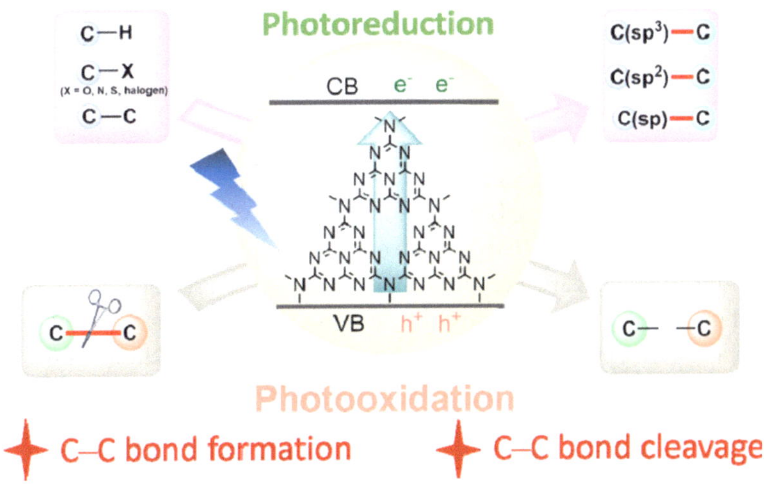

Figure 6.6. g-C_3N_4, with unique electronic properties and visible light response, is a promising photocatalyst in organic synthesis.

Source: https://ars.els-cdn.com/content/image/1-s2.0-S0040402025002078-ga1_lrg.jpg.

6.2.3. Green Chemistry and Sustainability in Photocatalytic Processes

The integration of green chemistry principles into photocatalytic reactions is crucial for enhancing sustainability in various chemical transformations, particularly in carbon–carbon (C–C) bond construction. Photocatalysis taps sunlight as a renewable energy to drive chemical reactions, therefore reducing the reliance on non-renewable energy sources. This process is in accordance with the general aims of green chemistry, which promote the use of safer solvents, reduce energy consumption, and prevent toxic waste. "Through the utilization of photocatalysts, researchers can achieve selective transformation under gentle conditions, which restricts the supply of energy and enhances overall process efficiency." The importance of optimizing reaction parameters such as light intensity and reactor design ensures that the reactions are efficient and sustainable, adhering to the principles of green chemistry. In addition to energy, photocatalytic processes offer primary advantages in the degradation of pollutants, thereby facilitating environmental sustainability. For example, the use of titanium dioxide (TiO_2) as a researched photocatalyst is helpful in tackling the issue of organic dye contamination rampant in textile industries. As described, organic dyes are known to be biodegradation-resistant, which contaminates water sources and poses drastic ecological risks. Scientists have confirmed, through the application of TiO_2 in photocatalytic degradation, its ability to degrade harmful dyes, such as indigo carmine, into harmless by-products.

This conversion not only minimizes the environmental footprint of textile waste but also enhances the sustainability of wastewater treatment processes, serving as a tangible example of green chemistry in photocatalytic systems (Saggioro et al., 2016). Moreover, developments of continuous-flow photocatalytic systems are a revolutionary step towards scalable and sustainable chemical processes. Such systems play very nicely with light-assisted reactions, offering promising possibilities for improved control of reaction conditions and enhanced photocatalytic activity. For instance, continuous flow reactors allow precise control of parameters including temperature, pressure, and residence time, leading to maximum C–C bond formation. Automation within such systems allows further seamless operation of research and development, allowing rapid screening of catalysts and reaction conditions. The modular design of such reactors allows easy access for researchers, thus democratizing green chemistry progress. The capability to produce intermediates, such as carbon monoxide, through photochemical processes, at will, is merely one of the many methods that new reactor designs can facilitate more sustainable chemical production (Bragato et al., 2024).

6.3. ADVANTAGES OF PHOTOCATALYTIC METHODS

Photocatalytic reactions have emerged as new routes to organic synthesis, especially in carbon-carbon (C–C) bond construction. One of the strengths of such methodologies is that they can be conducted under visible light as a readily available green energy source. Such reliance upon light not only reduces the need for harsh reaction conditions, e.g., high temperature and pressure, but also reduces the environmental impact of traditional chemical methods. In addition, solar energy utilization aligns with green chemistry principles, being efficient and sustainable in synthetic pathways. As recently shown by research, photocatalytic reactions can generate small organic radicals, which are important intermediates in the creation of C–C bonds, further enhancing the overall versatility of organic synthesis and minimizing waste (Antonietti et al., 2020).

Other than energy efficiency, photocatalytic processes also show high selectivity and functional group tolerance. Selective radical generation allows direct and effective modification of organic substrates, with the capability to construct sophisticated compounds in high yields. For instance, advances in photocatalysis have recently facilitated perfluoroalkylation of any organic molecule through controlled radical generation, with evidence of the methodology's effectiveness in manipulating C–C bonds (Vallejo et al., 2018). These tailored responses expand the range of chemistry accessible to chemists, offering diverse functionalized products that are otherwise not easily obtained through conventional means. Such high-level control over reaction pathways is one of the key advantages of photocatalytic approaches, as it allows interrogation of intricate molecular architectures without significant side reactions. Use of photocatalysis will also revolutionize industrial applications by streamlining the scalability of C–C bond-forming reactions.

Employing heterogeneous photocatalysts, such as potassium poly(heptazine imide) (K-PHI), under flow-reactor systems demonstrates how these methods could be more efficient compared to conventional batch-processing approaches. With the yields of γ,γ-dichloroketones ranging between 18% and 89% in a quasi-homogeneous system, such processes can be optimized for large-scale industrial synthesis (Antonietti et al., 2020). As more businesses adopt greener techniques, the promise of producing useful organic compounds with high efficiency is a major reason in favor of photocatalytic processes. Therefore, the advantages delivered by photocatalysis not only revolutionize research at an academic level but also drive substantial advancement in industrial organic chemistry.

6.3.1. Mild Reaction Conditions and Energy Efficiency

Employment of mild reaction conditions in photocatalysis is a significant advancement in C–C bond construction, if only because they are energy-efficient and environmentally friendly. Traditional C–C coupling processes typically require extremely high pressure or temperature, which accounts for a significant energy demand and increased cost of manufacture. Photocatalytic reactions, however, usually occur under ambient conditions, and utilization of light as a green source of energy is both efficient and selective. This not only decreases the energy impact of the reactions but also increases the feasibility of using these approaches in existing industrial processes.

The possibility of realizing huge savings in energy aligns with the growing interest in green chemistry, suggesting that conditions previously considered inefficient may actually be the optimal means for synthesizing complex organic molecules. In this regard, recent developments highlight the excellence of metal–organic frameworks (MOFs) in facilitating photocatalytic reactions under mild conditions. Incorporation of molecular catalysts such as $FeFe(CO)_6$ into robust Zr(IV)-based MOFs is an advanced approach to optimizing catalytic efficiency while being energy efficient. The structural stability of the catalytic site is preserved, allowing for improved stability and charge recombination avoidance compared to traditional alternatives.

Such a design promotes efficient hydrogen evolution reactions, demonstrating that MOFs are viable platforms for achieving enhanced photocatalytic transformation efficiencies while simultaneously conserving energy. Thus, this finding underscores the necessity of developing new materials to maximize reactions and align them with both economic and environmental imperatives. Moreover, donor-acceptor (D–A) fluorophore engineering in photocatalytic systems is instrumental in optimizing energy transfer processes under mild conditions. By adjusting solvent polarity and modulating D–A character, researchers can reliably access excited states, thereby facilitating energy transfer in photocatalytic reactions.

The findings that a superior fluorophore can achieve high efficiency in nickel-catalyzed cross-coupling reactions not only demonstrate the versatility of photocatalysis but also highlight energy-saving approaches for constructing C–C bonds. These approaches are vital as they ensure the sustainability of chemical synthesis processes and signal the promise of photocatalytic systems to redefine synthetic organic chemistry practice. The concurrent breakthroughs in these areas highlight the importance of adhering to mild reaction conditions as an advancement in the area of energy-efficient and sustainable synthesis (Figure 6.7).

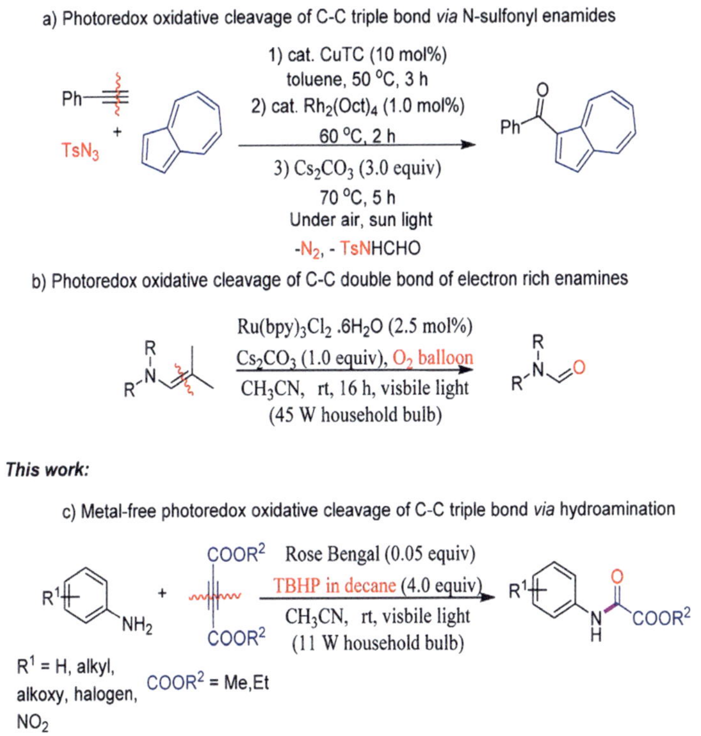

Figure 6.7. Selected examples of photocatalytic oxidative cleavage of C–C double and triple bonds.

Source: https://www.researchgate.net/figure/Selected-examples-for-photocatalysis-oxidative-cleavage-of-C-C-double-and-triple-bonds_fig1_340230499.

6.3.2. Selectivity and Functional Group Tolerance

For photocatalysis, selectivity and tolerance of functional groups are of the utmost significance in defining the efficacy and efficiency of C–C bond-forming reactions. Selective activation of a specific collection of substrates, while tolerating additional functional groups, is of the highest significance, especially when synthetic approaches involve complex molecules. Recent developments in photocatalytic methods have demonstrated that the application of donor–acceptor (D–A) fluorophores can easily modulate the reactivity of chemical systems, allowing for more efficient control of product formation. By adjusting the energy levels of these fluorophores, as demonstrated by the findings of Huang et al. (2018), chemists can enhance the selectivity of reactions, ensuring that desired products are achieved without undesired side products. Consequently, selectivity and tolerance towards functional groups not only control the success of a synthetic route but also serve as criteria in developing photocatalytic systems for specific applications. Furthermore, the tolerance to functional groups exhibited by photocatalytic reactions is an indication of the sophistication of this synthetic strategy. Photocatalysis provides the ability to transform diverse substrates, even with potentially reactive functional groups. This feature is significant because classical methodologies can struggle to maintain reactivity in the presence of numerous functional groups. The versatility of photocatalysis allows complex assemblies of molecules to be constructed without rigorous protective group tactics, which can be time and resource-consuming. The work presented by Kärkäs et al. (2016) highlights how photo-induced reactions can yield intricate polycyclic frameworks, with the ability to incorporate a variety of functional groups discreetly. By using photocatalytic systems that tolerate a wide range of functional groups, chemists can reach new levels of synthesis, achieving molecular frameworks previously out of reach. Ultimately, the combination of selectivity and functional group tolerance in photocatalytic C–C bond-forming reactions is a paradigm-bending approach to synthetic organic chemistry. By utilizing the unique features of excited states and their reactivity, researchers can design reactions that are not only efficient but also highly selective and compatible with a wide range of functional moieties. The new strategies that have emerged from this field have much potential in enhancing the synthesis of very complicated natural products and medicines because they allow them to follow a simple route in assembly. As shown in the above studies, advancements in D–A D-A fluorophores and mechanistic insights can significantly increase reaction yields, paving the way for future advances in this promising field of chemistry. Exploration of selectivity and tolerance to functional groups in photocatalysis will undoubtedly evoke new protocols and methodologies, further enriching the repertoire available to synthetic chemists.

6.3.3. Reduction of Waste and Environmental Impact

The philosophy of waste reduction and environmental impact minimization is inherent to contemporary chemical processes, particularly the creation of carbon-carbon (C–C) bonds. Photocatalysis is an innovative technology that employs light to activate chemical reactions, offering a new route to the synthesis of desirable organic products with reduced waste. The technique utilizes photocatalysts that facilitate the conversion of renewable feedstocks such as lignin into carbonaceous products. By invoking the principles of advanced oxidation processes (AOPs), photocatalysis not only the recovery of valuable constituents from wasteful materials but also offers a green alternative to traditional petrochemical-based products, whose manufacture often significantly contributes to environmental degradation. The ramifications of the use of photocatalytic methodologies in C–C bond formation extend far beyond the simple manufacture of materials; they reflect an added degree of commitment towards taking care of the environment and minimizing resource use. Furthermore, the merger of photocatalytic reactions with biocatalytic systems is a giant step towards making chemical synthesis more sustainable. The inclusion of these methods, in addition to expanding product yield, also virtually eliminates some of the issues of waste disposal and environmental pollution. For instance, when lignin-derived materials are processed by photocatalysis, they can potentially lead to aromatic products that could be suitable alternatives to fossil fuel-derived chemicals. This emerging synergy boasts promising directions for reducing the environmental footprint of the chemical industry. More specifically, studies indicate that employing such an integrated process can enhance the degradation of toxic organic pollutants, demonstrating the effectiveness of photocatalytic processes in minimizing the adverse environmental effects associated with traditional chemical processes (Busse et al., 2015). The environmental applications of photocatalysis can even be extended to solving larger problems like water pollution due to industrial processes. Conventional wastewater treatment techniques tend to fail to remove recalcitrant organic pollutants effectively, and therefore novel solutions are essential. The integration of photocatalysis in wastewater treatment offers the possibility of achieving substantial pollutant degradation while simultaneously providing energy-efficient processes. The technology uses light to convert hazardous chemicals into less harmful byproducts, thereby aligning with sustainable development goals. The incorporation of new technologies, including hybrid processing systems that combine photocatalysis with other processes for treatment, is an encouraging approach to achieving total waste minimization and improved cleaner production processes (Abou-Ghazala et al., 2014). Although further research must be undertaken to optimize these systems, their promise to redefine C–C bond formation and sustainability is profound and necessary for a sustainable future.

6.4. CHALLENGES AND LIMITATIONS

The pursuit of photocatalysis in C–C bond construction is confronted with significant challenges, primarily in the efficiency and stability of utilized photocatalysts. Most recent research attributes inefficiency to the low quantum yields that are inherent in most photocatalytic reactions. Common photocatalysts such as titanium dioxide are usually handicapped by their band gap, which restricts their activation by visible light and depresses overall reaction rates. The intrinsic photocatalytic activity of TiO_2, for example, is compromised by its electronic characteristics, which are not fully optimized under applicable conditions. Accordingly, while promising approaches such as doping have been developed to reduce such limitations, the pragmatics of incorporating such optimizations into effective technologies present lingering issues with optimizing reaction conditions and realizing consistently high yields (Abou-Ghazala et al., 2014). Besides, the scalability problems of photocatalytic reactions cannot be overlooked. Though laboratory experiments of C–C bond formation through photocatalysts have seen great success, scaling them up to industrial levels remains a problem. The financial feasibility of photocatalysis is often inhibited by the high cost of novel materials and the energy required for efficient light activation. In addition, their ability to regenerate and maintain longevity is also an important factor; several cycles of reaction might lead to a loss of activity and thus are not suitable for continuous operations. These limitations highlight the importance of future studies in developing stable long-term photocatalytic systems (Abou-Ghazala et al., 2014). Furthermore, the environmental effects of photocatalytic operations also have to be critically assessed. While photocatalysis offers a means of organic pollutant reduction via integrated technology such as sonophotocatalysis, problems with reaction by-products remain of concern. Degradation of the pollutants may lead to the formation of intermediate products complex to be removed and may be poisonous. This necessitates careful evaluation and monitoring of the whole process to ensure that it is founded on sustainability principles. In addition, the adaptation of photocatalysis to current wastewater treatment systems adds yet another level of complexity, requiring innovations to consider both economic models and environmental law requirements (Chiodo et al., 2010).

6.4.1. Stability and Reusability of Photocatalysts

The recyclability and stability of photocatalysts are of particular interest in enhancing the efficiency of photocatalytic reactions, particularly the construction of carbon-carbon (C–C) bonds. The ability of a photocatalyst to maintain its structure and catalyst activity through multiple cycles significantly contributes to cost savings and addresses environmental issues by minimizing waste production. For instance, titanium dioxide (TiO_2) is presently utilized extensively in heterogeneous photocatalysis due to its choice characteristics, including plentiful availability, good stability, and non-toxicity. Of late, it has been shown that TiO_2 is not only endowed with long-lived stability in performing C–C bond formation, such as decarboxylative fluorination of carboxylic acids with Selectfluor, but also achieves high turnover frequencies—even 1050 h^{-1}. This outcome points towards the possibility of utilizing stable photocatalysts to maximize the sustainability of chemical conversions in industrial operations (Hammond et al., 2018). Furthermore, incorporating modifications into photocatalytic systems can achieve maximum stability and reusability. Synthesis of nanohybrids, utilizing materials like carbon nanotubes (CNTs) and iron oxide (Fe_3O_4) into TiO_2, is a state-of-the-art approach for improving photocatalytic efficacy. These modifications not only facilitate structural stability of photocatalysts under operating conditions but also allow efficient photocatalyst separation and recovery from reaction mixtures, enhancing the practicality of photocatalytic systems in industrial applications.

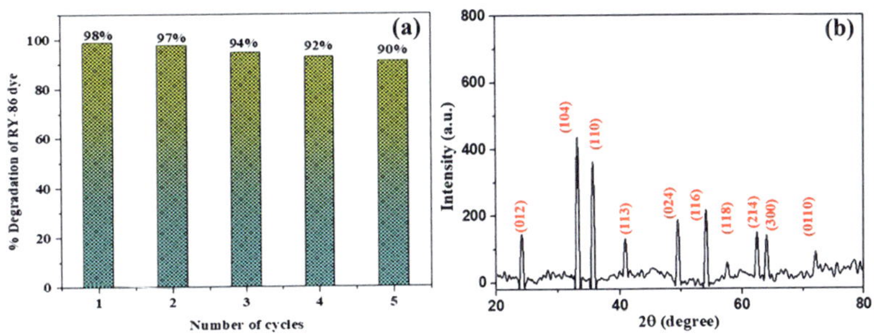

Figure 6.8. Reusability and photostability studies: (a) Degradation efficiency over five runs; and (b) XRD analysis after the photocatalytic process.

Source: https://www.researchgate.net/figure/Reusability-and-photostability-studies-a-degradation-efficiency-during-five-runs-and-b_fig4_344782691.

Redesign of the reactor and addition of innovative materials allow the efficiency of photocatalysts to be significantly enhanced, such that they are applicable for a great number of cycles. This development opens up possibilities for photocatalysis to be achieved as a practical solution for degradation in industry, thus making it indispensable in applications ranging from environmental remediation (Mason et al., 2018). The implications of reusability and stability extend to economic significance because they also address the environmental responsibility of photocatalytic systems when utilized in C–C bond formation. Waste minimization through efficient catalyst recycling inherently promotes chemical manufacturing sustainability. As increasingly stable photocatalytic systems are designed, it becomes more feasible to address the challenge of refractory pollutants in wastewater treatment. A photocatalyst capable of withstanding multiple cycles not only assure consistent performance but also diminishes the environmental footprint of industrial processes. Therefore, as the field of photocatalysis evolves toward stability and reusability, technological advancements in this direction imply a shift to greener, more sustainable chemical methodologies (Mason et al., 2018) (Figure 6.8).

6.4.2. Scalability of Photocatalytic Processes

The scalability of photocatalytic reactions significantly influences their applicability in industrial settings, particularly in C–C bond-forming processes. Scale-up from minor laboratory reactions to large batches presents unique challenges and opportunities. A primary advantage of photocatalytic processes is to utilize light as an energy source to facilitate reactions that otherwise would require too much heat. Enhanced photon efficiency demonstrates a basic scale factor, as demonstrated by the activity of catalysts like TiO2, which achieve high turnover frequencies, surpassing traditional processes in efficiency without stoichiometric heavy metals. Innovation of photocatalytic approaches paves the way for even more sustainable synthetic methods, reflecting a broadening trend towards sustainability in the chemical industry. Thus, understanding the scalability dynamics of photocatalytic processes is critical to facilitate the transfer of laboratory success into practical applications. A key component in realizing scalability for photocatalytic processes is providing steady improvement of catalyst activity under varying conditions. Improvements in material science and catalyst technology have provided heterogeneous photocatalysts with significant activity and stability. For example, advances in the photocatalytic use of TiO_2 as a selective catalyst for $C(sp^3)$–F bond formation demonstrate the potential of photocatalysis in evading deficiencies characteristic of traditional catalytic processes, which are usually restricted by activity and selectivity. Furthermore, the ease of recovery and recycling of these photocatalysts further contributes to their operational efficiency on a larger scale. Fluorinated molecule construction by heterogeneous photocatalysis is a compelling case for scalable manufacturing methods, emphasizing the requirement to create resilient and efficient systems as we move toward more sustainable chemical practice (Hammond et al., 2018). The diversity of photochemical transformations is another essential determinant of scalability for photocatalytic processes. Photocatalysis enables the rapid assembly of complex molecular structures from more primitive precursors through the assistance of the unique reactive intermediates generated following the absorption of photons. The process enables the preparation of highly elaborate polycyclic compounds that are challenging to prepare under conventional thermal processes. For example, the use of photochemical conversions for C–C bond assembly not only optimizes synthetic efficiency but also optimizes the diversity of products available from one operation (Kärkäs et al., 2016). As the chemical industry increasingly seeks opportunities to and maximize production while minimizing environmental impact, being able to scale up photochemical processes with no loss of selectivity or overall yield is essential. The continued exploration of photochemical methods in conjunction with improved catalyst engineering is directly aligned with growing commercial pressure for sustainable chemistry.

6.4.3. Competition with traditional methods of C–C bond formation

In organic synthesis, the competition between traditional carbon-carbon (C–C) bond-forming methods and photocatalytic approaches is increasingly critical. Traditional approaches, such as cross-coupling reactions under palladium or nickel, tend to be highly dependent on extensive heating and highly reactive intermediates, which limit the scope and efficiency of the reaction. However, these reactions have limitations, including decreased selectivity and substrate constraints, which lead to the formation of undesired by-products. Photocatalysis, on the other hand, has introduced a paradigm shift by using light to generate reactive species under a more flexible and controlled regime. The ability to efficiently produce intricate molecular scaffolds has made photocatalytic methods attractive, although they are still in their nascent stages of widespread adoption compared to traditional counterparts. The combination of photocatalysis with transition metal catalysis presents significant advantages over conventional ways. For instance, nickel or palladium photoredox catalysis reactions are now pivotal alternatives as they offer the ability to create carbon–carbon bonds with less input energy and less side reaction. The nickel photocatalysis has been proven effective by facilitating the coupling of less-activated substrates, bypassing the conventional limitations of C–C bond-forming reactions. These protocols simplify the process, offering good yields and selectivity, along with the promise of expanding the scope of the reaction. As revealed in recent studies, this new coupling combines traditional transition metal reactivity with radical-based methods, offering an appealing approach for synthetic chemists aiming to build C–C bonds (이근석, 2021). Even with the promise of photocatalysis, traditional methods of C–C bond formation remain prevalent, mainly due to their established strategies and acquired instincts from decades of research. Traditional methods are typically more reproducible for chemists experienced with their mechanisms. Furthermore, their scaling up and ease of transferability to industrial processes often hold a competitive edge. But as photocatalysis continues to develop, there is more recognition of the advantages this process can bring. Light-powered processes for transformation have the potential to reduce energy requirements and introduce novel reaction mechanisms, which could soon lead to a reevaluation of current standards in synthetic chemistry. Lastly, the transition from traditional methods to photocatalytic processes in C–C bond formation is an evolutionary step driven by ongoing advancements in the field (Kärkäs et al., 2016).

6.5. RECENT ADVANCES IN PHOTOCATALYTIC RESEARCH

The area of photocatalysis has also been a promising horizon for recent developments to facilitate the process of forming C–C bonds, thus making organic synthesis more efficient and valuable. One such development is visible light-driven photocatalysis, which has enriched the chemist's toolkit by enabling reactions that were previously hindered by high energy requirements. This approach is founded on the generation of reactive radicals that mediate carbon-carbon bond formation in a more environmentally friendly manner. Current research specifies the effectiveness of using perfluoroalkyl radicals in these reactions. It highlights their potential to engage various organic substrates through mechanisms such as visible-light transition-metal-mediated photocatalysis and electron donor-acceptor complexes. Taking advantage of these advancements, scientists can utilize light energy for selective transformations, contributing significantly to the growing interest in new and sustainable synthetic approaches to organic chemistry (Vallejo et al., 2018). Supplementing the promises of photocatalytic systems, the study of ligand-to-metal charge transfer (LMCT) mechanisms has gained prominence in recent reports. LMCT forms a novel mechanism for the visible light activation of photocatalysts like titanium dioxide (TiO2), which traditionally exhibit negligible activity under visible light. Surface adsorbates induce charge transfer complexes that facilitate efficient energy absorption and electron transfer under light exposure. This paradigm shift not only enhances the photocatalytic activity of TiO_2 but also broadens the substrate range that can be targeted for C–C bond construction. As such, the development of LMCT-mediated photocatalytic systems opens up new opportunities for the application of solar energy in chemical conversion reactions, and organic reaction flexibility, and advances a greener synthesis approach to synthetic chemistry (Choi et al., 2025). Also, continued progress in photocatalytic methods emphasizes the importance of interdisciplinary collaboration for advancing the field. By merging fundamental chemistry concepts with state-of-the-art materials science, researchers have developed hybrid photocatalysts with enhanced performance, stability, and selectivity for C–C bond formations. Researchers are increasingly investigating novel semiconductor materials with traditional catalysts to enhance electronic properties as well as sensitivity toward visible light. As demands for sustainable chemical processes continue to grow, the information provided by recent advances in the science of photocatalysis will be critical to future synthetic design. Lastly, the facilitation of light-promoted C–C bond construction is a milestone in the development of greener methodologies that satisfy global agendas for efficiency and sustainability in chemical manufacture (Figure 6.9).

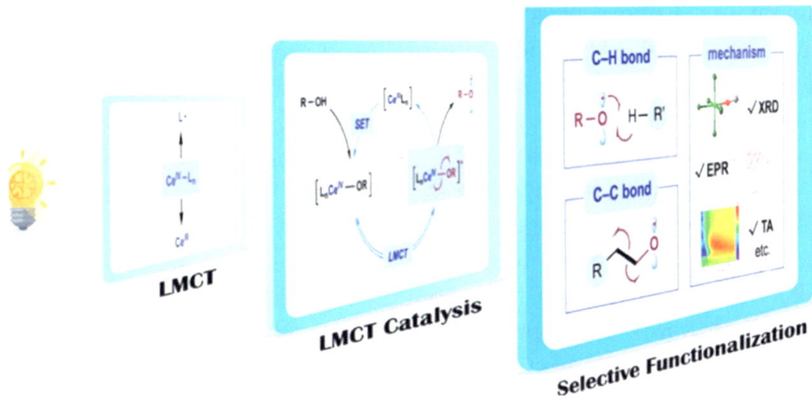

Figure 6.9. Ligand-to-metal charge transfer (LMCT) catalysis.

Source: https://pubs.acs.org/cms/10.1021/acs.accounts.4c00510/asset/images/large/ar4c00510_0012.jpeg.

6.5.1. Novel Photocatalysts and Their Properties

New photocatalyst discovery is a crucial process in the development of photocatalytic processes, particularly in C–C bond formation. These catalysts are designed to effectively harvest light energy and facilitate reactions that are not possible with traditional catalytic procedures. New heterogeneous photocatalysts like potassium poly(heptazine imide) (K-PHI) show the potential of new materials in generating small organic radicals upon irradiation with visible light. These radicals are productive intermediates for the preparation of γ,γ-dichloroketones from enones and chloroform, achieving yields of 18%–89% across different substrates. Notably, its sub-micrometer particle size and negative zeta-potential enhance the reactivity and stability of K-PHI, enabling significant productivity improvements with conventional methods, reaffirming anew the efficacy of new photocatalysts in mediating intricate organic transformations (Antonietti et al., 2020). Furthermore, the synthesis of fluorinated compounds also employs new photocatalysts, which are essential to numerous industries due to their particular properties. The addition of fluorine to organic molecules using traditional methods often results in environmental and efficiency issues, necessitating more environmentally friendly alternatives.

Using titanium dioxide (TiO2) as a photocatalyst, recent research has successfully decarboxylated various carboxylic acids via heterogeneous photocatalysis. The catalytic cycle boasts staggering turnover rates of greater than 1050 h^{-1}, coupled with unprecedented photon efficiency of greater than 1, highlighting the advances in photocatalytic technology. With its minimized requirement for heavy metals and stoichiometric reagents, the catalytic process is a testament to the transition towards green methods of organic synthesis (Hammond et al., 2018). The introduction of such novel photocatalysts not only maximizes reaction efficiency but also opens new avenues for the discovery of functionalized organic materials. The combination of K-PHI and TiO_2 in C–C bond formation shows the strength and versatility of photocatalytic systems to facilitate complex chemical reactions. As scientists increasingly study the mechanisms of photocatalysts, the lessons learnt from such catalysts are likely to carry over into forthcoming developments in organic synthesis, particularly towards developing methods that emphasize sustainability and lower environmental impact. Continued research into such photocatalytic systems guarantees the continued growth of synthetic methodology, supporting the applicability of new photocatalysts to C–C bond formation and other applications.

6.5.2. Innovations in Reaction Conditions and Setups

Advances in reaction conditions and geometries have greatly enhanced the efficiency of photocatalytic C–C bond formation, a significant step for synthetic organic chemistry. Innovations such as electron donor-acceptor (EDA) complexes have played a key role in the rationalization of reaction conditions. The process allows for effective interaction of electron-donor cycloalkanols with electron-acceptor alkenes, causing a proton-coupled electron transfer leading to the ring opening of cycloalkanols. Interestingly, the procedure sidesteps the utilization of an external photocatalyst, simplifying experimental designs and reducing material costs. The alkyl radicals generated effectively react with Michael acceptors through Giese reactions, enhancing the chemistry of selective C–C bond formations and optimizing reaction modes. The greater understanding of the process of electron transfer underscores avenues for future developments in rendering the practice of photocatalysis more sustainable and efficient (Alemán et al., 2024). Besides, improvements in reactor designs have contributed significantly to photocatalytic reaction efficiency and versatility. The advent of 3D-printed photoreactors that can be customized has afforded chemists unparalleled control over parameters, including geometry, light intensity, and fluid flow. In this approach, precise tuning is performed to obtain the best conditions for photocatalytic reactions, which are necessary for higher selectivity and yield. LEDs play a significant role in such reactors with their narrow emission spectra, which are tunable for optimal wavelength selection based on the photochemical conversion of interest. Using high-power LEDs with cooling devices to manage heat dissipation ensures steady reaction conditions and thorough probing of reaction dynamics. This versatility puts such imaginative arrangements at the center of photochemical C–C bond formation research (Bernin et al., 2022). Finally, ongoing research into large-scale photochemical reaction arrangements mirrors the overall transition to sustainable chemistry protocols. As chemical manufacturing faces increasing scrutiny over its environmental impact, efficient photocatalytic systems are set to reduce waste and minimize by-products. By utilizing modular reactors that can accommodate varying vessel sizes and distances, the scientists can efficiently compare the results of various configurations, providing valuable information regarding reaction efficiency. Estimating incident light intensity for various reactor configurations provides further improvement in optimization methods. With the combination of innovative photoreactor design and new photocatalytic methods, there is significant potential for greener C–C coupling reaction innovation, paving the way for future developments aligned with current sustainability targets in chemistry (Bernin et al., 2022).

6.5.3. Future Directions in Photocatalytic C–C Bond Formation

The future of photocatalytic C–C bond assembly lies in the synthesis of new materials and structures that will enable improved efficiency and selectivity in chemical reactions. Current trends indicate a growing use of inverse opal structures, which possess exotic optical properties that enable enhanced light-matter interactions necessary for photocatalysis. These ordered macroporous materials provide not only a high surface area but also plasmonic-photonic coupling when metal nanoparticles are filled into them. Such arrangements have shown great promise in improving photoelectrochemical reactions, so scientists need to focus on understanding the connection between photocatalyst activity and material design. With the potential for enhanced performance of such intricate structures, the science can achieve improved rates of reaction and product distributions in C–C bond formation reactions, opening doors to novel innovations in upcoming research (Armstrong et al., 2018). Besides structural innovation, the integration of machine learning and predictive modeling will also be capable of revolutionizing the landscape of photocatalytic C–C bond formation. As various systems of photocatalysis yield massive data sets, utilizing advanced computational methods can identify patterns and optimize conditions wherein C–C bond formations yield maximum efficiency. Machine learning models can forecast optimum outcomes as a function of various parameters such as reaction conditions and photocatalyst properties. This approach not only accelerates research but also improves understanding of complex reaction mechanisms underlying photocatalysis. Also, synergy between predictive modeling and experiment might lead to wiser tweaking of catalysis protocols with the development of a sound platform for designing tailor-made photocatalysts that address specific C–C bond-forming problems (Fapohunda et al., 2025). Interdisciplinary collaborative research has great potential for the field of photocatalytic C–C bond formation, particularly in environmental sustainability and practical applications. As worldwide interests in the and ecological consequences of synthetic methods grow, photocatalysis is ready to deliver more environmentally benign alternatives for traditional C–C bond-forming techniques. Through collaborative studies among chemists, materials scientists, and environmental engineers, creating highly efficient multifunctional photocatalysts with low waste production and high solar energy harvesting becomes more feasible. Moreover, integration of photocatalytic reactions into existing industrial processes can also drive large-scale application of this technology. Scaling up photocatalytic reactions to industrial levels and assessing their economic viability should be the focus of future studies, ultimately bridging the gap between laboratory success and application in green chemistry (Armstrong et al., 2018).

SUMMARY

In short, photocatalysis represents a key advancement in the field of C–C bond forming chemistry, bringing environmentally benign methodologies together with sophisticated chemical synthesis techniques. Using light as a source of power for chemical reactions reduces energy use and decreases waste generation, creating a more effective mechanism for organic conversion. As science evolves, the fine-tuning of photocatalyst parameters, such as TiO2, becomes increasingly useful for their applicability in C–C bond formation. For instance, recent research has revealed the influence of metal doping of catalysts with elements like nitrogen to alter their electronic nature, ultimately expanding their working spectral range and also improving catalytic activity (Atanelov et al., 2015). The combination of photocatalysis with high-tech materials like mesoporous silica opens new avenues for research and development in this field. Hybrid material formation with engineered structures enables increased photocatalyst loading and reaction rates. The close interaction between substrates and the photocatalyst has a tendency to create unique phenomena such as high adsorption capacity, as is the case with mesoporous synthesized materials (Asahi et al., 2015). This synergistic approach not only indicates the universality of photocatalytic systems in organic synthesis but also shows the importance of material design for improving photocatalytic efficiency in real applications. Lastly, the future of photocatalysis in the construction of C–C bonds lies in continuous innovation and interdisciplinary collaboration. With sciences like materials science, chemistry, and environmental sciences converging, new photocatalytic mechanisms and systems will likely accelerate developments in this field. Closing the theoretical-practical gap will be critical to unleashing the maximum potential of photocatalysis. Techniques to better utilize visible light, overcome limitations in current systems, and implement scalable processes will drive this field into the future, eventually creating greener and more sustainable chemical manufacturing processes.

In carbon–carbon (C–C) bond synthesis photocatalysis, band gap engineering of the semiconductor materials plays a crucial role. Emerging developments have examined how modifying electronic states in material systems like $SrTiO_3$ can enhance photocatalytic activity. By replacing transition metal ions like Fe into the material lattice, researchers can manage the valence band effectively. This process, particularly when coupled with co-doping with La ions, has led to the stabilization of low-spin electronic configurations, generating new optical transitions in the near-infrared. The ability to regulate defect formation during this doping process has significant implications for the efficiency of these materials as photocatalysts. This demonstrates how structurally engineered semiconductors can open new doors to promoting C–C bond formation in organic synthesis (Bowden et al., 2015).

The utilization of two-dimensional materials as photocatalysts has received considerable attention, particularly following the discovery of phosphorene from black phosphorus. Due to its unique physical properties and pronounced chemical reactivity, phosphorene presents an intriguing opportunity for numerous applications, such as serving as a photocatalytic reaction substrate. Its layer structure allows for increased interfacial interaction with photocatalysts, promoting faster rates of reaction to form C–C bonds. Recent studies into the synthesis and functionalization of phosphorene recognize its worth as a new class of metal ligand, thereby facilitating the development of better catalysts. Recent studies into encapsulation methods and nanoparticles integration further attest to the versatility and utility of phosphorene in catalyzing varied chemical reactions, making it the leading material in the area of photocatalysis (Benito et al., 2019). In summary, the controversies around photocatalysis in C–C bond formation emphasize material breakthrough and controlled tailoring of electronic features. The findings in the case of $SrTiO_3$ shed light on vital aspects of band gap engineering, providing a better understanding of how localized electronic states can enable photochemical reactions. Concurrently, the study of phosphorene reaffirms the significance of two-dimensional materials and their potential to optimize the performance of photocatalysts for organic synthesis. With the trend set to continue, these factors underscore the promise of integrating next-generation materials into photocatalytic systems and pave the way for cleaner methods in building C–C bonds, with far-reaching impacts on the future of organic chemistry and materials science.

REFERENCES

1. Tijani, J. O., Fatoba, O. O., Madzivire, G., & Petrik, L. F. (2014b). A Review of Combined Advanced Oxidation Technologies for the Removal of Organic Pollutants from Water. *Water Air & Soil Pollution*, *225*(9). https://doi.org/10.1007/s11270-014-2102-yAlemán Lara, J. J., Carli, B., Goicuría, M., Martínez Fernández, L., et al. (2024). Proton-coupled electron transfer ring opening of cycloalkanols followed by a Giese radical addition enabled by an electron donor–acceptor complex. https://core.ac.uk/download/619330064.pdf (accessed on 28th September 2025).
2. Angnes, R., Bagal, D., Brachet, E., Bräse, S., Chen, Y., Chen, J., Cismesia, M., et al. (2017). Sulfonylative and azidosulfonylative cyclizations by visible-light-photosensitization of sulfonyl azides in THF. https://core.ac.uk/download/96621313.pdf (accessed on 28th September 2025).
3. Antonietti, M., Kurpil, B., Mazzanti, S., Pieber, B., et al. (2020). Dichloromethylation of enones by carbon nitride photocatalysis.
4. Armstrong, E., Collins, G., Geaney, H., McNulty, D., et al. (2018). 2D and 3D photonic crystal materials for photocatalysis and electrochemical energy storage and conversion. https://core.ac.uk/download/153478439.pdf (accessed on 28th September 2025).
5. Asahi, R., Chang, H., Cong, Y., Fang, W., Frank, B., Frank, D., Fujishima, A., et al. (2015). Carbon-dot-sensitized, nitrogen-doped TiO_2 in mesoporous silica for water decontamination through nonhydrophobic enrichment-degradation mode. https://core.ac.uk/download/42138215.pdf (accessed on 28th September 2025).
6. Atanelov, J., Gruber, C., & Mohn, P. (2015). The electronic and magnetic structure of p-element (C, N) doped rutile-TiO_2: A hybrid DFT study. http://arxiv.org/abs/1501.05233 (accessed on 28th September 2025).
7. Barata Vallejo, S., & Postigo, J. A. (2019). Photocatalytic difluoro methylation reactions of aromatic compounds and aliphatic multiple C–C bonds. https://core.ac.uk/download/343675310.pdf (accessed on 28th September 2025).
8. Barata Vallejo, S., Cooke, M. V., & Postigo, A. (2018). Radical fluor alkylation reactions. https://core.ac.uk/download/287880745.pdf (accessed on 28th September 2025).
9. Benito, I. I., Bini, R., Bolognesi, M., Caporali, S., et al. (2019). A perspective on recent advances in phosphorene functionalization and its application in devices. https://core.ac.uk/download/301577585.pdf (accessed on 28th September 2025).

10. Bernin, D., Kvist, P., & Riddell, A. M. (2022). A 3D printed photoreactor for investigating variable reaction geometry, wavelength, and fluid flow. https://core.ac.uk/download/535015224.pdf (accessed on 28th September 2025).
11. Bowden, M. E., Chambers, S. A., Comes, R. B., Heald, S. M., et al. (2015). Infrared optical absorption in low-spin Fe^{2+}-doped $SrTiO_3$. http://arxiv.org/abs/1510.05965 (accessed on 28th September 2025).
12. Bragato, N., Buskens, P., Claros, M., den Hartog, T., et al. (2024). Light-assisted carbon dioxide reduction in an automated photoreactor system coupled to carbonylation chemistry. https://core.ac.uk/download/624275548.pdf (accessed on 28th September 2025).
13. Busse, N., Czermak, P., Herrenbauer, M., Lekelefac, C. A., et al. (2015). Photocatalytic based degradation processes of lignin derivatives. https://core.ac.uk/download/33372036.pdf (accessed on 28th September 2025).
14. Chiodo, L., García-Lastra, J. M., Iacomino, A., Mowbray, D. J., et al. (2010). Tailoring electronic and optical properties of TiO_2: Nano structuring, doping and molecular-oxide interactions. https://core.ac.uk/download/36054376.pdf (accessed on 28th September 2025).
15. Choi, W., Kim, G., & Zhang, G. (2025). Visible light driven photocatalysis mediated via ligand-to-metal charge transfer (LMCT): An alternative approach to solar activation of titania. https://core.ac.uk/download/148797921.pdf (accessed on 28th September 2025).
16. Cohen, S. M., Fei, H., Orthaber, A., Ott, S., et al. (2013). Enhanced photochemical hydrogen production by a molecular diiron catalyst incorporated into a metal-organic framework. https://core.ac.uk/download/323064774.pdf (accessed on 28th September 2025).
17. De Vos, A., Hendrickx, K., Lejaeghere, K., Van Der Voort, P., et al. (2017). Missing linkers: An alternative pathway to UiO-66 electronic structure engineering. https://core.ac.uk/download/84045313.pdf (accessed on 28th September 2025).
18. Di Valentin, C., Fazio, G., & Selli, D. (2017). Using density functional theory to model realistic TiO_2 nanoparticles, their photoactivation and interaction with water. http://arxiv.org/abs/1809.02336 (accessed on 28th September 2025).
19. Fapohunda, O., Fapohunda, O., Ige, A. O., Ige, K., et al. (2025). Predictive modeling of PFAS behavior and degradation in novel treatment scenarios: A review. https://core.ac.uk/download/647986103.pdf (accessed on 28th September 2025).

20. Gherardi, F., Goidanich, S., Roveri, M., Toniolo, L., et al. (2018). Photocatalytic nanocomposites for the protection of European architectural heritage. https://core.ac.uk/download/151431903.pdf (accessed on 28th September 2025).
21. Hammond, C., & Tarantino, G. (2018). Catalytic formation of $C(sp^3)$-F bonds via heterogeneous photocatalysis. https://core.ac.uk/download/287579966.pdf (accessed on 28th September 2025).
22. Howe, R. F., & Rhydderch, S. (2015). *In-situ* EPR studies of reaction pathways in titania photocatalyst-promoted alkylation of alkenes. https://core.ac.uk/download/29151174.pdf (accessed on 28th September 2025).
23. Huang, J., Li, S., Liu, Q., Lu, Y., et al. (2018). Donor–acceptor fluorophores for energy-transfer-mediated photocatalysis. https://core.ac.uk/download/231843482.pdf (accessed on 28th September 2025).
24. Kärkäs, M. D., Porco, J. A., & Stephenson, C. R. J. (2016). Photochemical approaches to complex chemotypes: Applications in natural product synthesis. https://open.bu.edu/bitstream/2144/26416/1/Photochemical%20Approaches%20to%20Complex%20Chemotypes%3a%20Applications%20in%20Natural%20Product%20Synthesis.pdf (accessed on 28th September 2025).
25. Kudoahor, E. (2024). Two Studies: Aromatization-Driven Ring Opening Functionalization of Unstrained Cycloalkanones Through Visible Light Catalysis, and Synthesis of Amino Endoperoxides Using Self-Doped Titanium Dioxide (Ti^{3+}@TiO_2) as a Photocatalyst. https://core.ac.uk/download/619597405.pdf (accessed on 28th September 2025).
26. Mason, E. (2018). Synthesis and Photocatalytic Performance of TiO_2-CNT and Magnetized Fe_3O_4-TiO_2-CNT Multifunctional Hybrids: A Pickering Emulsion Platform for Organic Degradation. https://core.ac.uk/download/211346833.pdf (accessed on 28th September 2025).
27. Mendes Saggioro, E., Moreira, J. C., & Sousa Oliveira, A. (2016). Heterogeneous photocatalysis remediation of wastewater polluted by indigoid dyes. https://core.ac.uk/download/322426764.pdf (accessed on 28th September 2025).
28. Prats Luján, A. (2024). Discovery, Characterization and Application of Nitro reductases for Pharmaceutical Synthesis. https://core.ac.uk/download/620146778.pdf (accessed on 28th September 2025).

CHAPTER 7

ENERGY TRANSFER AND DUAL CATALYSIS

Contents

INTRODUCTION

The field of dual catalysis and energy transfer is a new and developing field with fresh methods of energy capture and utilization more effectively. At the center of such research lies the all-important phenomenon of photocatalysis, which can utilize light energy to energize chemical reactions and thus serve as an environmentally friendly substitute for traditional light-hungry processes. During recent years, utilization of visible light in such catalytic conversions has picked up a tremendous pace, which precisely adheres to the principles of green chemistry. By using low-cost and abundant solar light, scientists can convert this into clean energy carriers such as hydrogen and thus help address vital issues concerning the use of fossil fuels. As mentioned, particulate photocatalyst-based water splitting is a simple and inexpensive technology to convert renewable and clean solar energy into hydrogen energy on a large scale. Dual catalysis, where two different catalytic entities simultaneously interact, also increases the efficiency and selectivity of the reaction. This technique takes advantage of the complementary capabilities of each catalyst to achieve novel mechanisms that might not be found in conventional single-catalyst systems. Recent advancements in visible-light photocatalysis have highlighted the potential of combining energy and electron transfer mechanisms for sophisticated organic conversions. As highlighted, Visible-light photocatalysis is a rapidly expanding and powerful means of promoting organic transformation, and the discovery of energy-transfer (EnT) reactions, along with electron transfer pathways, opens unparalleled possibilities for synthetic chemists. Not only do these dual-catalytic systems accelerate reaction rates, but they also broaden the substrate range that can be effectively converted, thus making significant contributions to sustainable chemistry science. The implications of energy transfer and dual catalysis extend far beyond bare efficiency and cost savings; they also present new paradigms for overcoming the barrier of global energy. As scientists continue to improve these methods, the possibility of a saleable application in carbon-free technology becomes increasingly viable. The synergy of dual catalysis with visible-light-driven reactions can potentially trigger paradigm shifts in the production of sustainable energy resources, emphasizing out-of-the-box solutions to combat greenhouse gas emissions. Through optimal design and optimization of such systems, the chemistry community is well-positioned to play a significant role in the energy transition. By applying concepts in dual catalysis and cutting-edge research in energy-transfer methods, the field stands to contribute greatly to clean energy production and consumption at an unprecedented level.

Definition of Energy Transfer

Energy transfer is a fundamental principle of physical and chemical processes, reflecting the transfer of energy from one system to another. It is a unifying idea to comprehend all types of processes, from mechanical work to thermal conduction, and is particularly significant in catalysis. In catalysis, energy transfer occurs when reactants absorb energy, resulting in bond-breaking and product formation, thereby lowering the activation energy needed for a reaction to take place. The efficiency of energy transfer has an important role in the overall efficiency of catalytic processes, where different channels of energy transfer may lead to variation in reaction rates and yields. Understanding these dynamics enables scientists to tailor catalytic processes to specific uses, particularly in the instance of dual catalysis, where two catalytic processes are interrelated in order to enhance reaction yields. In dual catalysis, energy transfer is more complicated because several catalytic species and routes are involved simultaneously. The interplay between different catalysts may lead to enhanced energy transfer, making reactants transform more effectively into desired products. For instance, through the use of advanced catalytic materials, including graphene-based architectures, researchers have found that the incorporation of non-metallic heteroatoms into such materials can create catalytically active sites that enable efficient energy transfer during electrochemical reactions, like oxygen reduction (Alammari et al., 2018). In addition to improving the kinetics of the reactions, this also has a profound influence on the stability and lifetime of the catalysts employed in the dual catalytic systems. Thus, understanding the nuances in energy transfer enables more effective catalytic processes that can perform under varied conditions. Apart from that, studies on energy transfer in catalytic processes incorporate intricate factors based on the mechanisms of the reactions involved. For example, zeolites, extremely popular in chemical catalysis, possess characteristic topological and acid properties that significantly influence energy transfer processes. Advanced modeling techniques, such as ab initio molecular dynamics, shed light on how temperature changes and alterations within the catalyst structure can affect reaction kinetics, such as methanol conversion to hydrocarbons. This kind of modeling assists in explaining how energy transfer may influence not only the rates of reactions but also mechanisms along which they occur, including concerted or stepwise processes (Bailleul et al., 2016). By explaining such mechanisms at the molecular level, researchers can optimize catalytic processes and design innovative technologies that enhance the efficacy of energy usage in related applications. Understanding energy transfer in this multidimensional context is fundamental to developing catalytic science.

Overview of Dual Catalysis

Dual catalysis is a novel approach that has emerged as a game-altering mechanism in the realm of synthetic chemistry by combining two catalytic processes to enhance reaction specificity and efficiency. The mechanism utilizes the synergistic action of two distinct catalytic species that can act in parallel or sequential modes within the same reaction environment. In the process, dual catalysis not only accelerates reaction rates but also enables the transformation of stubborn substrates that would otherwise be difficult to change through traditional procedures. The versatility intrinsic in dual catalytic systems provides chemists with unprecedented liberty to maximize synthetic pathways, thereby allowing the production of more environmentally friendly and economically viable chemical processes. Among the critical uses of dual catalysis is in renewable energy conversion and storage technology, such as electrocatalysis and photocatalysis. These two processes are fundamental to the advancement of renewable energy technologies such as fuel cells and solar-powered reactions. Dual catalytic systems utilize the unique properties of both catalysts, ensuring maximum energy transfer mechanisms required for overall performance. For instance, employing metallic catalysts in combination with graphene-type materials has garnered significant attention due to the latter's ability to stabilize and support catalytic sites while promoting electrical conductivity (Alammari et al., 2018) (Figure 7.1).

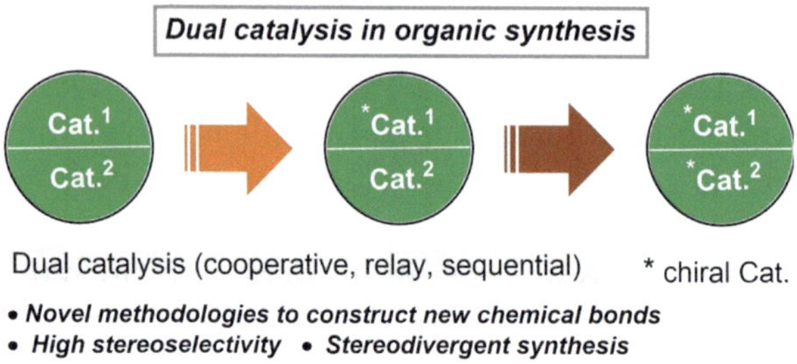

Figure 7.1. Dual catalysis in organic synthesis.

Source: https://external-content.duckduckgo.com/iu/?u= https%3 A%2F%2 Fchemi stry- europe.onlinelibrary.wiley.com%2 Fcms% 2Fasset% 2F6da376df-4f19-478a-8ebd-6697 a5e2 5b9e% 2Fejoc20 2201114-toc-0001-m.jpg&f=1&nofb=1&ipt=f6d63f7a4f01d7fd72eca0b70fdd7d57641616de11941e553700dc8d4339ffeb.

Hence, these composite systems exhibit considerably increased electrocatalytic activity, a requirement for efficient energy conversion reactions. Long-term stability of dual catalytic systems is also essential, particularly with the use of these systems on the rise in industrial processes. The integration of various catalytic materials needs to deliver not only short-term performance enhancement but also sustainability under working conditions. Efforts have focused more on designing hierarchical nanostructured catalysts that resist degradation over time, thereby maintaining efficiency over extended periods of use (Alammari et al., 2018). This evolution highlights the sustained effort required in material science and catalysis to enhance the strength of dual catalytic methods. As technology in this field improves, application of such systems has a great potential to assist in lessening energy consumption and maximizing chemical conversions, in line with overall objectives of sustainable development.

Importance of Studying Energy Transfer in Catalysis

The understanding of energy transfer during catalysis is essential for optimizing reaction efficiency and conditions critical to the practical application. Mechanisms of energy transfer dictate how the reactants intermingle and transform into products, dictating the overall kinetic rate of catalytic reactions. By understanding these mechanisms, researchers can unveil the pathways traversed by energy in the catalytic cycle, driving the creation of catalysts that work under less severe conditions with lower energetic inputs. For instance, the role of bases in facilitating catalytic processes, as demonstrated by Deydier et al. (2015), illustrates how a stable base can create active species that promote high reaction rates through energy transfer. Ultimately, such knowledge is applied to design more sustainable and cost-effective catalytic processes that minimize resource consumption and environmental impact. Furthermore, the interaction of energy transfer with dual catalysis presents fascinating potential for increasing catalytic effectiveness. Two or more catalytic entities in dual catalysis systems act synergistically to catalyze a reaction, typically enhancing energy transfer that can lead to higher product selectivity or reaction rates.

By understanding how these energy transfer processes operate in dual catalytic systems, scientists can deliberately design catalysts that interact complementarily. For instance, the intercalated oxide $Na_{0{\cdot}67}CoO_2$ exhibits high oxygen evolution reaction activity due to its unique electronic character. This character is attributed to effective energy transfer between active sites, which is favorable for peroxo ion-involved reaction pathways, as detailed by Dolocan et al. (2019). This discovery underscores the contribution of energy transfer analyses in the optimization of catalyst design with significant implications both for fundamental research and application. In addition, the study of energy transfer in catalysis can result in novel ideas in catalyst design from the ground of physical chemistry. Investigating the thermodynamic landscape of catalytic systems reveals energy barriers and transitions that play a central role in determining reaction pathways. Mathematical models derived from energy profiles can guide the optimization of the optimum reaction conditions and allow predicting and achieving new catalytic behavior. As graphically illustrated by recent computational calculations, energy landscapes can reveal routes for outer sphere transfer mechanism-based hydrogenation reactions, paving the way for catalytic innovation as demonstrated by Deydier et al. (2015). Therefore, investigations of energy transfer not only illuminate fundamental catalytic processes but also set the stage for the next generation of catalysts that have the potential to cause a paradigm shift in energy efficiency and sustainability in chemical manufacture.

7.1. FUNDAMENTALS OF ENERGY TRANSFER

It is critical to understand the concepts of energy transfer to maximize dual catalysis processes, which utilize the synergy of two or more catalytic processes to achieve maximum reaction efficiency. Energy transfer refers to the transfer of energy from one system to another, a concept crucial to various fields of science, including physics and chemistry. In catalysis, energy transfer is a critical process in electron transfer and other processes that can significantly affect reaction kinetics. The distance between active sites and the nature of interaction between catalysts dictate the facilitation of energy transfer, and this influences the dual catalyst's overall performance. By studying the energy transfer mechanisms, scientists can find means of designing more efficient catalytic systems that not only speed up reactions but also enhance selectivity and yield of chemical processes. The progress in the field of materials science, specifically the creation of graphene-type materials, has provided new opportunities to improve energy transfer mechanisms in catalysis.

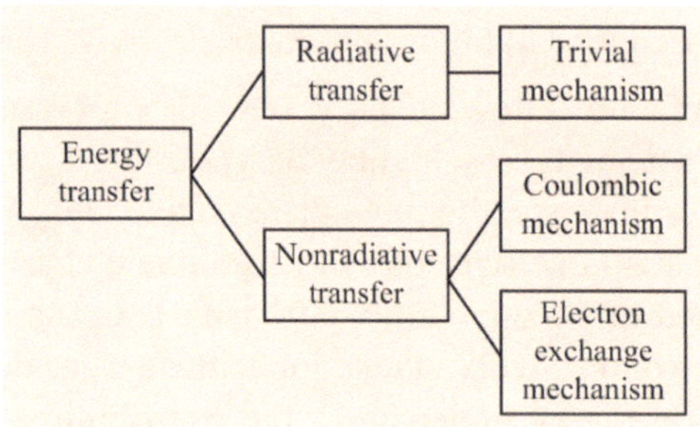

Figure 7.2. Energy transfer.

Source: https://media.springernature.com/lw685/springer-static/image/chp%3A10.1007%2F978-3-030-63713-2_2/MediaObjects/468596_1_En_2_Fig10_HTML.png.

The unique features of graphene, including high surface area and conductivity, render it a relevant substrate for electrocatalytic applications. The recent literature highlights the importance of incorporating non-metallic heteroatoms into graphene structures, which create catalytically active sites with enhanced energy transfer during reactions. Such structures not only facilitate electron transportation but also aid in the stability and longevity of the catalyst under working conditions. As such, it is essential to understand the mechanisms of energy transfer in advanced materials for constructing next-generation dual catalysts capable of operating effectively under varied environmental conditions, enhancing both efficiency and sustainability in energy applications (Alammari et al., 2018). The relationship between energy transfer and dual catalysis can also be understood in terms of hierarchical nanostructures. These frameworks can be structured to allow multiple catalysis processes to be performed in parallel, speeding up the overall reaction process. By placing and interacting multiple catalysts optimally, researchers can facilitate enhanced energy transfer, effectively lower activation barriers, and accelerate reaction rates. The dual catalysis process facilitates synergistic processes, where the strengths of one catalyst can counterbalance the weaknesses of another, allowing unprecedented reactivity. Here, the understanding of energy transfer not only serves as the foundation for maximizing an individual catalytic site but also offers insight into how composite materials can be optimized for real-world applications (Boettcher et al., 2010) (Figure 7.2).

7.1.1. Mechanisms of Energy Transfer

The energy transfer mechanisms form the focal point of the field of dual catalysis, where alternate routes of energy enhance the efficiency of chemical reactions. Molecular species-catalyst surface interaction lies at the core of such mechanisms that facilitate energy transfer in a myriad of processes, including electron transfer and energy level matching. Interestingly, integration of advanced materials such as graphene-type catalysts has been found to generate optimal energy transfer channels. Advanced materials enhance the efficiency of catalysis through enhanced charge transfer and reactive species stability, leading to improved energy conversion to chemical form. The development of hierarchical nanostructured electrocatalysts highlights the role of structural design in promoting effective energy transfer, thereby enabling multiple electrochemical reactions, particularly oxygen electroreduction reactions important for fuel cell applications (Alammari et al., 2018). Dual catalysis approaches take advantage of the complementary mechanisms of two catalysts to provide efficient energy transfer, an essential consideration in the synthesis of valuable chemicals from low-cost feedstocks. For instance, the integration of high-performance metal catalysts and acid-base catalysts can lead to synergistic effects that enhance reaction rates and product selectivity. One such typical example is the conversion of CO_2 to value-added hydrocarbons, where the dual catalysis by potassium superoxide-doped iron oxide and zeolites (e.g., ZSM-5 and MOR) optimizes energy transfer processes. These catalysts not only facilitate the cleavage of stable C=O bonds in CO_2 but also allow the formation of crucial intermediates with the assistance of intermediaries such as surface formate species. This level of control over the energy transfer mechanism results in considerably lower CO selectivity and better yields of desired products, pointing to the utility of dual catalyst systems (Abou-Hamad et al.,

2019). Energy transfer reactions in dual catalysis also include the stabilization of reaction intermediates, which play a crucial role in propelling the process with high efficiency and selectivity. Correct catalyst design has an essential influence on the intermediates' interaction with the catalytic surface, directing the global energy profile of the reaction. For example, in the conversion of CO_2 to hydrocarbons, the formation of carbenium ion intermediates is a decisive factor in achieving the required product selectivity. Research has established that first-principles calculations can provide insights into these interactions in order to construct a more elaborate picture of how distinct zeolite topologies, such as MOR and ZSM-5, selectively stabilize distinct intermediates. As researchers delve deeper into complex interactions, the development of sophisticated dual catalytic systems will continue to influence the mechanisms of energy transfer, supporting innovative routes in green chemical production.

7.1.2. Types of Energy Transfer Processes

Energy transfer phenomena play a critical role in numerous scientific disciplines, particularly chemistry and catalysis. Energetic to these phenomena are mechanisms of energy conversion and transmission that entail energy, irrespective of form, to pass from one system to another. Heat conduction, convection, and radiation are the most significant types of energy transfer, each having unique properties and applications.

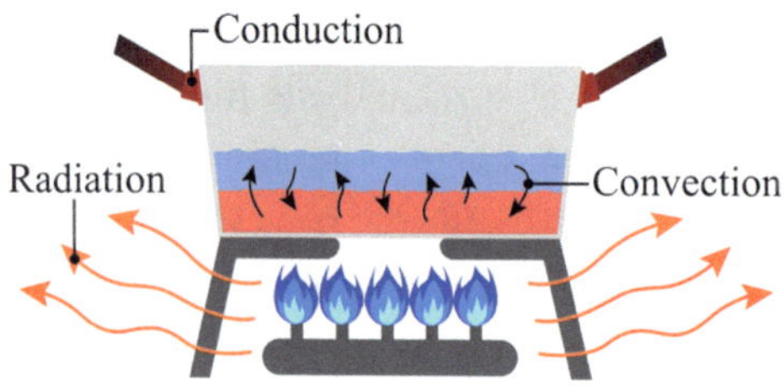

Figure 7.3. Three modes of heat transfer.

Source: https://www.shalom-education.com/wp-content/uploads/2021/03/image-19.png.

For instance, conduction of heat is direct thermal energy transfer by contact, which is of vital significance in catalysis, where reactants must collide and interact effectively to facilitate a reaction. Understanding these mechanisms allows scientists to fine-tune conditions of catalysis for maximum efficiency and yield in chemical reactions, particularly in energy applications. In dual catalysis, the interaction between two or more catalytic agents tends to enhance the efficiency of energy transfer. This method involves the integration of different catalysts to improve multiple reaction routes simultaneously, enabling maximum energy utilization in a chemical reaction. The cooperation of such catalytic systems can significantly enhance reaction rate and selectivity, as well as promote the conversion of renewable energy sources. For instance, the application of graphene-type materials in bimetallic catalysis is a manifestation of this tenet, as they not only facilitate efficient charge and energy transfer in reactions such as oxygen reduction but also stabilize the entire system. This breakthrough follows from the increased scientific interest in the optimization of energy transfer processes for improving catalytic activity in clean energy technologies (Alammari et al., 2018) (Figure 7.3).

Besides, the study of energy transfer in bimetallic catalysis is closely associated with the advancement of materials science, particularly through the introduction of novel nanostructured catalysts. These novel materials possess particular properties that allow for simultaneous modes of energy transfer, such as enhanced electron mobility and improved interaction with the reactants. Designing catalytically active sites through doping with non-metallic heteroatoms signifies the ongoing endeavor to maximize the efficacy of such processes. As researchers delve deeper into how material structure and properties contribute to energy transfer mechanisms, they pave the way for new applications that have the potential to transform the field of electrocatalysis, particularly when applied to fuel cells. Lastly, the development of energy transfer processes in dual catalysis research is an excellent step towards achieving sustainable energy solutions (Boettcher et al., 2010).

7.1.3. Role of Energy Transfer in Chemical Reactions

Energy transfer is of utmost significance in the study of chemical reactions because it is a fundamental process that maintains the transformation of reactants to products. It is a process involving the transfer of kinetic, potential, or thermal energy between molecules and atoms to break the activation energy barrier. When energy is absorbed, the reactants move to a higher energy level, making the breaking and formation of chemical bonds necessary to initiate the reaction. Efficiency in energy transfer can regulate the rate of reaction, making the effective regulation of energy flow essential in chemical kinetics. Understanding of these dynamics is fundamental, especially in catalysis, where catalysts can lower the activation energy required, thereby enhancing the efficiency of energy transfer and accelerating reaction rates without being expended in the process. In dual catalysis, attention to energy transfer is even more emphasized. Dual catalytic systems not only provide alternative redox pathways to the reaction but also generate synergistic effects that optimize energy transfer between involved catalysts. The systems often utilize various catalytic mechanisms or activate multiple chemical species, thus increasing efficiency and selectivity. For instance, graphene-type materials that are introduced to electrocatalytic systems have the potential to significantly improve energy transfer due to their extremely high electrical conductivity and large surface area. Such properties allow for faster electron transfer rates, thereby enhancing the energy management of catalytic reactions for better transformations. This new paradigm of utilizing heterogeneous structures enhances the efficiency of catalysis twofold and highlights the importance of optimized energy transfer for better performance in advanced reactions (Alammari et al., 2018). Moreover, knowledge of energy transfer mechanisms is also used in the explanation of longevity and stability of catalytic materials. Investigating how energy is dissipated or conserved in bidirectional catalytic systems can provide insights into more sustainable catalyst design. For instance, introducing non-metallic heteroatoms into graphene matrices is recognized as a viable method for designing catalytically active sites that can enhance energy transfer while maintaining structural stability during reactions. This is particularly important in oxygen reduction reactions, where energy transfer dynamics not only influence the reaction rate but also affect the durability of the catalyst under operational conditions. Hence, the intricacies of energy transfer during chemical reactions inform the development of highly efficient and resilient catalytic systems that echo the inherent synergy between chemical catalysis and energy control (Boettcher et al., 2010).

7.2. PRINCIPLES OF CATALYSIS

The fundamentum of catalysis lies in the inherent ability of a catalyst to lower the activation energy of a reaction and thereby facilitate the velocity at which the reactants are converted into products. This enables the acceleration of chemical reactions under less rigorous conditions, typically yielding greater energy efficiency. In the field of energy transfer, this concept assumes particular significance, as catalysis can enable the use of plentiful but underutilized resources such as carbon dioxide and renewable feedstocks to produce proper forms of energy. Through the use of catalytic systems, the energy required for transformations can be simplified, accelerating processes and rendering them more environmentally friendly. By optimizing molecular constituent interactions, researchers are exploring dual-catalytic modes that integrate multiple catalytic functions to improve selectivity and yield while simultaneously minimizing by-product formation. The most intriguing advance in the art of catalysis is the development of hybrid catalyst systems, which synergistically blend different catalytic functionalities. For instance, the integration of metal catalysts with zeolitic material has been an exciting area in direct CO_2 conversion to hydrocarbons. According to recent studies, some hybrid compounds of potassium superoxide-doped iron oxide and zeolite structures, including ZSM-5 and MOR, have been established to convert CO2 to valuable light olefins and aromatics effectively. More significantly, the hybrid catalysts exhibit extremely low selectivity towards undesired products such as carbon monoxide, a characteristic issue in conventional reaction processes. Their ability to maintain high space-time yields and reduced side reactions emphasizes the mechanisms of dual catalysis in energy transfer reactions, providing a strong potential for future carbon-neutral fuel synthesis (Abou-Hamad et al., 2019). In addition, research on new materials, such as graphene-type structures, has introduced new opportunities in catalytic applications, particularly in electrocatalysis for energy conversion processes. Their application in oxygen electroreduction is merely one illustration of their potential as efficient catalysts, owing to their tunable electronic states and large surface area. Careful incorporation of non-metallic heteroatoms allows them to act as active catalysts by creating novel active sites, facilitating more efficient interaction with reaction intermediates. Creating hierarchical nanostructured electrocatalysts is even superior for enhancing stability and performance under working conditions. While the demand for renewable energy continues to increase, the role of catalysis, particularly as demonstrated by binary catalytic strategies, will also remain a pivotal factor in sustainable activities and the minimization of environmental impacts (Alammari et al., 2018).

7.2.1. Definition and Types of Catalysts

Catalysts play a vital role in chemical reactions by increasing their rate without being altered permanently in the process. Their primary function is to provide an alternative reaction pathway with reduced activation energy, helping reactants transform into products more efficiently. There are two general classes of catalysts: heterogeneous and homogeneous catalysts. Heterogeneous catalysts are in a phase distinct from the reactants, typically solids when the reactants are gases or liquids, making recovery and separation simpler after the reaction. Homogeneous catalysts are in the same phase as the reactants, producing more uniform reactivity but perhaps making it harder to isolate the product. The role of catalysts extends beyond merely speeding up reaction rates; they also influence selectivity, determining the products obtained in a given reaction. Acid-base catalysts, metal catalysts, and enzyme-based biocatalysts are typical examples of heterogeneous and homogeneous catalysts, respectively. Metal catalysts feature prominently in most industrial operations, e.g., in hydrogenation reactions, where noble metals catalyze the transformation of alkenes into alkanes. Enzyme catalysts are paradigms of outstanding selectivity and efficiency in biological processes, showcasing nature's optimization of catalytic reactions by natural evolution. They are of high promise for sustainable chemistry strategies in line with present-day environmental and economic imperatives (Figure 7.4).

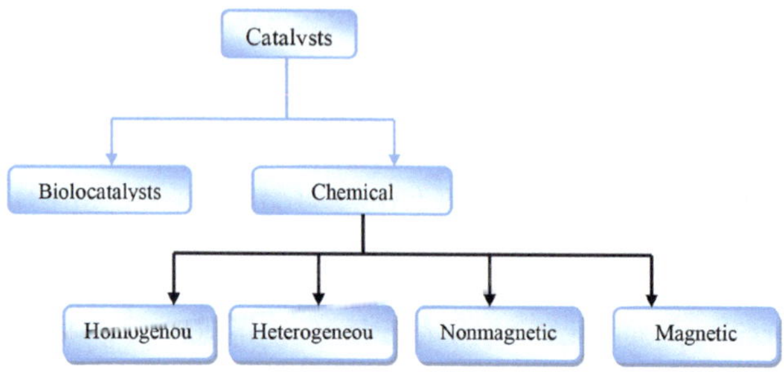

Figure 7.4. The various types of catalysts.

Source: https://www.researchgate.net/figure/The-various-types-of-catalysts_fig3_368463199.

Recent advancements in catalysis, particularly through nanostructured catalyst synthesis, have shed light on novel directions in the energy sector, particularly in fuel cell technology. Graphene-like substances, for instance, have demonstrated substantial potential as electrocatalysts, specifically in oxygen electroreduction. This research highlights the importance of upgrading catalytic activity through careful doping with non-metallic heteroatoms, which creates novel active sites. These findings not only result in improved long-term catalyst stability but also enable the development of hierarchical structures that optimize reactivity and energy transfer. The results of such research highlight the central position of catalysts in the mediation of energy transfer mechanisms, validating their intrinsic value in chemical synthesis and renewable energy solutions (Brak et al., 2014; Alammari et al., 2018).

7.2.2. Mechanisms of Catalytic Action

The mechanisms of catalytic action are the key to understanding energy transfer in all kinds of chemical reactions. Catalysts facilitate by providing an alternate route with lower activation energy, thereby accelerating the reaction rate without being consumed in the process. This is especially important in systems with more than one pathway, e.g., in dual catalysis, where two separate catalytic processes operate in parallel to each other. A good example of this is the electrochemical oxygen reduction reaction (ORR), where the slow kinetics create significant problems in microbial fuel cells (MFCs) (Bergel et al., 2012). The addition of some catalysts may lead to enhanced working efficiency, making the study of various catalytic mechanisms crucial for advancing optimized energy transfer processes across a broad scope of applications. In microbial fuel cells, microbial-level catalytic processes have been researched for insights into optimizing ORR. Microbial catalysis studies reveal that it occurs spontaneously on metallic material surfaces, facilitating reactions crucial for energy conversion with high efficiency. This occurrence demonstrates how biological catalysts contrast with their conventional chemical counterparts, particularly in their reactions to fluctuations in environmental conditions, such as neutral pH and progressive oxygen mass transfer (Bergel et al., 2012). Through these novel catalytic processes, researchers anticipate finding new ways to improve the performance of MFCs, which will in turn translate to more sustainable energy systems that harness biological processes to produce energy (Figure 7.5).

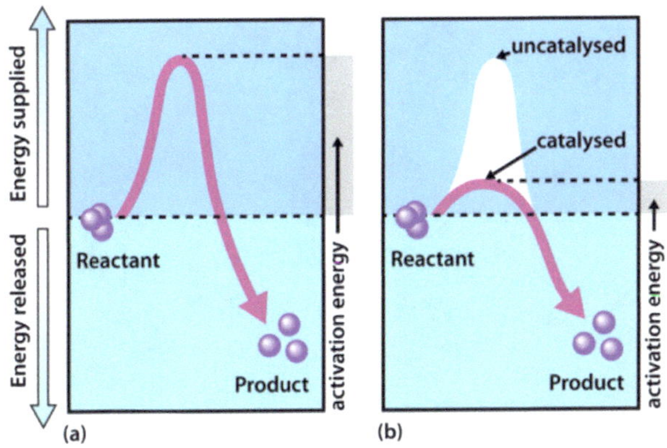

Figure 7.5. Mechanisms of catalytic action in energy transfer in various chemical reactions.

Source: https://cdn.wou.edu/chemistry/files/2019/10/transition-state-energy.jpg.

Moreover, the action of reductive and oxidative mechanisms in biological systems reveals a further area of complexity in catalytic mechanisms. In cellular environments, maintaining redox equilibrium is essential for the proper function of various pathways, particularly those involved in protein folding and managing oxidative stress. Processes like thioredoxin and glutaredoxin are key players in managing such redox reactions, dictating cellular health and metabolic proficiency. Disruption of these processes has cascading effects, leading to complications such as neurodegenerative diseases and cancer. Explanation of these mechanisms not only sheds light on the fundamental biological processes but also provides insight into the design of catalysts that can operate effectively in various biochemical contexts, thereby facilitating energy transfer by dual catalytic systems.

7.2.3. Factors Affecting Catalytic Efficiency

The efficiency of catalysts is greatly influenced by numerous intrinsic and extrinsic factors, all of which are significant in the case of energy transfer and bimolecular catalysis. First off, the inherent nature of the catalyst—comprised of its metal, support, and structure—influences how it operates. For instance, the surface area of a catalyst will determine its reactive sites employed in reaction, thus having a direct correlation with the rate of chemical conversions. High surface area catalysts increase interaction with the reactants, leading to improved catalytic efficiency.

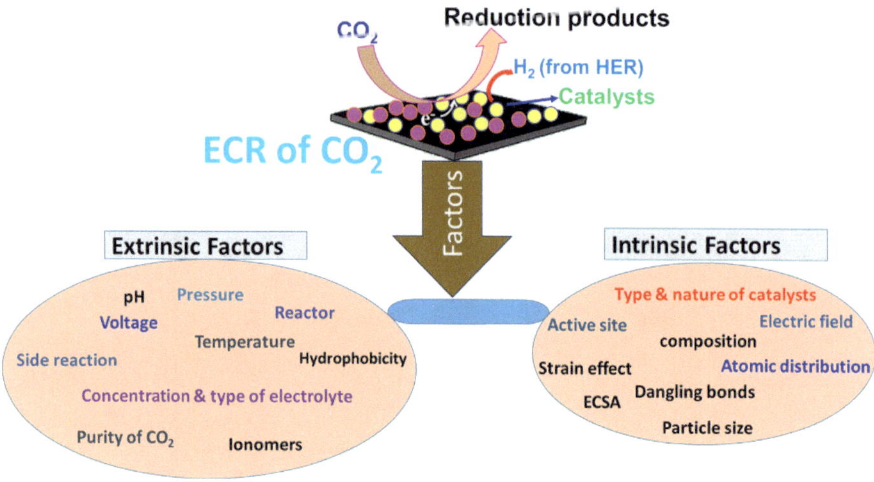

Figure 7.6. Intrinsic and extrinsic factors affecting catalytic efficiency.

Source: https://pubs.rsc.org/image/article/2025/CY/d4cy01091d/d4cy01091d-f6_hi-res.gif.

Finally, the presence of various solvents can have a profound impact on catalytic efficiency and selectivity, creating a complex landscape for catalytic performance. The solvent will affect the solubility of reactants and products and shift the reaction pathway due to solvation effects. Polar solvents, for instance, can stabilize charge states and transition states in a reaction and thus can reorganize the energy barriers that accompany catalysis. The choice of solvent often tailors the dynamic balance between reactants and catalysts, allowing the catalytic reaction to be optimized. In bidirectional catalysis, solvents are also crucial in differentially engaging the two catalytic environments further to optimize overall reaction kinematics and energy transfer mechanisms (Koumpouras et al., 2008; Xi et al., 2025).

7.3. DUAL CATALYSIS EXPLAINED

The concept of dual catalysis, particularly energy transfer catalysis, has received significant interest due to its potential to enhance reaction efficiency and selectivity in chemical reactions. This technique involves the simultaneous use of two distinct catalytic systems working in harmony to enhance substrate conversion to the desired product. By virtue of the complementary functions of two disparate catalysts, dual catalysis is able to bypass the limitations encountered in typical single-catalyst systems effectively. Specifically, this mechanism facilitates the selective conversion of recalcitrant feedstocks, such as carbon dioxide, to high-value-added hydrocarbons, highlighting its crucial role in sustainable energy solutions. The combination of two catalysts encourages greater energy coupling, which is the secret to optimal productivity with minimum waste. This highlights the revolutionary impact the application of dual catalysis has on future energy solutions. In its pursuit of better efficiency in the conversion of CO_2, recent developments have revealed the strong potential of hybridizing metal catalysts and Brønsted acid catalysis. An obvious example is the formation of a hybrid system comprised of potassium superoxide-doped iron oxide and acidic zeolites such as ZSM-5 and MOR (Figure 7.7).

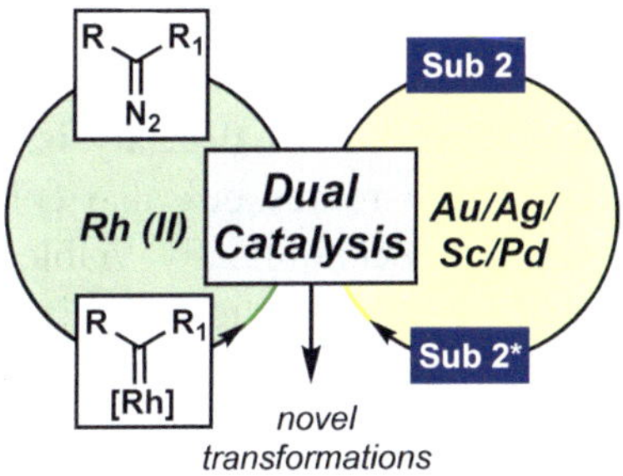

Figure 7.7. Dual catalysis.

Source: https://chemistry-europe.onlinelibrary.wiley.com/cms/asset/627f0e26-5ab8-48d9-9e37-8d62eee2b7e8/ejoc202101419-toc-0001-m.jpg.

This two-phase catalytic system allows direct CO_2 conversion to valuable olefins and aromatics in excellent space-time yields, representing a visionary carbon recycling strategy. Evidence suggests that the efficacy of dual catalysis is demonstrated by the formation of surface formate species over the zeolite network, which significantly reduces the selectivity towards the formation of undesirable carbon monoxide. These new strategies highlight the potential of dual catalysis as a way of not just reducing CO_2 emissions but also generating economically viable products from this greenhouse gas and thereby nourishing the circular economy. Advanced materials' role in facilitating dual catalysis cannot be overstated, particularly in electrocatalysis applications. Graphene-like materials are viewed as new promises for enhancing catalytic performance, especially in reactions like oxygen reduction, which are a pillar of fuel cell technologies. These materials bring new dimensions of functionality to conventional catalysts by introducing heteroatoms, hierarchical nanostructures, and improvements in long-term stability, thereby creating an ideal environment for efficient catalysis. Further, the potential to act as supports for metal catalysts scattered across surfaces implies an intrinsic synergy that can facilitate dual catalytic reactions. With complex design and material optimization aided by graphene, researchers not only achieve efficiency thresholds but also create opportunities for new approaches to reduce energy demand in catalytic processes. This further demonstrates the essential intersection of material science and catalysis in the development of energy technologies today.

7.3.1. Definition and Significance of Dual Catalysis

The concept of dual catalysis involves an interaction between two catalytic processes that co-occur in one reaction system. Through the individual abilities of the two catalysts, this method enhances the reaction efficiency and selectivity. For instance, in energy transfer reactions, one may be used for substrate activation and another for consecutive transformations, yielding a synergistic effect that cannot be realized under normal catalysis. The dual role not only facilitates more complex reaction routes but also enhances overall reaction rates, rendering processes more economically viable and sustainable. Such subtle coordination among varying catalytic activities leads to new synthetic routes, rendering dual catalysis a valuable tool in both organic synthesis and energy conversion processes. The impact of dual catalysis pervades broad areas, most prominently within the realm of renewable energy applications. By allowing dual catalysis to activate and transform simultaneously, it provides a means of higher efficiency light-harvesting and energy transfer, which is vital in photochemical systems. The technique is crucial when exploring complex materials for solar energy conversion, as it allows the building of advanced molecular architectures necessary for efficient energy transfer. Furthermore, the synergistic coordination of two catalytic processes can stabilize reactive intermediates to an even greater extent, ultimately leading to maximum product yield and minimizing byproduct formation. The coordination among different catalytic sites not only enhances the efficiency of synthetic operations but also overall chemical reaction sustainability. Experimental research demonstrates the application of dual catalysis in real-world situations and validates its potential to change most industrial processes. For example, investigation work has that certain gold complexes can effectively facilitate dual catalytic reactions with organometallic compounds, resulting in high reaction rates and enhanced selectivity for specific products. Advanced characterizations reveal that bond dissociation energies and reaction mechanisms are significantly affected by the interaction of catalyst types. By judiciously selecting catalysts with complementary behavior, scientists can leverage the virtues of dual catalysis to enhance reaction conditions and achieve maximum performance. As evidenced in recent studies, this approach has significant implications for creating more efficient and sustainable energy conversion methods, highlighting the importance of investigating and optimizing dual catalytic systems in contemporary chemical science (Gonz et al., 2018; N/A, 2025).

7.3.2. Comparison with Traditional Single-Catalyst Systems

The history of catalysis has seen a dramatic change from the traditional single-catalyst systems to the more elaborate dual catalysis paradigms. In conventional systems, one catalyst operates in an established mechanism that contributes to reaction pathway and product selectivity control to a minimal extent. This tends to lead to wasteful substrate transformation and suboptimized reaction conditions.

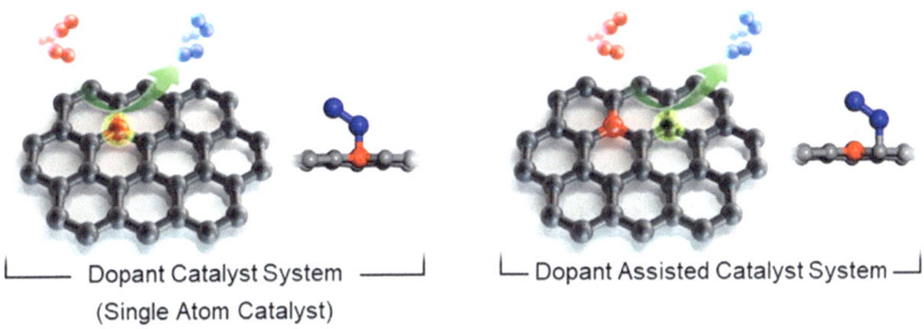

Figure 7.8. Single-atomic catalyst (SAC).

Source: https://pubs.acs.org/cms/10.1021/accountsmr.1c00016/asset/images/accountsmr.1c00016.social.jpeg_v03.

Their one-dimensionality limits their potential to utilize multiple sources of energy efficiently, which is a constraint in areas such as renewable energy technologies. Hence, breakthroughs from single-catalyst systems open up new possibilities for more complex catalytic interactions, leading to higher reaction rates and enhanced efficiency of energy conversions in chemical reactions. In contrast to traditional single-catalyst methodologies, dual catalysis exploits the twofold abilities of two catalytic species to produce synergetic interactions for enhanced reactivity and selectivity. By providing an arena for the interaction of disparate catalysts, it becomes feasible to utilize multiple catalytic processes in one homogenous reaction system. This provides opportunities for enhanced energy transfer processes because multimodal energy channels would be optimized to prefer specific reaction channels (Figure 7.8).

Furthermore, the use of dual catalytic systems can significantly expand the range of reactants and products designed for controlled transformations and thus make up for some inefficiencies of traditional approaches. The novel paradigms made possible by dual catalytic frameworks are not an incremental progress but rather a fundamental enhancement of catalytic capability in synthetic chemistry. Recent advances in photoelectrochemical systems provide a good illustration of the conceivable benefit of dual catalysis over the standard single-catalyst systems. As established in the current literature, semiconductor nanowires have emerged as the new trend in the development of this area, offering improved surface area and improved light absorption characteristics (Deng et al., 2019). Dual catalysis takes advantage of these properties through simultaneous reaction steps that are feasible, allowing light energy and chemical conversion to be harvested in tandem. This opens up new applications, particularly in the simulation of natural photosynthesis, wherein the transfer of energy is a primary necessity. Lastly, the integration of bimetallic catalytic systems in the energy sector is a very flexible approach that, besides enhancing efficiency and sustainability, aligns closely with current environmental and technological needs and maximizes the social contribution of catalytic processes (Boettcher et al., 2010).

7.3.3. Examples of Dual Catalysis in Practice

In dual catalysis, a great example is the production of graphene-based compounds for electrocatalytic processes, including oxygen reduction reaction (ORR). The incorporation of graphene-type structures enhances catalysis due to their different electronic properties and high surface area. This duality is observed when heteroatoms, which are non-metallic in nature like nitrogen or boron, are incorporated into the graphene structure to create active sites. These designed sites not only facilitate the ORR but also improve the overall stability and performance of fuel cells, thus making it a significant innovation in energy transfer processes. These findings are just a few examples of how the interaction between different catalytic elements can enhance performance and longevity in practical applications of dual catalysis, a crucial breakthrough in sustainable energy technologies (Alammari et al., 2018). But another interesting instance of dual catalysis is found in the use of metal-organic framework (MOF) materials that contain both organic linkers and metal catalysts. In this arrangement, the dynamic catalytic environment supports the catalysis of several chemical reactions, including nitrogen fixation and CO_2 reduction. The dual functionality arises from the synergistic coupling between the organic components and the metal centers, which can stabilize reactive intermediates and enhance the selectivity of the target products. Recent research has indicated that through optimizing such frameworks, scientists can obtain excellent rates of reaction while avoiding the formation of by-products. This practical application defines the potential of dual catalytic systems since they not only focus on efficiency but also ensure that energy conversion processes are environmentally sustainable, a basic tenet in creating alternative energy solutions (Boettcher et al., 2010). Furthermore, advancements in photocatalytic systems based on dual catalysis have been very promising in addressing energy transfer problems. For instance, semiconductor materials combined with noble metals allow efficient harvesting of solar energy and the simultaneous powering of chemical transformations. Light is captured by the semiconductor, which forms electron-hole pairs, while the noble metal acts as a catalyst to accelerate reduction and oxidation reactions. This dual catalytic process explores new opportunities for converting solar energy into chemical energy, particularly in hydrogen generation from water splitting or CO_2 reduction to organic fuels. The integration of these two catalytic processes demonstrates the innovative approaches being researched to achieve maximum energy efficiency and set new benchmarks for clean energy technology (Alammari et al., 2018).

7.4. ENERGY TRANSFER IN DUAL CATALYSIS

The mechanism of energy transfer in dual catalysis is an area of research of particular emphasis that combines theory and technology. Dual catalysis promotes simultaneous activation of two substrates, enhancing the efficiency and selectivity of the reaction. The process typically involves a synergistic effect between various catalytic species, where energy absorbed or emitted by one step of a reaction directly influences the control of subsequent steps. In this regard, such materials as graphene-type catalysts have been emphasized as potential candidates. Their ability to form catalytically active sites by introducing non-metallic heteroatoms into the material improves their performance during energy transfer processes. This property enables more efficient reactant conversion, opening the doors to advances in technologies such as fuel cells and electrocatalysis, thereby reiterating the enormous significance of energy interactions in optimizing catalytic performance (Alammari et al., 2018). The conversion of carbon dioxide (CO2) into valuable hydrocarbons is an excellent example for illustrating the principles of energy transfer in dual catalysis. Sustaining such high productivities through the cost of unwanted side reactions is a big challenge. New methods involving a mixture of catalysts, for example, potassium superoxide-doped iron oxide with zeolites like ZSM-5 and MOR, are typical of the art in this field. Such catalysts directly catalyze CO_2 to olefins and aromatics with high space–time yields and very low CO selectivities. The success of this process is founded on the formation of surface formate species over the zeolite framework, illustrating how energy transfer within catalytic species can regulate product distributions and reaction kinetics in general. By explaining these mechanisms, scientists unlock doors to more sophisticated strategies in chemical synthesis (Abou-Hamad et al., 2019). In summary, understanding energy transfer phenomena under dual catalysis involves the complex interaction between multiple catalytic entities and their corresponding environments. The study of energy dynamics not only facilitates catalyst selection and synthesis but also enhances their operational stability and effectiveness. With additional advances in the pipeline, the marriage of theoretical insight and experimental methodologies is likely to propel the discipline forward, facilitating innovations capable of addressing current challenges in energy conversion and green chemistry. Emphasizing the interrelatedness between catalysis and energy transfer is essential for unveiling new fronts in chemical transformation and serves as a foundation for future studies aimed at achieving superior efficacy in catalysis across various applications. Consequently, this study remains essential as the scientific community continues to strive towards utilizing dual catalysis for the betterment of society (Figure 7.9).

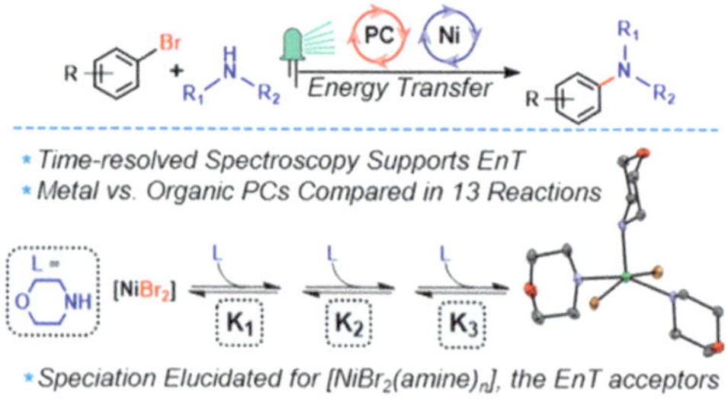

Figure 7.9. Dual catalytic light-driven cross-coupling methodologies utilizing a Ni(II) salt with a photocatalyst (PC) have emerged as promising approaches to forge aryl C–N bonds under mild conditions.

Source: https://pubs.acs.org/cms/10.1021/jacs.9b11049/asset/images/medium/ja9b11049_0006.gif.

7.4.1. Mechanisms of Energy Transfer in Dual Systems

The energy transfer mechanisms in dual systems represent a critical research theme within the broad subject of energy transfer and dual catalysis. Two active sites in such systems synergistically interact to facilitate optimization of energy transfer processes. This synergism has the potential to enhance reaction kinetics and efficiency, significantly impacting overall performance in energy-related applications. For instance, understanding how electron and vibrational energy is transferred between the two halves of a dual catalysis setup can lead to new design rules for optimal energy use. Experiments have shown that altering the nature and distribution of active sites can change energy transfer pathways, thereby influencing reaction pathways and sustainability in applications ranging from fuel cells to solar energy conversion. Furthermore, the incorporation of nanomaterials, such as graphene-type structures, into dual systems has revolutionized energy transfer mechanisms. These materials possess excellent electronic characteristics and surface properties that allow for efficient charge transfer and energy conversion. The potential applications of such materials in electrocatalysis, particularly in oxygen reduction reactions, are a clear indication of how energy transfer mechanisms can be designed to increase system performance (Alammari et al., 2018). For example, doping graphene with non-metallic heteroatoms has been reported to create catalytically active sites that enhance the interaction of the catalytic components in order to ensure more efficient energy transfer. This discovery presents a prospective way of constructing sustainable energy technologies by regulating the interactions in dual catalysis systems. Finally, long-term stability of dual catalytic systems is also a crucial consideration in the mechanisms of energy transfer. Given that energy conversion reactions are usually afflicted with catalyst deactivation, longevity causes are critical to acquire. Hierarchical nanostructured electrocatalysts built around graphene-type material offer a robust support system that can endure long-term catalytic activity. This durability results from the structural stability of the catalysts, which enables continuous and effective energy transfer channels (Alammari et al., 2018). By investigating these mechanisms, researchers are able to make inferences towards the optimization of catalyst design that can lead to an increase in efficiency and service life. As the topic develops, it is increasingly evident that advanced knowledge in energy transfer mechanisms in double systems will contribute to the advancement of scientific theory and daily practice in the energy technologies field.

7.4.2. Impact of Energy Transfer on Reaction Pathways

Energy transfer plays a vital role in deciding reaction pathways, particularly in catalysis, where efficiency and selectivity are most desirable. By facilitating energy transfer between the reactants and catalysts, we can significantly redefine the mechanism by which a reaction occurs. Dual catalysis systems are a case in point where two distinct catalytic functions synergistically collaborate to enhance product formation. For instance, a combination of different catalysts can optimize energy states whereby more favorable pathways in the reaction can be attained with the least possible by-product formation. This concept is essential in reactions such as the conversion of CO_2 into hydrocarbons, where tuning the reaction conditions can lead to significant alterations in selectivity and product profile, as detailed in the synthesis of aromatics and olefins from CO_2 with extremely low CO selectivity (Abou-Hamad et al., 2019). In addition, the interfacial energy dynamics between various catalytic materials significantly contribute to the efficiency of the entire reaction. The design of high-performance catalysts, such as MXenes hybridized with carbons, is a classical illustration of this philosophy. These possess very low overpotentials for energy-consuming reactions due to efficient energy transfer between their heterostructures. Electronic coupling between constituents assists in charge transfer and, hence, enhances kinetics and optimal reaction conditions. Such configurations enhance overall reaction pathways and are thus critical in applications towards catalysis and energy storage. The structural properties of the material not only improve catalytic performance but also highlight the intricate interdependence of energy transfer and reaction kinetics (Liu et al., 2018) (Figure 7.10).

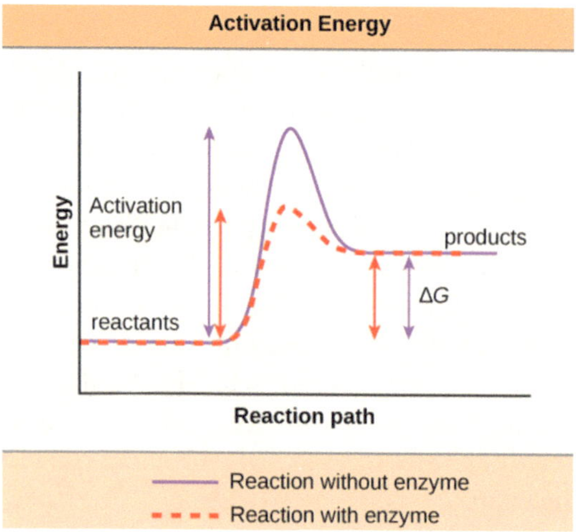

Figure 7.10. Impact of energy transfer on reaction pathways.

Source: https://opentextbc.ca/biology/wp-content/uploads/sites/96/2015/02/Figure_04_01_06.jpg.

In addition to optimizing reaction pathways through energy transfer, it is also essential to consider the mechanisms that control these interactions. The surface intermediates mechanism, as in CO_2 conversion reactions, is essential for determining selectivity and yield. According to studies, the nature of such intermediates may be influenced by energy input and choice of catalyst and, consequently, the entire reaction outcome. For instance, the presence of formate species within zeolite frameworks significantly affects the reaction landscape, highlighting the influence of energy dynamics on catalytic pathways. With the application of advanced characterization techniques and computational modeling, scientists are better positioned to predict and account for such behavior, paving the way for improved catalysis approaches. Hence, understanding the impact of energy transfer is not a speculative notion; it has deep implications for the development of sustainable chemical processes (Abou-Hamad et al., 2019).

7.4.3. Case Studies Demonstrating Energy Transfer in Dual Catalysis

The study of energy transfer processes in dual catalysis has been helpful in gaining insight into the efficiency and effectiveness of catalytic systems. A primary case study is the use of graphene-type materials for electrocatalytic processes, particularly oxygen reduction reactions. The study highlights the potential of these materials in designing catalytically active sites through the incorporation of non-metallic heteroatoms as dopants. This modification leads to hierarchical nanostructured electrocatalysts that are not only effective for energy transfer but also exhibit long-term stability in working conditions. These advancements convincingly demonstrate how designed materials can facilitate energy transfer processes between catalytic centers, improving reaction pathways and overall catalytic efficiency in fuel cell technologies (Alammari et al., 2018). In addition to graphene-based systems, carbon dots (CDs) represent another frontier in delving into energy transfer in dual catalysis. Recent advances in surface modification and the synthesis of CDs have revealed their tunable photoluminescent and photocatalytic properties. Modulation of CD size and functionalization allows one to modulate energy transfer dynamics as well as photocatalytic efficiency. Synergistic investigations have proven that combining CDs with metal oxides or transition metal chalcogenides can significantly enhance their photocatalytic activities, such as water splitting and hydrogen evolution. The synergy reflects an incredibly intricate interaction among different catalysis modalities, emphasizing the potential of engineered materials to be employed for enhancing energy transfer efficiencies within a bi-catalytic regime (Boonsith et al., 2020). In summary, these case studies emphasize the importance of catalytic material design innovation to optimize energy transfer efficiencies in dual catalysis systems. The use of advanced materials such as graphene-type structures and carbon dots not only expands the scope of catalytic applications but also creates pathways toward sustainable energy solutions. By focusing on the fundamental aspects of energy transfer in dual catalysis, researchers can pave the way for more effective catalytic reactions that are essential to solving contemporary energy challenges. With future advancements in the field, further study on the synergistic effects of employing varying catalytic units will hopefully lead to breakthroughs and highlight the need for multi-disciplinary research in this area.

7.5. APPLICATIONS OF DUAL CATALYSIS

In the dual catalysis study, numerous applications are found, particularly in energy conversion technology. The technique enhances the efficiency of catalytic processes via the synergistic combination of different catalytic species, frequently allowing more selective conversion. For instance, when applied to fuel cell technology, dual catalytic systems have superb performance for oxygen reduction reaction. Recent advances have emphasized the role of graphene-type materials because, in addition to their ability to increase conductivity, they also offer the possibility of creating active catalytic sites through doping with heteroatoms. This specific capability is critical in optimizing the performance of catalysts in energy systems, leading to vast advancements in fuel cell durability and efficiency. Thus, the incorporation of dual catalysis into energy transfer reactions is an important area of research for applications in sustainable energy. Integration of dual catalysis has also shown impressive potential in organic synthesis reactions, which require highly selective product formation. Through dual activation of several catalytic pathways, researchers are discovering new means of synthesizing complex organic molecules with efficiency. For instance, the use of dual catalysts can enable concurrent substrate conversion under milder conditions, conserving energy and enhancing yield. The two-component strategy enables fine-tuning of reaction parameters to eventually develop conditions favorable for desired outcomes. The use of graphene-based catalytic materials enhances the process because it provides support and stability for efficient coupling reactions. These applications highlight the revolutionary potential of dual catalysis not only in terms of maximizing chemical efficiency but also in facilitating more sustainable practices within the realm of synthesis. Lastly, the potential of dual catalysis to impact the production of renewable energy should not be underestimated. This catalytic method is up-and-coming towards the development of more efficient solar-to-fuel schemes. By utilizing the special properties of the dual catalysts — in which one is responsible for absorbing light and another for inducing chemical change — the overall efficiency of the energy conversion can be improved. Recent investigations into the synergy between heterogeneous and homogeneous catalysts have demonstrated significant progress in solar-driven processes. A combination of different catalytic mechanisms is more efficient in converting solar energy into storable chemical fuels that can also meet the world's pressing energy needs.

Hence, research into dual catalysis for use in renewable energy has promising avenues that are consistent with global aspirations for sustainable development and energy transition.

7.5.1. Industrial Applications and Benefits

The industrial context is shifting towards a paradigm of energy transfer and dual catalysis as the main technologies, particularly in areas striving for sustainability. Dual catalysis systems can catalysis, which significantly increases reaction efficiencies by integrating catalytic processes, reducing the total energy requirement and the time needed for chemical conversion. Among the major applications is electrochemical reactions, where dual catalysis accelerates oxygen reduction reactions, a requirement for fuel cell technology. Advanced materials, including graphene-type catalysts, not only improve the performance of catalysts but also allow for scalable production, which is an industrial requirement.

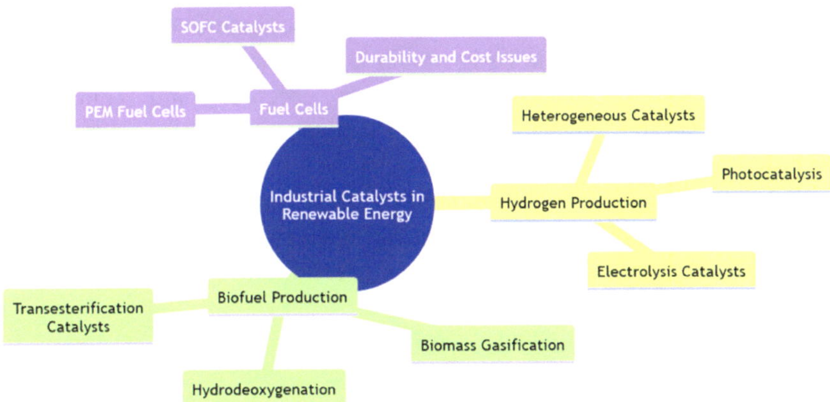

Figure 7.11. The use of industrial catalysts in renewable energy production systems.

Source: https://www.mdpi.com/chemproc/chemproc-17-00006/article_deploy/html/images/chemproc-17-00006-g001.png.

Besides, new materials are also promoting stability and durability, addressing some of the long-term challenges experienced with typical catalysts in industrial applications. This can eventually lead to cost reduction in operations and enhanced productivity (Alammari et al., 2018). Energy transfer and dual catalysis benefits extend beyond improved efficiency, having a significant impact on the economic and environmental sustainability of industrial processes. One of the most critical areas in this is the use of CO_2 valorization processes, which have gained pace due to increasing regulatory pressure to curb greenhouse gas emissions. The processes tend to utilize new catalytic approaches, such as the Sabatier reaction, which are capable of transforming CO_2 into commercial products, essentially turning waste into a commodity. By using waste CO2, industries can contribute to circular economy models, which prioritize resource efficiency and waste minimization. The use of designed reactors also enhances the viability of these processes, providing the optimized thermal management necessary to maximize production, thereby linking profitability with sustainability efforts (Gallego et al., 2018) (Figure 7.11).

In addition to their environmental benefits, dual catalysis and energy transfer technologies also hold significant economic opportunities in most industry sectors. Conversion to cleaner production methods can lead to new business models based on waste recycling and resource recovery, and gain competitive positioning in increasingly sustainability-conscious markets. While nations impose stricter environmental regulations, companies that embrace dual catalysis systems stand to gain early mover advantages, becoming innovation champions. In addition, by improving energy efficiency and reducing raw material usage through advanced catalytic systems, industries can potentially experience lower production costs. Thus, such innovations not only benefit the environment but also spur economic progress through the creation of job opportunities in development, manufacturing, and servicing of new technologies, leading eventually towards a sustainable industrial future (Gallego et al., 2018).

7.5.2. Role in Sustainable Chemistry

The quest for green chemistry is increasingly focused on addressing environmental concerns arising from resource use and pollution. An area of promising research in this respect is the novel application of nanowire technology within photoelectrochemistry, mimicking the natural process of photosynthesis. Using sunlight, water, and carbon dioxide, semiconductor nanowires may power chemical conversions to generate renewable fuels as well as valuable chemicals. This green process not only aids in the mitigation of greenhouse gases but also promotes a closed-loop process with low waste products. The use of nanowire technology in green chemistry presents the potential to enhance the efficiency of energy and catalysis processes, transfer, and incorporate ideally with the energy transition goals and sustainable techniques in industrial processes (Deng et al., 2019). In addition, green chemistry is crucial in transforming waste chemicals into valuable commodities, such as converting waste cooking oil (WCO) into fatty acid methyl esters (FAME). This is achieved through the application of transesterification, esterification, and hydrolysis, through innovative process equipment for the efficient transformation of waste to biodiesel. Application of heterogeneous catalysts on the newest equipment models, such as microwave reactors and oscillatory baffled reactors, has demonstrated significant increases in reaction performance and overall sustainability (Figure 7.12).

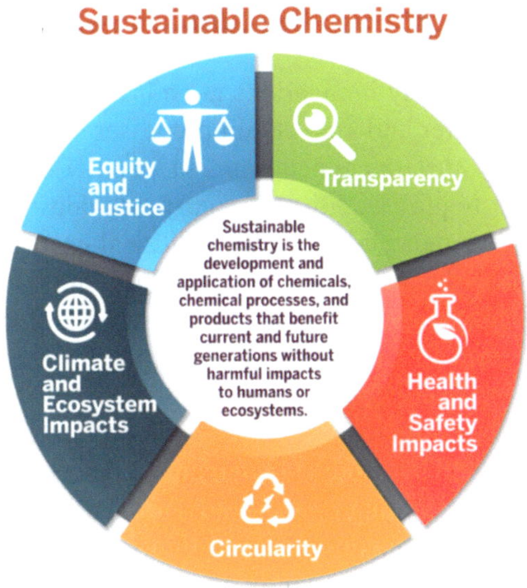

Figure 7.12. Sustainable chemistry.

Source: https://external-content.duckduckgo.com/iu/?u= https% 3A%2 F% 2Fbio-engineer.org%2Fwp-content%2Fuploads%2F2023%2F02%2FSustainable-chemistry-experts-create-blueprint-for-safer-future.png&f=1&nofb=1&ipt=90701fee59 0bb8b54d4e0312fb2a3b0811e6eaab8db0e760eb7e1c96adea82d8.

The continuous flow method, which integrates the reaction and separation steps, simplifies the process of FAME synthesis, enhancing efficiency and reducing energy consumption. As industries seek more sustainable alternatives, the efficient deployment of waste-to-energy principles is crucial in limiting environmental impact as well as in promoting the circular economy (Aubin et al., 2013). The integration of energy transfer and dual catalysis presents exciting prospects for the development of sustainable chemistry. Dual catalyst systems that utilize both light and chemical catalysts can actually speed up reaction rates and selectivity, thereby enhancing the efficiency of complex chemical reactions. By optimizing the energy transfer mechanisms in these systems, researchers can achieve higher yields for transformations that have traditionally been limited by slow kinetics. These developments not only increase the level of sustainable chemical production but also emphasize the importance of green materials and processes. Finally, these innovations are critical in developing scalable technologies that meet growing global energy demands while respecting sustainability principles. This enables a shift towards green chemistry, shaping the future of chemical manufacturing (Aubin et al., 2013; Deng et al., 2019).

7.5.3. Innovations in Dual Catalysis Technology

Bilichnal catalysis technology innovations are a breakthrough in the synthesis of chemical energy transfer and conversion processes. The technique involves two different catalytic systems functioning in parallel to catalyze reactions that would otherwise require additional energy or additional time. The new integration of light-powered catalysts with traditional heat catalysts has shown the ability to enhance the rates of reaction while simultaneously reducing the use of energy. This innovation is not only favorable for the efficient use of energy but also serves as a vehicle for chemical processes aligned with sustainable principles. Secondly, the ability to take advantage of the dual catalysis process allows for increased selectivity in product generation, which is key to waste reduction and by-product management, hence being a greener way to synthesize chemicals. Building on these advances, researchers have recently focused more intensely on maximizing reaction conditions in dual catalysis systems. By manipulating various parameters such as temperature, pressure, and reactant concentration, researchers have effectively optimized the efficiency and robustness of such catalytic reactions. Notably, the advances witnessed in this context can be attributed to multidisciplinary research collaborations that integrate knowledge from chemistry, material science, and biology. For instance, recent research on engineered biomolecules has highlighted how the operating properties of catalysts can be tailored for better performance. Not only does the use of biomolecules create potential for innovative catalytic design, but it also creates a 'green' alternative by reducing reliance on harmful chemicals and the facilitation of sustainable catalytic systems, as attested by methods that integrate biological interactions to ensnare nanoparticles successfully (Agustin et al., 2015). In addition, the finding of self-organizing catalytic materials opens a new ground in the development of dual catalysis technology. Such catalytic materials are designed to adapt and respond dynamically to diverse reaction conditions to enhance catalytic performance through feedback and self-regulation control. The concepts of ecological resilience established in the context of sustainable economic conduct (Richardward et al., 2012) also apply here, encouraging systems that are not only high-performing but are also able to reinvent their processes based on stimuli from the external environment. Such groundbreaking approaches play a significant role in developing adaptive catalytic systems that emphasize both performance and sustainability. As these technologies develop, they hold the potential to transform energy transfer processes, aligning them more closely with ecological laws and addressing the urgency of climate change adaptation and resource preservation.

SUMMARY

In summarizing the critical analysis of energy transfer and dual catalysis, it is essential to underscore the pivotal role of advanced materials to achieve improved catalytic efficiency. The study of graphene-type materials has opened broad prospects for electrocatalytic use, particularly for oxygen reduction reactions. The addition of heteroatoms as non-metallic dopants enhances not only the formation of catalytically active sites but also the long-term stability of the materials. This suggests the necessity of optimizing catalytic components to improve energy transfer efficiency in actual applications. The novel designs and methods discussed emphasize the synergy between theoretical models and experimental evidence, placing graphene-based materials at the forefront of energy conversion devices (Alammari et al., 2018). In addition, research on dual catalysis systems helps advance knowledge of photocatalytic processes, particularly in hydrogen production from biomass. Comparative analysis of various catalysts, like Pt/TiO_2 and Pd/TiO_2, proves that successful thermal processing can actually contribute to photocatalytic activity. The strong metal-support interaction (SMSI) effect enables effective electron transport from the metal particles to the titanium dioxide support, thus inhibiting electron-hole recombination. The connectivity suggests a significant path to increasing catalytic activity, which ultimately leads to more efficient energy transfer and increased hydrogen production. Such findings point to the green potential of dual catalysis as a means to mitigate global energy challenges (Aramendia et al., 2011). Lastly, the combination of knowledge from dual catalysis and energy transfer shows a promising future for more research and technological innovations. The precise building of catalysts through methods such as hierarchical nanostructuring and wise dopant incorporation not only enhances performance but also sheds light on the knowledge of ground-state interactions in catalysis. As evident through a variety of experimental techniques, it is crucial to optimize these interactions for the transition to more sustainable energy systems to occur. This integration of advanced materials and highly evolved catalytic mechanisms is crucial for developing effective energy solutions needed for the future, thereby endorsing the need for continued research within this vibrant discipline. The general advancement in dual catalysis can ultimately have a revolutionary impact on the development of sustainable energy practices and the reduction of environmental problems.

In research into the subtleties of dual catalysis energy transfer, several salient points underscore the significance and promise of this burgeoning field. First, the direct conversion of carbon dioxide (CO_2) to value-added hydrocarbons, as olefins and aromatics, is a central scientific issue. As much as has been researched, high productivity and low unwanted carbon monoxide (CO) selectivity remain elusive. The combination of new catalysts, such as potassium superoxide-doped iron oxide and zeolites ZSM-5 and MOR, demonstrates promising outcomes. These catalysts enable the utilization of CO_2 to form useful products with remarkable space-time yields, demonstrating their potential for industrial use and a reduced environmental footprint (Abou-Hamad et al., 2019). Furthermore, the zeolite's function in such a catalytic reaction also illuminates the inherent mechanisms that significantly control reaction selectivity. For instance, the formation of surface formate groups on the zeolite matrix is found to be an essential parameter in achieving low selectivity for CO in CO_2 conversion. Correct solid-state nuclear magnetic resonance characterization enhances our understanding of the catalytic environment, revealing differences in MOR and ZSM-5 reactivity. Such evidence promotes a tailored route in catalyst development, making it possible to tailor performance for various reactions. Scientists can further untangle the extraordinary behavior of carbenium ion intermediates in various zeolites through first-principles simulations to provide new avenues in energy transfer technologies (Abou-Hamad et al., 2019). In dual-catalysis applications, the use of advanced ligands with catalytic systems can significantly enhance reaction efficiency and selectivity. Interestingly, ferrocenyl phosphine-thioether ligands have been found helpful in enabling iridium-catalyzed hydrogenation of ketones, particularly with the help of a strong base co-catalyst. This hybrid system not only generates an anionic hydride complex but also allows mechanistic investigation of outer-sphere hydrogenation processes. Computational studies foretell a plausible operating cycle with an energy gap that indicates the viability of such complex systems. Interestingly, the firm base allows for twin atomics in complex formation and functioning as an atomically perfect Lewis acid in ketone activation, highlighting the delicate balance between catalyst design and reaction conditions for the best energy transfer.

REFERENCES

1. Ramirez, A., Chowdhury, A. D., Dokania, A., Cnudde, P., Caglayan, M., Yarulina, I., Abou-Hamad, E., Gevers, L., Ould-Chikh, S., De Wispelaere, K., Van Speybroeck, V., & Gascon, J. (2019). Effect of zeolite topology and reactor configuration on the direct conversion of CO2 to light olefins and aromatics. *ACS Catalysis*, *9*(7), 6320–6334. https://doi.org/10.1021/acscatal.9b01466Agustin, Y. E., & Tsai, S. L. (2015). The effect of storage temperature for the detection of silver nanoparticles via engineered biomolecules. https://core.ac.uk/download/42962818.pdf (accessed on 28th September 2025).
2. Alammari, M., Allen, J., Cai, J., Chen, L., Conner, T., Farooque, M., Gong, Y., et al. (2018). Graphene-based nanostructures in electrocatalytic oxygen reduction. http://arxiv.org/abs/1805.03149 (accessed on 28th September 2025).
3. Aramendia, M. A., Colmenares, J. C., Magdziarz, A., Marinas, A., et al. (2011). Influence of the strong metal support interaction effect (SMSI) of Pt/TiO_2 and Pd/TiO_2 systems in the photocatalytic biohydrogen production from glucose solution. https://core.ac.uk/download/187420267.pdf (accessed on 28th September 2025).
4. Aubin, J., Mazubert, A., & Poux, M. (2013). Intensified processes for FAME production from waste cooking oil: A technological review. https://core.ac.uk/download/18540047.pdf (accessed on 28th September 2025).
5. Bailleul, S., De Wispelaere, K., & Van Speybroeck, V. (2016). Towards molecular control of elementary reactions in zeolite catalysis by advanced molecular simulations mimicking operating conditions. https://core.ac.uk/download/55855326.pdf (accessed on 28th September 2025).
6. Bergel, A., Erable, B., & Féron, D. (2012). Microbial catalysis of the oxygen reduction reaction for microbial fuel cells: A review. https://core.ac.uk/download/12043853.pdf (accessed on 28th September 2025).
7. Boettcher, S. W., Lewis, N. S., McKone, J. R., Mi, Q., et al. (2010). Solar water splitting cells. https://core.ac.uk/download/4887222.pdf (accessed on 28th September 2025).
8. Boonsith, S., Kim, D. C., Sakdaronnarong, C., Sangjan, S., et al. (2020). Recent developments in synthesis and photocatalytic applications of carbon dots. https://core.ac.uk/download/323383933.pdf (accessed on 28th September 2025).
9. Brak, K., & Jacobsen, E. N. (2014). Asymmetric ion-pairing catalysis. https://core.ac.uk/download/28945734.pdf (accessed on 28th September 2025).

10. Cardenas-Rodriguez, M., & Tokatlidis, K. (2017). Cytosolic redox components regulate protein homeostasis via additional localization in the mitochondrial intermembrane space. https://core.ac.uk/download/84591464.pdf (accessed on 28th September 2025).

11. Casares González, J. Á., Espinet Rubio, P., & Pérez Iglesias, M. (2018). Comparing protonolysis and trans metalation reactions: Microcalorimetric studies of C–Au(I) bonds in [AuRL] complexes. https://core.ac.uk/download/225143515.pdf (accessed on 28th September 2025).

12. Centeno Gallego, M. Á., Laguna Espitia, O. H., Navarro de Miguel, J. C., Odriozola Gordón, M., et al. (2018). Policies and motivations for the CO_2 valorization through the Sabatier reaction using structured catalysts: A review of the most recent advances. https://core.ac.uk/download/188008301.pdf (accessed on 28th September 2025).

13. Deng, J., Liu, B., Liu, C., Liu, Y., et al. (2019). Nanowire photo electrochemistry. https://core.ac.uk/download/287623654.pdf (accessed on 28th September 2025).

14. Deydier, E., Hayes, J. M., Lledós, A., Malacea-Kabbara, L., et al. (2015). Ketone hydrogenation with iridium complexes with "non N–H" ligands: The key role of the strong base. https://core.ac.uk/download/42137987.pdf (accessed on 28th September 2025).

15. Dolocan, A., Goodenough, J. B., Lei, M., Li, Q., et al. (2019). Short O–O separation in layered oxide $Na_{0.67}CoO_2$ enables an ultrafast oxygen evolution reaction. https://core.ac.uk/download/287623779.pdf (accessed on 28th September 2025).

16. Fujishima, A., Paracchino, A., Song, A., Tamirat, A. G., Pinaud, B. A., Chen, C., Chen, C., et al. (2020). Benchmark performance of low-cost Sb_2Se_3 photocathodes for unassisted solar overall water splitting. https://core.ac.uk/download/304590791.pdf (accessed on 28th September 2025).

17. Koumpouras, G. (2008). Mathematical Modelling of a Low-Temperature Hydrogen Production Process with In Situ CO_2 Capture.

18. Liu, N., Pei, W., Yang, X., Zhao, Y., et al. (2018). Heterostructures of MXenes and N-doped graphene as highly active bifunctional electrocatalysts. http://arxiv.org/abs/1805.04213 (accessed on 28th September 2025).

19. N/A. (2025). Aerospace Medicine and Biology: A Continuing Bibliography (Supplement 160). https://core.ac.uk/download/pdf/42879782.pdf (accessed on 28th September 2025).

20. Richardson, J. (2012). Economic Resilience: Including a Case Study of the Global Transition Network. https://commons.ln.edu.hk/cgi/viewcontent.

cgi?article=1005&context=southsouthforum (accessed on 28th September 2025).

21. Saldien, J., & Vandevelde, C. (2014). Ono: An open platform for social robotics. https://core.ac.uk/download/55690833.pdf (accessed on 28th September 2025).

22. Wang, Z., Li, C., & Domen, K. (2018). Recent developments in heterogeneous photocatalysts for solar-driven overall water splitting. Chemical Society Reviews, 48, 2109–2125. https://doi.org/10.1039/c8cs00542g (accessed on 28th September 2025).

23. Zhou, Q. Q., Zou, Y. Q., Lu, L. Q., & Xiao, W. J. (2018). Visible-light-induced organic photochemical reactions through energy-transfer pathways. Angewandte Chemie International Edition, 58, 1586–1604. https://doi.org/10.1002/anie.201803102 (accessed on 28th September 2025).

CHAPTER 8

PHOTOCATALYSIS IN FLOW CHEMISTRY

Contents

INTRODUCTION

The evolution of photocatalysis has transformed the character of chemical synthesis, particularly with its confluence with flow chemistry. This significant advancement is primarily attributed to the ability of photocatalysis to harness light energy for use in catalyzing chemical reactions, offering a green alternative to traditional chemical synthesis. Flow chemistry, in which reactants are continuously fed through a reaction zone, is the counterpart of photocatalysis in that it enables more efficient mixing and heat transfer. This synergy is highly relevant for optimizing reaction rates and selectivity, thereby further increasing the overall productivity of chemical processes. By shortening the reactant exposure time to light and maximizing the reaction conditions, these two methodologies unveil an interesting future for greener synthetic pathways in chemical engineering. Further research indicates that photocatalytic reactions may be employed in many fields, from environmental pollution management to sustainable energy production. Combining material science and photocatalysis, such as using titanium dioxide (TiO_2) thin films, enhances photocatalytic efficiency by improving surface characteristics and the ability of catalysts to absorb light. Techniques such as sol-gel synthesis and chemical vapor deposition, which aim to utilize the maximum structural integrity of the materials, are essential in achieving efficient photocatalytic performance (Králová et al., 2012). This subsequently results in enhanced contaminant degradation rates or effective conversion of solar energy into chemical energy, thus representing the objectives of process intensification and sustainability in flow chemistry applications. Moreover, the application of photocatalysis in flow systems has also been utilized to form microfluidic devices, showing innovative ways of solving real-life issues. The application of anatase TiO_2 thin films as microfluidic channels, for instance, shows how photocatalysis can be employed for specific applications, such as blood oxygenation in medical devices (MNF et al., 2009). This ability to engineer targeted microenvironments for chemical reactions facilitates photocatalysis within flow chemistry platforms. With ongoing developments in research, the integration of photocatalytic systems with flow strategies is poised to revolutionize solutions to pivotal challenges in energy and environmental sustainability, advancing the boundaries of green chemistry.

Overview of Flow Chemistry

Flow chemistry, with continuous reaction processes compared to traditional batch approaches, has revolutionized the synthetic approach in the chemical industry. The emerging technology can provide enhanced scalability, improved efficiency, and enhanced safety. Flow chemistry, by enabling exact control over reaction parameters such as temperature, pressure, and residence time, provides chemists with the ability to improve reactions with high reproducibility and low waste. This systematic environment is particularly beneficial in multi-complex reactions, where the carefully controlled conditions influence product yield and structure. The versatility of flow reactors to accept numerous techniques, including temperature control and photonic interactions, represents a gigantic leap toward more sustainable chemical processes. Compatibility with photocatalytic processes, one of the spectacular strengths of flow chemistry, uses light to catalyze chemical transformations. The uninterrupted flow of reactants creates a more uniform exposure to light, which favors the efficiency of photocatalytic reactions. This helps increase reaction rates and yields, which are vital for advancing industrial applications.

The incorporation of photocatalysts, such as titanium dioxide (TiO_2), into flow systems has garnered much research focus, as the self-aligned structures of TiO_2 nanotubes can significantly improve catalytic efficiency. Specifically, the potential of these nanotubes to enhance electron mobility and support selective reactions can be tailored through designed growth approaches, highlighting the synergy between flow chemistry and materials science in ensuring optimized photocatalytic systems (Nguyen et al., 2016). Moreover, combining photocatalysis with flow chemistry presents a groundbreaking mechanism for solving environmental problems such as pollutant degradation and energy conversion. The continuity of flow processes allows for extended light exposure and efficient use of the reactant, which is required for processes like wastewater treatment or carbon dioxide reduction. Recent studies comparing model selection criteria for photocatalytic systems emphasize the necessity of robust methodologies for evaluating process efficiency (Abdullah et al., 2018). As researchers continue to explore the vast potential of this connection, the promise of enhancing sustainable and efficient technologies in flow chemistry is revealed, paving the way for novel solutions in chemical synthesis and environmental remediation (Figure 8.1).

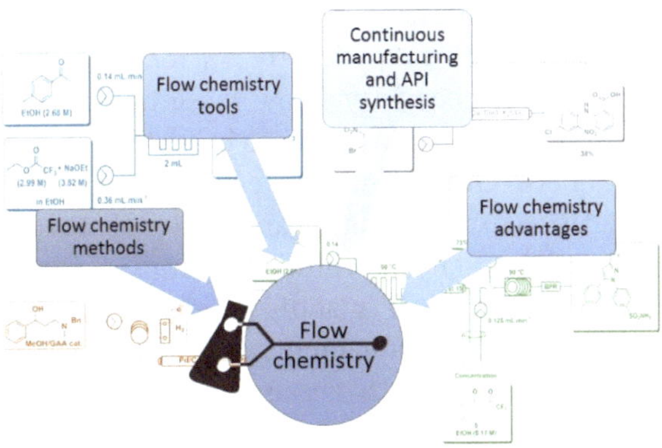

Figure 8.1. Flow chemistry.

Source: https://external-content.duckduckgo.com/iu/?u= https% 3A% 2F% 2 Fwww.up.ac.za%2Fmedia%2Fshared%2F104%2FZP_Images% 2FResearch% 2F flowimage1.zp132706.jpg&f=1&nofb=1&ipt=aab65fa6b12ffcced038e7a8e4882e de7dd5f42de35e3e405a74df2093db0f95.

Importance of Integrating Photocatalysis in Flow Chemistry

The incorporation of photocatalysis into flow chemistry is a seminal development in the synthesis of complex chemical products. The idea of flow chemistry allows continuous processing, which is advantageous for enhancing reaction efficiency and decreasing byproduct formation. Such efficiency is critical in photocatalysis, where reaction conditions such as light illumination and concentrations of reactants can be accurately managed in real time. Photocatalytic reactions are enhanced by effective mass transfer and greater heat removal that flow systems offer, hence reducing the occurrence of undesirable side reactions so prevalent in batch processes. Through the smart, innovative design of photocatalysts, such as metal-organic frameworks, researchers have established that optimally designed electronic characteristics significantly enhance photocatalytic activity. In this case, the findings on the performance of NH_2-MIL-125(Ti) demonstrate highly optimized catalysts with long-lived excitons and increased efficiency in charge transfer, which are necessary for effective photocatalysis (Fedin et al., 2016). The use of nanoporous carbons in photocatalytic systems has also opened up a new level of complexity and potential in flow chemistry. By mixing these materials with standard photocatalysts such as tungsten oxide, researchers have described drastic increases in pollutant degradation, demonstrating that the chemistry of the additives is key to photocatalytic activity. For instance, enhanced interaction of hydrophobic nanoporous carbons with phenol illustrates how modification of the support material can alter the photocatalyst's selectivity and efficiency. This systematic architecture allows researchers to design the photocatalytic process for particular contaminants, and the advantages of flow chemistry allow these tailored approaches to be achieved at an industrial scale. These advances bring forward the importance of synergistic processes between photocatalyst components, as seen in improved photocatalytic degradation of rhodamine B and phenol (Fernández et al., 2016). Moreover, combining photocatalysis with flow chemistry maximizes chemical reactions while making chemical manufacturing more sustainable. Flow processes are continuous, thus eliminating waste generation and maximizing the utilization efficiency of materials. These qualities are extremely crucial in an era becoming increasingly environmentally conscious, in the bid to keep the footprint of industrial activities to a minimum. Photocatalytic systems operate under mild conditions, aligning with the principles of green chemistry and reducing energy consumption and the use of hazardous reagents. Along with the development of the field, the development of hybrid photocatalysts that are effective in harnessing solar energy and also operate in flow chemistry environments will be of prime significance. This synergy not only has potential for enhanced efficiency and sustainability but also presents opportunities for new uses in environmental clean-up and clean energy production, solidifying the position of photocatalysis as a cornerstone of future chemical process innovation.

8.1. FLOW CHEMISTRY FUNDAMENTALS

The integration of photocatalysis with flow chemistry represents a significant advancement in the efficiency and effectiveness of chemical synthesis. In traditional batch processing, fixed reaction conditions can lead to issues concerning reproducibility and scalability. Flow chemistry, however, offers continuous processing possibilities with fine control over reaction conditions such as temperature, pressure, and concentration. This control is particularly beneficial for photocatalytic reactions, where continuous irradiation of the reactants with light can enhance reaction rates and yields of the product. Flow reactors tailored for photocatalysis can optimize reaction routes and reduce waste, which allows chemists to adopt more sustainable chemical production practices. Photocatalyst choice is crucial in flow chemistry use, particularly with reactants such as titanium dioxide (TiO_2). TiO_2 has evoked significant interest due to its photochemical stability, non-toxic nature, and ability to generate reactive species upon irradiation with light. There is proof that the rate of photocatalytic reactions can be significantly accelerated by adjusting factors like reaction medium pH and concentration of the photocatalyst. For instance, in the degradation of anthraquinonic dye Acid Blue 25, the highest rates of degradation were observed at specific pH levels and optimum concentrations of TiO_2, indicating that the synergy of photocatalysis with flow chemistry can lead to superior outcomes compared to conventional methods (Durme et al., 2014). This indicates that these parameters should be precisely adjusted to achieve the greatest benefits from flow chemistry in photocatalytic reactions. More importantly, the structure and geometry of TiO_2 photocatalysts play a role in their performance when applied to flow processes. Self-organization methods in recent developments in the fabrication of TiO_2 nanotubes indicate potential for improved photocatalyst activity. Individual properties of such nanotubes, including morphologies, may influence electron mobility and stability, two factors key to achieving high-efficiency photocatalytic processes. Through research of many structuring approaches, researchers have demonstrated that specially designed TiO_2 nanotube structures are able to offer superior properties such as improved surface area and light-absorption (Nguyen et al., 2016). Eventually, the adoption of these novel photocatalysts into flow chemistry processes can give rise to the next generation of sustainable chemical synthesis and water purification methods (Figure 8.2).

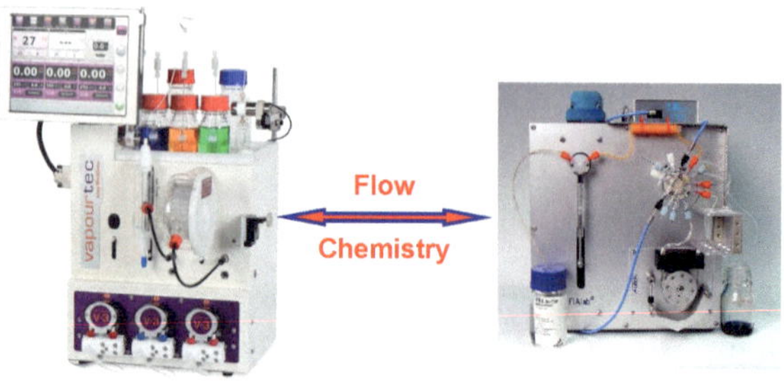

Figure 8.2. The integration of photocatalysis into flow chemistry represents a significant advancement in the efficiency and effectiveness of chemical synthesis.

Source: https://pub.mdpi-res.com/molecules/molecules-25-01434/article_deploy/html/images/molecules-25-01434-ag.png?1585744757.

8.1.1. Definition and Advantages of Flow Chemistry

Flow chemistry, in which reactants are pumped continuously through a reactor, is a paradigm shift in synthesis chemistry. Unlike typical batch operations, where reactants are combined and allowed to react within a static vessel, flow chemistry enables instantaneous mixing and control of reaction conditions. This innovation not only enhances reaction kinetics but also improves heat and mass transfer, which is essential for optimum performance in photochemical reactions. For photocatalysis, where good absorption and dissemination of light are necessary, flow systems can facilitate unmatched reaction efficiencies by ensuring even light exposure for all reactants, thereby maximizing the photocatalyst's potential (Aillet et al., 2013). The advantages of flow chemistry are diverse, particularly in scalability and safety in operation. The continuous flow of systems allows reaction conditions to be detected and controlled precisely, significantly reducing the time needed to optimize and develop chemical processes. Moreover, flow chemistry automatically minimizes the risk of hazardous accidents with large quantities of reactive chemicals used in batch processing. In photocatalysis, the controlled environment provided by flow reactors allows safer handling of photoactive materials, and hence, useful utilization is possible with reduced connected risks of photochemical reactions (Aillet et al., 2013). A second significant advantage of flow chemistry is its potential to reduce chemical waste and enhance overall process efficacy. By promoting proper stoichiometry and inhibiting side reactions, flow systems are able to lessen the environmental footprint of chemical synthesis significantly. This aligns with the growing interest in environmentally friendly processes for chemical manufacturing that consider yield and ecological concerns.

Additionally, the incorporation of nanotechnology into applications for flow chemistry introduces new solutions for the decontamination of pollutants, enhancing agricultural methods with low environmental impact (Bacchini et al., 2013). These advances show the tremendous synergy between photocatalysis and flow chemistry and pave the way for novel methods towards sustainable chemical synthesis (Figure 8.3).

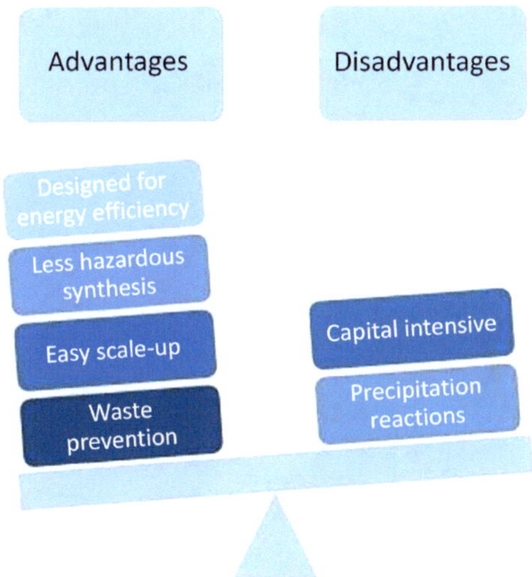

Figure 8.3. Advantages and disadvantages of flow chemistry.

Source: https://www.up.ac.za/chemistry/article/2581564/flow-chemistry.

8.1.2. Comparison with Batch Processing

Conversion from batch operation to flow chemistry is the main advancement in the field of photocatalysis. In traditional batch reactors, under stationary conditions over a long duration, the reactions occur. This creates inefficient mass and energy transfer, resulting in less efficient reactions. But in flow systems, continuous processing is enabled where reactants are pumped continuously into the reactor, allowing better control over reaction conditions such as residence time and temperature. This regulation also significantly enhances the efficiency of photocatalytic processes, as comparative studies show that microreactors operate better than batch systems, achieving increased chemical conversions with less irradiation time (Aillet et al., 2013). All this ultimately results in flow chemistry enabling not only faster reaction rates but also greater sustainability by minimizing waste and better utilizing resources. Also, the operational character of flow systems allows for more stable handling of photochemical reactions, i.e., in dealing with complex substrates. For example, studies have established the enhanced degradation of recalcitrant azo dyes in continuous photocatalytic systems, with high levels of efficiency in pollution removal (Falco et al., 2004). It is feasible to manage parameters such as pressure in membrane cells and initial substrate concentration, which could be very important in controlling reaction outcomes.

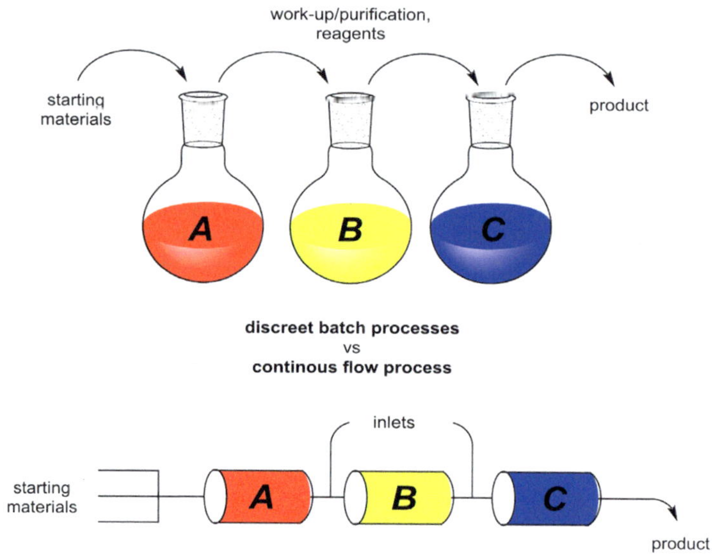

Figure 8.4. A basic depiction of batch versus flow.

Source: https://www.researchgate.net/figure/A-basic-depiction-of-batch-vs-flow_fig4_351682193.

During batch processing, substrate concentration variations could lead to oscillatory reaction kinetics, which typically make the process challenging. Unlike flow systems, where conditions are more stable and operating parameters can be dynamically changed in real-time, the latter can provide steady-state operation, which is favorable for achieving maximum degradation of the pollutant. Apart from increasing the efficiency of photocatalytic reactions, flow chemistry also eliminates scalability concerns typically troublesome in batch processing. The ability to handle highly concentrated reaction solutions—such as in the case of azo-dye degradation, where solution concentrations can reach up to 500 mg/L—is the key to the practical use of flow systems (Falco et al., 2004).

The continuous nature lends itself to easy mixing with established industrial processes, rather than the bottle-necking action of batch reactors that is likely to limit output. As photocatalysis is directed towards macro-scale applications, the benefits of adopting flow chemistry become more apparent, including better operational stability, faster processing times, and higher product yields through a more scalable and flexible production process (Figure 8.4).

8.1.3. Applications of Flow Chemistry in Various Fields

Flow chemistry has been an extremely versatile technique in most areas of application, particularly the synthesis of functional materials. One notable example is the fabrication of titanium dioxide (TiO2) nanotubes, which are crucial for use in photocatalytic reactions. The self-aligned TiO_2 nanotubes are endowed with unique properties for energy conversion and purification of the environment, as reported in recent studies. The ability to control their morphology—anything from smooth tubes to more complicated structures—should make them invaluable for applications in solar energy and air purification technologies, among others. On top of that, the reproducible and controlled environment of flow chemistry makes it possible to precisely engineer these nanotubes, which is significant to maximize their physical and chemical properties. By applying flow chemistry, researchers can develop more effective photocatalyst systems to address contemporary issues in environmental health and energy sustainability, as demonstrated by Nguyen et al. (2016). Apart from materials science, flow chemistry is also key to addressing global issues on water safety through the use of photocatalytic disinfection technology. The need for safe drinking water, especially with predicted looming shortages, necessitates innovative solutions. Flow chemistry allows for integration with solar disinfection technologies and photocatalytic processes to enhance the removal of pathogenic microorganisms, such as *Cryptosporidium* and *E. coli*. The hybrid approach utilizes flow systems to catalyze the rate of photocatalytic reactions, significantly reducing effective disinfection times. The use of immobilized catalysts in such systems facilitates better mass transfer, ultimately leading to a more stable system for viable application. The incorporation of flow chemistry into these water purification technologies highlights its potential to facilitate safer drinking water solutions worldwide, as explained by Byrne et al. (2014). Furthermore, the extent of flow chemistry extends to the pharmaceutical industry, where it aids in the continuous synthesis of complex organic compounds. Batch reactions have traditionally suffered from inadequacies such as lengthy reaction times and variable yields. In contrast, flow chemistry enhances the efficiency and scalability required for drug discovery. Photocatalysis in the flow system optimizes reactions that might otherwise be slowed by limitations in light penetration and catalyst deactivation. By a simplified process, flow chemistry allows for a more extensive study of the reaction conditions, leading to improved yields and quality of drug precursors. The use of photocatalytic approaches in flow systems not only accelerates the process of identifying and manufacturing new drugs but also exemplifies the potential of this approach to advance medical research and public health activities.

8.2. INTEGRATION OF PHOTOCATALYSIS IN FLOW CHEMISTRY

The integration of photocatalysis with flow chemistry is a significant advancement in the conversion and synthesis of chemical compounds. This partnership offers a controlled system to optimize the photocatalytic reaction conditions, thus maximizing reaction kinetics and selectivity. Flow chemistry allows continuous processing, which counteracts the disadvantages of batch reaction, such as inadequate mixing and temperature gradients. By combining these methods, researchers can leverage the unique properties of photocatalysts, i.e., titanium dioxide, within an active system, leading to improved product yields and conversion efficiency. As seen in recent studies, microfluidic technologies enable the precise manipulation of reaction conditions, making it feasible to efficiently harvest light energy to drive photocatalytic reactions. These applications not only make the processes more efficient but also contribute to sustainable chemistry by minimizing waste and improving energy use. The development of nanostructured photocatalytic materials has also spurred the use of photocatalysis in flow chemistry applications. Recent advances in new photocatalyst synthesis have recorded significant boosts in activity, surpassing traditional materials frequently used in batch processes. The evolution of composite materials and sophisticated optofluidic architectures has enabled researchers to synthesize photocatalysts with tailored characteristics, promoting improved light absorption and faster electron transfer reactions. As outlined in recent reports, these developments have made the photocatalyst process more efficient and cost-effective, overcoming the long-existing drawbacks of traditional photocatalysts, which are incapable of scale-up for industrial purposes. Such progress significantly boosts the field of application for photocatalysis in flow chemistry, and the resulting revolutionary approaches to solar energy conversion and environmental purification open the way to sustainable chemical synthesis and energy generation (Jang et al., 2015). Moreover, practical applications of photocatalytic flow chemistry are continually growing, substantiating the viability and effectiveness of this integrated approach. Notably, microfluidic channels have been modified for novel applications such as blood oxygenation, where photocatalytic reactions can generate dissolved oxygen from flowing blood, emphasizing applications to everyday health. The structure allows for effective integration of photocatalytic cells, utilizing titanium dioxide thin films and appropriate electrodes for facilitating such reactions. The potential for photocatalytic technology to be used in medical and biotechnological processes demonstrates its capability to generate oxygen at substantial rates. The methodology also displays the ability of flow chemistry to provide a stable and controlled platform, which is vital for precision- and reliability-seeking applications (MNF et al., 2009). The creation of these integrations showcases the revolutionary significance of combining photocatalysis with flow chemistry, unveiling new avenues for study and practical implementation (Figure 8.5).

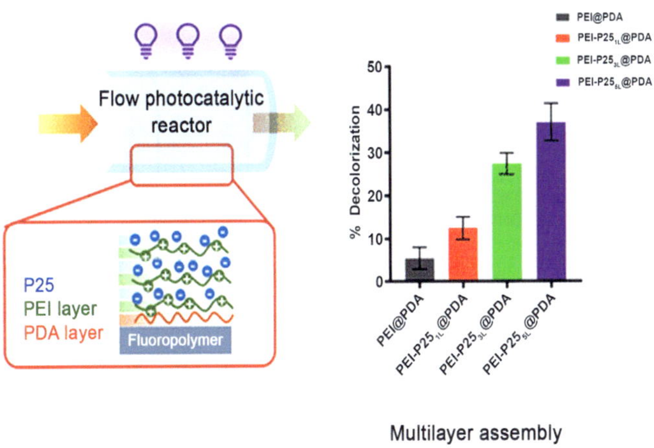

Figure 8.5. A flow reactor is advantageous for photocatalytic reactions due to enhanced transport phenomena and photon transfer.

Source: https://pubs.acs.org/doi/10.1021/acs.iecr.1c04303.

8.2.1. Techniques for Combining Photocatalysis and Flow Chemistry

The flow chemistry-photo-catalysis merger is a promising area of synthetic chemistry characterized by greater efficiency and improved control of the reaction. Among the remarkable methods is the use of thin TiO_2 films, which may be immobilized on substrates using techniques like material printing or chemical vapor deposition. These substrates offer a consistent flow of reactants over the photocatalytic surface, promoting efficient light capture and catalysis.

Of critical note, the structure and morphology of the TiO_2 films can be tuned by using sol-gel and hydrothermal synthesis routes, enhancing their photocatalysis as well. Their physical properties directly relate to their performance. This illustration is evidenced by analyses demonstrating how changes in the conditions of preparation affect the rates of degradation for such impurities as 2,6-dichloroindophenol and formic acid, thereby demonstrating the utility of the technology in practical applications in flow chemistry (Králová et al., 2012). Flow chemistry enables continuous reaction flow to improve reaction efficiency, safety, and scale-up. Scientists use photocatalytic flow reactors supported by TiO_2 films to achieve higher turnover and shorter reaction times compared to traditional batch techniques.

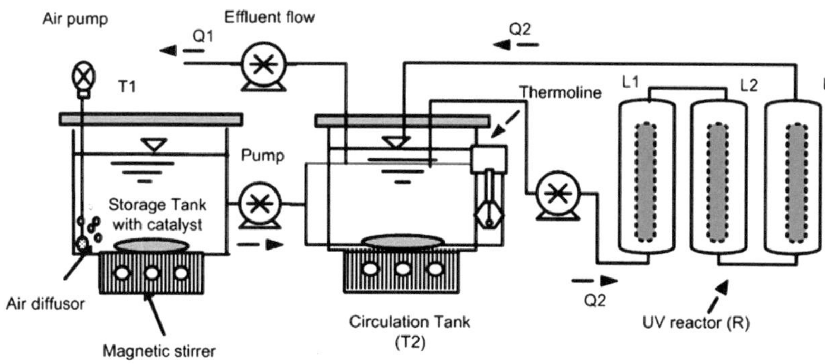

Figure 8.6. Schematic of the continuous-flow photocatalytic reactor with the catalyst.

Source: https://www.researchgate.net/profile/Atul-Wankhade/publication/281631151/figure/fig3/AS:284614506631173@1444868794691/Schematic-of-the-continuous-flow-photocatalytic-reactor-with-the-catalyst.png.

The influence of various operational parameters, such as flow rate, light intensity, and temperature, can be systematically investigated to improve reaction conditions for specific photocatalytic reactions. These parameters play a critical role in optimizing the light-photocatalyst interaction, which is responsible for the conversion and selectivity of the reaction. In addition, flow systems can also be automated to offer real-time monitoring and control of such parameters to provide reproducible reaction outcomes. In contrast, batch processes are riddled with uncertainty and variability (Abdullah et al., 2018). Also, the combination of advanced characterization techniques is necessary for monitoring and evaluating the performance of photocatalytic flow chemistry systems. Scanning electron microscopy and X-ray diffraction techniques can be employed to assess the structural properties of photocatalytic films, while photodegradation tests can be used to determine quantitatively how active the movie are in degrading the pollutants. With this understanding of the dynamics of light-photocatalyst-reactant flow interaction, researchers can make informed decisions on material choice and reactor setup to realize increased efficiencies in photocatalytic reactions. This harmonious interaction between flow chemistry and photocatalysis not only enhances the versatility of photocatalytic materials but also promises to usher in new avenues for designing reactors and processes, which can significantly aid the field of sustainable chemistry (Králová et al., 2012) (Figure 8.6).

8.2.2. Benefits of Using Photocatalysis in Flow Systems

The integration of photocatalysis with flow systems has revolutionized the field of synthetic chemistry by making the reactions more efficient and selective. Continuous operation is one of the critical advantages as it provides a higher throughput of products compared to batch methods. The continuous operation provides a constant photon flux, which makes it easier to control the environment of the reaction accurately. Therefore, the stability of the photocatalyst is improved with fewer chances of deactivation, resulting in more reproducible outcomes. Further, flow systems also use less reagent, which not only saves on cost but also avoids waste, conforming to green chemistry strategies. The ability of the flow systems to offer optimized reaction conditions has the potential to lead to enhanced reaction kinetics, and therefore, this method is very valuable for the synthesis of complex molecules in which controlling reaction parameters strictly is imperative. The second key advantage of employing photocatalysis under flow conditions is its flexibility. The photocatalytic reactions can be tailored to deliver multiple functional groups and reaction modes, a key consideration in modern organic synthesis. The ability to perform reactions under mild conditions, such as at room temperature and with light, offers possibilities for reactions that may otherwise prove difficult. One classic illustration is recent developments specified in the literature, which highlight the use of visible-light-induced photocatalysis to facilitate a variety of transformations, such as radical reactions utilizing alternatives like perfluoroalkylation, without harsh chemicals or conditions (Vallejo et al., 2018). This versatility not only broadens the scope of potential applications but also stimulates imagination in the design of new synthetic methods, further reinforcing the role of photocatalysis as a central technique in flow chemistry. Additionally, the safety enhancements that are associated with the use of photocatalysis in flow systems cannot be exaggerated. Continuous flow systems inherently reduce the risk posed by harmful materials by limiting exposure duration and reagent quantity at any given moment. The environment can be better regulated to handle exothermic reactions, which lead to runaway situations under traditional batch techniques. By integrating high-technology monitoring and control tools, researchers are now able to perform real-time adjustment of parameters for safe and efficient operation. Also, the fact that light is used as a catalyst means that many reactions are now accessible without using toxic reagents, which is particularly meaningful for attaining regulatory compliance and minimizing environmental footprint. Overall, these safety and sustainability benefits put photocatalysis in flow systems on the list of key developments of modern chemical processes (Hakki et al., 2017).

8.2.3. Challenges Faced in Integration

The integration of photocatalysis within flow chemistry presents complex challenges that hinder its application on a large scale in daily life. A specific challenge is the need for reactors that are specifically designed to suit the precise needs of photocatalytic processes. Although some progress has been achieved, as illustrated in multiscale-designed optofluidic reactors with enhanced performance, such reactors must be tailored to maximize efficiency in various applications. For instance, the established rate constant of reaction of 1.32 s^{-1} for photocatalytic decolorization of methylene blue shows the significance of reactor design in achieving higher performance. However, despite such advancements, the expense and complexity of adopting new reactor geometries deter researchers from using new photocatalytic solutions, preventing the technological advancement necessary for integration into flow chemistry applications (Leung et al., 2014). Further to this, the synthesis of appropriate photocatalytic materials poses additional challenges that directly impact integration processes. The complexity of material science is reflected in the need for synthesis processes that are not merely efficient but also compatible with specific application requirements. For photocatalysis, this is reflected in the studies conducted on thin film materials, whose composition and processing conditions must be under tight control. For example, the development of La–Bi–Cu oxysulfide photocathodes relies on a thorough understanding of composition, processing, and structure relationships. Reactive sputtering and thermal synthesis processes utilized to produce such materials need to be finely tuned, as minor variations can lead to drastic differences in performance parameters. These challenges in material synthesis underscore the need for sophisticated methodologies to facilitate easier integration into flow systems and meet the demands of photocatalytic reactions (Gregoire et al., 2020). Lastly, the dynamic equilibrium between photocatalytic performance and working conditions in flow chemistry presents challenges that can be addressed by adaptive solutions. Because photocatalytic reactions are usually variable, dependent on factors such as light, flow rate, and temperature, it would be a challenge to optimize conditions. High-temperature processing needs for certain materials would also conflict with other materials that significantly contribute to degradation mechanisms, complicating the outcome of reactions. This interaction necessitates monitoring and adjustments during the operational process, which is resource and technically demanding. Hence, scientists need to develop robust systems that not only respond to changes in reaction conditions but also operate independently for enhanced reliability and efficiency. With advances in dynamic processing capabilities emerging, overcoming such operation challenges will be the key to realizing the potential of photocatalysis in flow chemistry applications (Leung et al., 2014).

8.3. CASE STUDIES OF PHOTOCATALYSIS IN FLOW CHEMISTRY

Certain groundbreaking case studies on the subject of photocatalysis highlight the benefits of integrating this technology into flow chemistry platforms. One of the leading examples involved the study of immobilizing titanium dioxide (TiO_2) thin films onto various substrates, including soda-lime glass and glassy carbon. Material printing and chemical vapor deposition were utilized as preparation techniques, providing a fair idea of the adaptability of these methods in achieving the desired photocatalytic activity. The structure and morphology of the TiO_2 films were thoroughly analyzed using scanning electron microscopy and X-ray diffraction to gain critical insights into their crystal phases. Notably, the photocatalytic activity of the fabricated films was validated by degrading pollutants such as 2,6-dichlorindophenol and formic acid, showing promising results justifying the application of TiO_2 in environmental clean-up procedures (Králová et al., 2012). More research into the real-world application of photocatalysis is demonstrated by a groundbreaking study that utilizes microfluidic technology to oxygenate blood. Microfluidic channels molded from polydimethylsiloxane (PDMS) here facilitated the construction of photocatalytic cells that could generate dissolved oxygen from circulating blood with high efficiency. By integrating anatase titanium dioxide thin films with platinum electrodes, researchers were able to make electrically assisted photocatalytic reactions feasible, generating oxygen at significant rates. The proof of concept proved the system's ability to elevate venous blood to arterial oxygen saturation levels, proving the versatility of photocatalysis beyond traditional environmental contexts. This integration also demonstrates the potential of flow chemistry innovation in delivering innovative therapeutic technology that enhances patient care and medical practice (MNF et al., 2009).

The comparative analysis of different photocatalytic methods verifies that material property optimization and synthesis conditions are critical for optimal performance in flow chemistry applications. By trial and error with sol-gel concentration adjustment, hydrothermal synthesis temperature adjustment, and deposition techniques, researchers can tune up the photocatalytic activity of TiO_2 thin films. For instance, parameters such as the radiation intensity and time of exposure during chemical vapor deposition were found to play a significant role in controlling the efficiency of the photocatalysts. It is worth noting that the analysis of performance by various degradation assays indicates the significance of systematic analysis to establish best practices in the discipline. These findings not only pave the way for improved photocatalysts in flow systems but also highlight the value of interdisciplinary approaches to solving daunting scientific challenges in chemical engineering (Králová et al., 2012) (Figure 8.7).

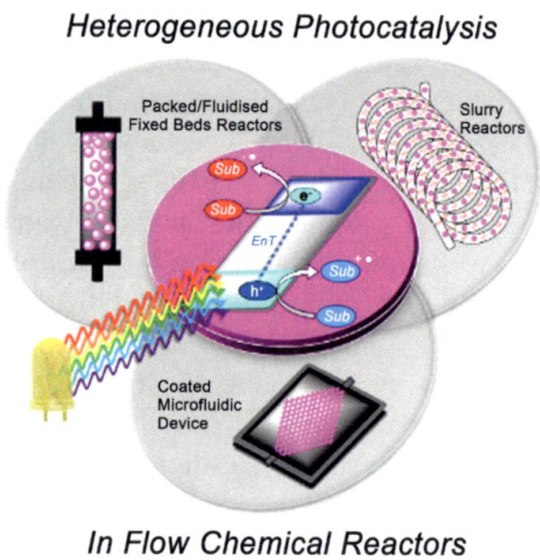

Figure 8.7. Heterogeneous photocatalysis in flow chemical reactors.

Source: https://external-content.duckduckgo.com/iu/?u=https%3A%2F%2Fwww.beilstein-journals.org%2Fbjoc%2Fcontent%2Ffigures%2F1860-5397-16-125-presentation.png&f=1&nofb=1&ipt=dabb542694613b8e2745749c04aee78a160e21200c6bfdc8cec2ca3d2b268818.

8.3.1. Successful Applications in Organic Synthesis

The marriage between flow chemistry and photocatalysis has revolutionized organic synthesis by bringing in new methods for the synthesis of complex molecules. Photocatalytic methods utilize light energy to drive chemical reactions, which is very useful in flow systems because they provide better control over reaction conditions and reaction rates. The technique has been successfully employed for various organic conversions, such as C–C coupling reactions and functional group interconversions. The use of photocatalysts, such as TiO_2 and CuS nanocrystals, significantly enhances the efficiency of this reaction, resulting in less by-product formation and high selectivity toward desired products.

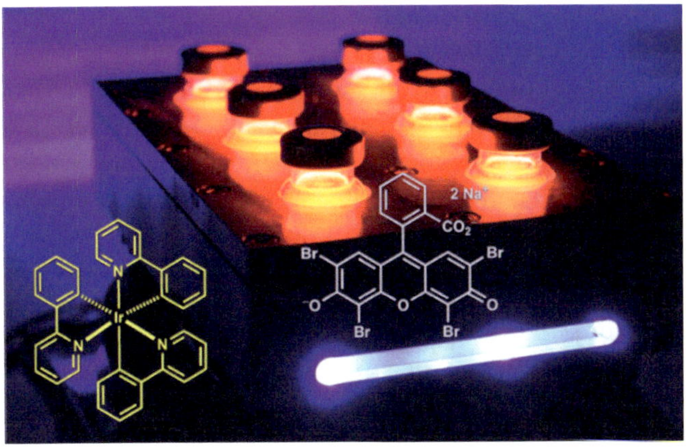

Figure 8.8. Photocatalysis with flow chemistry has revolutionized organic synthesis.

Source: https://external-content.duckduckgo.com/iu/?u=https%3A%2F%2Fwww.thieme.de%2Fstatics%2Fbilder%2Fthieme%2Ffinal%2Fen%2Fbilder%2Ftw_chemistry%2FSOS-D-workbench-photocatalysis-in-organic-synthesis.jpg&f=1&nofb=1&ipt=48d65e607f5a2aa9db85cf5d5729c0c67d264290e1456d7587714c9f75febc3f.

The combination of photocatalytic reactions and continuous flow processing maximizes the collision between substrates and light, leading to higher yields and greater potential for scaling up to commercial purposes. In photocatalytic applications, CuS nanocrystals are among the emerging stars in sustainable synthesis due to their unique properties. Recent work suggests the photoinduced carrier charge dynamics in CuS nanodisks and demonstrates that carrier recombination time scale manipulation is critical in maximizing photocatalytic efficiency. The fact that they can control hole dynamics by modifying surface chemistry is one of their noteworthy features. These findings emphasize the importance of understanding the principles behind the effects that can dominate the pathways of organic reactions. The integration of advanced photocatalysts into flow chemistry platforms leads to more efficient and sustainable synthetic methods, impacting the overall goals of sustainable chemistry in industrial processes, as outlined in research documents (Aiga et al., 2015). Additionally, the synthesis of self-standing anatase TiO_2 nanoparticle chains represents a significant advancement in photocatalytic materials. These chains, synthesized from carbon nanotube substrates using atomic layer deposition, provide an incredibly high surface area that enhances photocatalysis. In regions such as degradation of organic pollutants, e.g., acetaldehyde, the highly crystallized anatase phase of TiO_2 has been reported to be extremely efficient when exposed to UV light. By using the properties of these nanoparticle chains on flow chemistry platforms, researchers have been able to achieve rapid degradation rates and adapt them to practical organic synthesis and environmental remediation. The continued development of these materials in photocatalysis highlights the potential for significant innovation in organic syntheses, promising cleaner and greener methods in the chemical sector (Bals et al., 2014) (Figure 8.8).

8.3.2. Environmental Applications and Sustainability

The implications of photocatalysis in environmental processes are increasingly regarded as a critical advancement in sustainable technology. The method operates through photocatalytic degradation of environmental pollutants, such as organic dyes and pathogenic microbes within water reservoirs. Emerging trends indicate that integrating photocatalysis with flow chemistry maximizes reaction efficiency and improves scalability for practical applications. For instance, systems based on TiO_2 using photocatalysis have a massive potential for solar disinfection of water for drinking, particularly where clean water is not accessible or is only accessible with difficulty. Current reports indicate that implementing photocatalytic processes can significantly reduce the time required for conventional solar disinfection, thereby transforming water treatment in both developing and developed nations. These approaches highlight the core social importance of advances in photocatalytic technologies in the context of environmental sustainability (Byrne et al., 2014). The sophistication of photocatalytic materials and reactor engineering also alludes to their contribution to sustaining ecological outcomes. Of particular interest is the use of advances in nanotechnology to develop composite materials that enhance the efficiency of photocatalysts. Research on nanocomposite films, for example, halloysite clay and TiO_2, has proven to degrade organic pollutants, such as the azo dye Basic Blue 41. The composite films do not only have high stability in photocatalytic performance but also the potential for scalability in case they are immobilized onto substrates like glass. The combination of halloysite with TiO_2 forms a structured environment that enhances surface reactions, thus facilitating a more sustainable method of wastewater treatment. The use of such novel materials in photocatalytic systems illustrates the broader impact of sustainable environmental methods that exploit enhanced materials for effective pollutant degradation (Kaplani et al., 2014). Moreover, the economic viability of photocatalytic technologies remains a top issue as society moves toward sustainable means. While promising results in laboratory tests, there are barriers to the large-scale production and standardization of photocatalytic systems. System efficiency in larger scales will require overcoming challenges such as mass transfer limitations and the need for uniform test protocols to evaluate photocatalytic activity. Continuous research to optimize reactor design and the absorption capabilities of photocatalytic materials is required to tap the true potential of such technologies. The union of photocatalysis and flow chemistry provides a broad avenue for innovation in environmental treatments, backed up and supported by continuous research and development efforts aimed at sustainability and resource efficiency. Ultimately, the translation of laboratory success into real-world uses will be key to rendering photocatalysis an anchor of sustainable environmental technologies (Byrne et al., 2014; Kaplani et al., 2014).

8.3.3. Industrial Scale Implementations

The scaling-up of photocatalysis in flow chemistry has numerous advantages, particularly in the efficient conversion of organic molecules. Through continuous flow reactors, the reactions are capable of achieving high throughput and scalability, the two primary requisites for industrial manufacturing. Traditional batch reactions are likely to have limitations with respect to reaction time, reagent handling, and homogeneity of the end product. On the other hand, flow photocatalysis can enable greater mass transfer and uniform exposure to light, and thus significantly improve reaction yields and kinetics. Moreover, the ability to operate under mild conditions, utilizing visible light as a source of energy, aligns well with the principles of green chemistry, ultimately reducing the environmental impact generally associated with bulk chemical synthesis. One of the key features that facilitates industrial advancement in photocatalytic flow processes is the development of multidomain photocatalysts, with the ability to effectively activate various types of organic bonds. Photocatalysts like aromatic ketones and metal-oxo porphyrins have significant potential in promoting selective C-H bond functionalization through direct hydrogen atom transfer mechanisms. The careful selection of reaction conditions and additives enables these processes to be designed for particular industrial applications, enabling complex molecules to be synthesized with increased efficiency compared to conventional methods. By leveraging the distinctive features of different photocatalysts under the philosophy of reticular chemistry, stable and tunable frameworks for chemical reactions can be created. These frameworks can be utilized in a range of applications, thus positioning photocatalysis at the forefront of future sustainable production (Capaldo et al., 2021). Despite the promising developments, several challenges must be addressed to achieve the full exploitation of photocatalysis in an industrial setting. Stability and regeneration problems of the photocatalysts, reactor design scale-up, and incorporation into existing production lines are significant challenges. Interdisciplinary cooperation, which is triggered by advancements in reticular chemistry, is necessary to engineer the materials and structures involved in such processes (Freund et al., 2021). Furthermore, a significant investigation into the long-term economic viability and regulatory implications of novel photocatalytic technologies is essential for the successful deployment of the same. Overcoming these barriers, flow chemistry photocatalysis can be scaled from laboratory success to widespread industrial application, bringing in a new era of sustainable chemical production.

8.4. RECENT ADVANCES IN PHOTOCATALYTIC MATERIALS

Current advances in photocatalytic materials have significantly enhanced the efficiency and applicability of photocatalysis in flow chemistry. The synthesis of highly crystallized titania nanoparticles, deposited through low-temperature procedures, demonstrates a viable route to developing photocatalytic activity. This technique not only facilitates the creation of self-cleaning surfaces but also maintains the structural integrity of heat-sensitive materials like polymers. As described in the research, the inclusion of silica buffer layers enhances the compatibility of inorganic coatings and results in higher photocatalytic activity when exposed to organic impurities such as methylene blue under UV illumination. The results indicate that titania-coated silica-polymer systems can offer improved degradation rates upon use. There is a need for material engineering in photocatalytic processes, with the potential for convergence with flow chemistry processes (Buysser et al., 2018) (Figure 8.9).

Moreover, the nanomaterial research has not just revolutionized photocatalysis but also energy conversion devices such as dye-sensitized solar cells (DSSCs). The use of nanostructured materials in photoelectrode engineering has improved the efficiency and stability of the devices to suit all low-power applications, especially under conditions with restricted light. Building upon the pioneering work of ORegan and Grätzel, advancements in the development of both *p*- and *n*-type semiconductors have consistently enhanced the efficiency of DSSCs, demonstrating the robust synergy between the advances made in materials and the efficiency of energy conversion. The judicious application of nanomaterials highlights broader potential in advancing photocatalytic systems for incorporation into energy harvesting technology, providing a synergistic approach in flow chemistry (Bonomo et al., 2017).

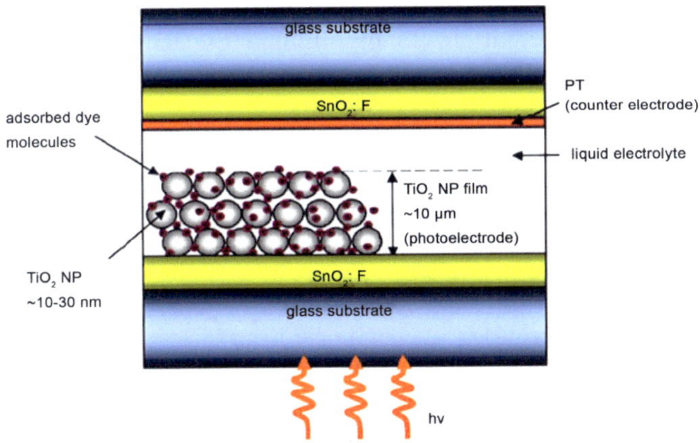

Figure 8.9. Structure of a dye-sensitized solar cell.

Source: https://www.researchgate.net/figure/Structure-of-a-dye-sensitized-solar-cell_fig1_235940468.

These recent advances not only display the potential for enhanced photocatalytic activity but also present new avenues in clean energy production and environmental cleaning. The integration of photocatalytic systems into flow chemistry allows continuous operation, which is inherently more efficient compared to batch operation. With innovation in materials science, scientists can look forward to witnessing more developments that integrate cutting-edge photocatalytic and hybrid materials, tuned for enhanced performance under varied operating conditions. The research effort in efficient photocatalytic materials is, thus, central to solving the world's energy and environmental sustainability problems, justifying the significance of the role of material innovations in charting the future of photocatalysis in flow chemistry.

8.4.1. Development of Novel Photocatalysts

The synthesis of new photocatalysts is an essential area of research within the broader field of photocatalysis in flow chemistry. The traditional photocatalysts are less efficient and affordable compared to other traditional sources of energy. As a means of addressing such a challenge, significant research into new nanostructured materials has increased. There has been a quest for alternative methods of producing and enhancing these materials to enhance photocatalytic efficiency from sunlight. Of special concern, new studies are aimed at composing materials and designing mobile photocatalysts able to adapt to some reaction conditions. Each step is designed to optimize efficiency and broaden the applicability range of photocatalysis in clean energy conversion and environmental remediation, as highlighted by Jang et al. (2015). The development of advanced photocatalytic systems on a flow chemistry platform has been a vital change in synthetic protocols.

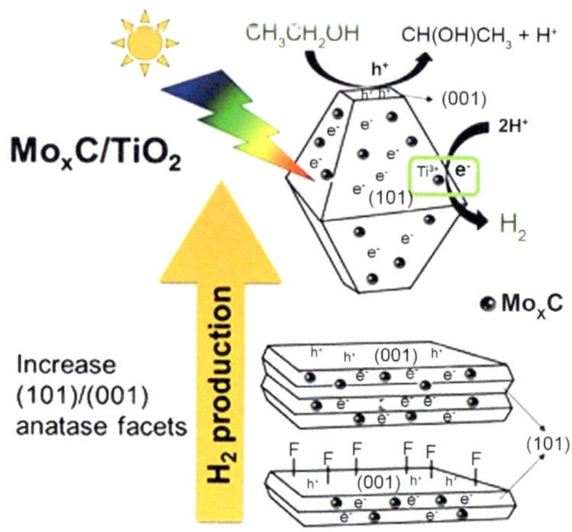

Figure 8.10. Development of a novel, efficient, noble-metal-free photocatalyst.

Source: https://external-content.duckduckgo.com/iu/?u=https%3A%2F%2Fwww.ub.edu%2Fin2ub%2Fwp-content%2Fuploads%2F2022%2F09%2FEnvironmental_22.jpg&f=1&nofb=1&ipt=47b57f4104ea163a10752eeee101b481d2e97628ad63307b99ca34a68b0eac11.

New optofluidic platforms are in the pipeline that exploit the benefits of continuous flow, hence allowing for better control of reaction parameters such as temperature and light exposure. The process encourages faster rates of mass transfer and accommodates quick scaling of reactions, which is critical in industry. For instance, heterogeneous photocatalysts have been shown to significantly enhance the yields of responses when combined with continuous flow processes. Such systems can efficiently harness solar power, demonstrate improved productivity, and reduce the overall environmental footprint of chemical synthesis. This interplay between photocatalysis and flow chemistry continues to challenge creative designs in the development of photocatalysts, as also documented in the literature (Jang et al., 2015). The production and application of small organic radicals are critical to expanding the functional capability of photocatalysis. By creating methods that yield strongly disparate radical intermediates, researchers have the ability to enhance the functionality of photocatalytic reactions by inserting varying functional groups into target molecules. One recent example toward this end is the employment of potassium poly(heptazine imide) (K-PHI) as a photocatalyst for dichloromethyl radical formation from chloroform. The process is highly effective in extending C1-enones to yield γ,γ-dichloroketones at impressive rates. The novel quasi-homogeneous flow photoreactor design featuring K-PHI has been shown to significantly increase productivity compared to traditional batch methods. The integration of such a kind suggests the formation of value-added organic compounds through radical-based transformations in the development of novel synthetic applications in flow chemistry (Figure 8.10).

8.4.2. Nanomaterials and Their Role in Photocatalysis

The addition of nanomaterials to the application of photocatalysis revolutionized conventional approaches to chemical reactions, particularly flow chemistry. The nanomaterials' size-dependent characteristics possess greater surface area and higher reactivity, which are desirable for effective photocatalytic reactions. Nanostructured titanium dioxide (TiO_2), one of the widely used materials in photocatalysis, illustrates this effect strongly. Previous research has shown that numerous synthesis methods, such as sol-gel and hydrothermal processes, can have a significant impact on TiO_2 layer morphology and crystallinity. The produced nanostructures, thoroughly analyzed via scanning electron microscopy methods, display distinct photocatalytic behavior, particularly against the degradation of organic pollutants such as formic acid and 2,6-dichlorophenol, when embedded in a flow chemistry platform (Králová et al., 2012). It is this enhanced performance that renders nanomaterials the central components in contemporary photocatalytic systems. Yet another significant development in the field of photocatalytic technology is the utilization of graphitic carbon nitride (g-C_3N_4). This compound is recognized for its tunability of electronic properties, which can be established through sulfur doping and post-synthesis treatments. With well-controlled heating protocols, the different forms of g-C_3N_4, such as exfoliated and mesyl-functionalized variants, have exhibited remarkable photocatalytic characteristics. For instance, studies indicate that mesyl-functionalized g-C_3N_4 is better at the photodegradation of organic dyes under ultraviolet A light than its non-functionalized counterparts (Cvejn et al., 2020). Such advances in photocatalyst performance can be explained by enhanced light harvesting and molecular-level interactions, underlining the significance of functionality in nanomaterials.

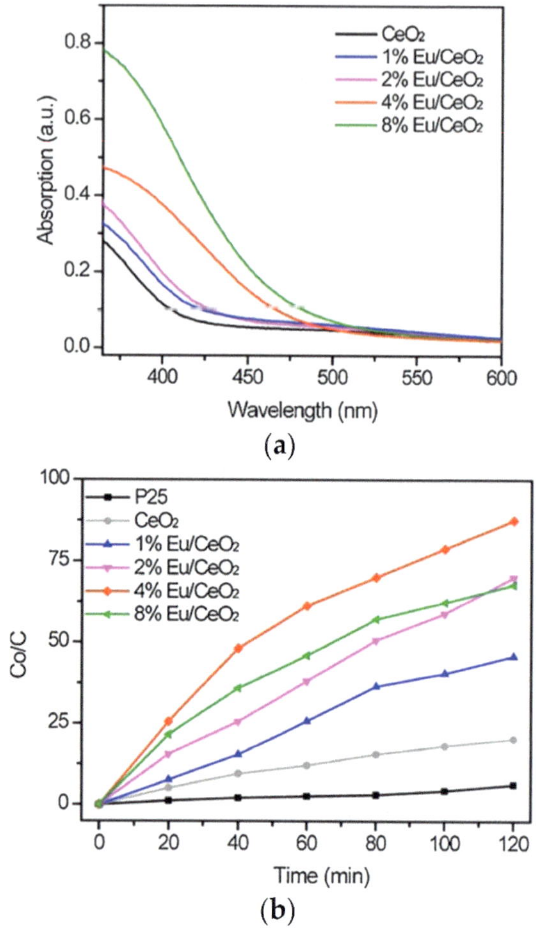

Figure 8.11. (a) UV-vis spectra of CeO_2 and Eu/CeO_2 containing different amounts of Eu; and (b) photocatalytic degradation of formaldehyde using P25, CeO_2, and Eu/CeO_2 containing different amounts of Eu.

Source: https://www.mdpi.com/2076-3417/15/3/1629.

This ability to tailor nanomaterials for photocatalytic applications demonstrates their versatility and utility in flow chemistry systems. Furthermore, the role of nanomaterials in photocatalysis extends beyond improving efficiency in traditional reactions; they also allow for new mechanisms and pathways. For example, in photocatalytic degradation of pollutants, nanomaterials are capable of producing reactive species upon irradiation with light, which in turn are involved in degradation reactions. The unique surface properties of nanomaterials increase the rates of adsorption of reactants, which is important in flow systems where constant input and output of materials is a requirement. Furthermore, the ability to modify the properties of nanomaterials offers prospects for developing hybrid systems that can work synergistically with light to achieve net reaction rates. Thus, the role of nanomaterials in photocatalysis extends beyond that of catalysts but to that of centerpieces that redefine the abilities and possibilities in flow chemistry applications, an exciting field of research in materials science and environmental remediation (Figure 8.11).

8.4.3. Innovations in Catalyst Design for Flow Systems

Recent developments in catalyst design have revolutionized the landscape of flow chemistry, particularly in photocatalysis. Dynamic flow systems foster consistent reaction processes, thus enhancing efficiency and scalability compared to the traditional batch process. Of critical significance is the use of nanomaterials, that is, titanium dioxide (TiO_2) and zinc oxide (ZnO), which are highly active as photocatalysts. These semiconducting materials are usable for numerous applications, ranging from wastewater treatment to self-cleaning surfaces, indicating their potential and versatility towards sustainability. Nevertheless, as the use of such nanoparticles expands, they must inevitably undergo thorough life cycle and environmental assessments, as identified in the need for novel frameworks within nanotoxicology. Understanding how these materials behave when interacting with biological systems and ecosystems is essential to ensure that the shift towards such innovation is compatible with environmental safety measures (Bahnemann et al., 2011). Also, sophisticated structural designs, such as inverse opal catalysts, have emerged as a core research focus in the search for improved photocatalytic performance. Such ordered materials have unique optical properties that facilitate enhanced light-matter interactions, essential for photocatalysis. The structured macroporosity of inverse opals allows for enhanced light absorption and optimized charge carrier dynamics, crucial for efficient energy conversion in photocatalysts. Recent studies demonstrate that the integration of these structures with plasmonic elements allows for even increased enhancement of catalysis through plasmonic-photonic structure coupling. The synergistic approach not only optimizes rates of reaction but also increases the applicability of these catalysts in a broad range of electrochemical and photocatalytic reactions. These innovations mark a significant step in the evolution of catalysts that are not only efficient but also sensitive to the specific needs of flow-based chemical processes (Armstrong et al., 2018). In light of these innovations, ongoing exploration of catalyst design for flow systems indicates a broader shift toward sustainable and efficient chemical synthesis. The ability to harness light and novel catalytic frameworks creates the door to new photochemical routes that can result in more environmentally friendly chemical alternatives. Researchers are increasingly focused on perfecting flow conditions, using real-time monitoring devices, and constructing more robust catalysts to push these systems further. As greener practices become imperative, catalyst design innovations will be the game-changers to shape the future of flow chemistry. In resolving questions of efficiency and safety concerns, the innovations can reshape the standard of photocatalytic reactions towards a greener chemical industry (Bahnemann et al., 2011; Armstrong et al., 2018).

8.5. FUTURE PERSPECTIVES

With continued advances in photocatalysis in flow chemistry, future directions focus on enhancing efficiency and effectiveness in applications. Integration of nanotechnology into traditional photocatalytic processes is tremendously promising for accelerating reaction rates and enhancing reaction selectivity. Through the use of nanomaterials, scientists have been able to develop innovative methods that can significantly reduce the environmental impact associated with routine chemical processes. This is of significant application in agricultural production, where nanotechnology can improve food production while minimizing risks of contamination in soil and water systems. This kind of innovation shows the importance of science in addressing real-world problems by improving productivity as well as sustainability in agri-industries (Bacchini et al., 2013). Besides, the future of flow chemistry's photocatalysis is very bright in water treatment, driven by the need for cleaner, more sustainable methods. Sophisticated oxidation processes like photocatalysis have proved to be one of the leading solutions to wastewater problems in urban and industrial settings. As researchers continue to improve the methods, the focus will be on aligning the photocatalytic systems with conventional treatment systems to enhance the degradation of contaminants. Most notably, countries like China, Spain, and the USA lead research in this area, and their collaborative model can potentially result in breakthrough solutions to a vital global resources shortage issue. The advantages gained through synergy-based means through advanced oxidation not only improve water quality but can also contribute to resource sustainability (Agüera et al., 2019). Overall, the current work on photocatalysis through flow chemistry is sure to unveil significant developments that address environmental as well as resource management problems. As researchers discuss the efficiency of different photocatalysts and optimize reaction conditions, the science will further develop, leading to wider, more diverse applications. From enhancing crop yields to developing efficient water treatment systems, these enhancements hold significant importance in building a more sustainable world. Integration of research efforts from various geographic and scientific communities will facilitate the sharing of knowledge and innovation. Eventually, the integration of these future visions will make photocatalysis a frontline strategy in addressing global issues in environmental conservation and resource management.

8.5.1. Potential Advancements in Photocatalytic Technologies

Advances in photocatalytic systems drive the improvement of efficiency and effectiveness in the use of flow chemistry. Typical methods of contaminant elimination and pathogen annihilation are often marked by long processing times and low efficiency, particularly in water filtering. New technologies show that the integration of photocatalytic materials can significantly reduce such limitations. For instance, the photocatalytic application of titanium dioxide (TiO_2) has immense potential to trigger solar disinfection mechanisms. By leveraging the synergistic application of solar power and photocatalytic mechanisms, researchers have been able to effectively enhance disinfection processes, thereby facilitating faster and more reliable access to clean water. This integration of photocatalysis and flow systems highlights not only the potential for more efficient designs in photo-reactor technology but also the challenges of developing innovative methods for mass transfer and catalyst immobilization, which are crucial for achieving the optimum reaction efficiencies observed in solar-driven operations (Byrne et al., 2014). Further, these structural advancements in photocatalytic materials, such as inverse opal structures, have opened up new avenues for improving light-matter interactions. These three-dimensional ordered macroporous materials, with their unique optical properties, enhance the activity of photocatalysts in applications for flow chemistry. More recent work has identified that the integration of plasmonic-photonic coupling and metal nanoparticles can significantly improve the functionality of catalysis, particularly in solar-to-fuel and photoelectrochemical systems. The utilization of such intricate material designs enhances not just light absorption but also charge carrier mobility, which is essential for efficient photocatalytic reactions. This strongly indicates a shift towards material architecture optimization in photocatalytic technology, leading to enhanced applications in various chemical processes. In addition, future studies on photocatalytic technologies will be focused on transcending current challenges in terms of catalyst efficacy and sustainability. There has been a recent focus on emphasizing the need to standardize test methods and enhance reactor designs to facilitate more extensive commercial uses of photocatalytic systems. In addition, there is a need to develop alternative catalysts that are more effective at absorbing visible light than traditional TiO_2. Developments in composite material and heterojunction systems are already showing the way, suggesting that future photocatalytic technologies will be able to take advantage of a combination of enhanced efficiencies and reduced operating costs. By enhancing interdisciplinarity among materials science, chemical engineering, and environmental science, the development of stable photocatalytic systems can be hastened to become pivotal in addressing environmental challenges such as water quality and energy sustainability (Armstrong et al., 2018) (Figure 8.12).

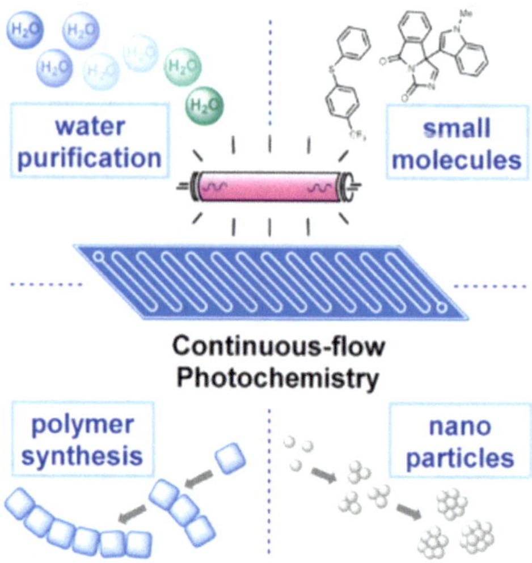

Figure 8.12. Continuous-flow photochemistry in microreactors has received significant attention from researchers in academia and industry, as this technology offers reduced reaction times, higher selectivities, straightforward scalability, and the ability to safely use hazardous intermediates and gaseous reactants.

Source: https://pubs.acs.org/doi/abs/10.1021/acs.chemrev.5b00707.

8.5.2. Role of Artificial Intelligence (AI) in Optimizing Processes

The marriage of artificial intelligence (AI) in process optimization within photocatalysis and flow chemistry has been a revolutionary factor. The technology applies machine learning algorithms to read volumes of complex data sets generated during the process of chemical reaction. By predicting optimal conditions for various reaction parameters, AI minimizes the trial-and-error approach traditionally applied in chemical synthesis, hence enhancing efficiency and curbing waste. For instance, the latest breakthrough in machine learning enables one to model multivariable reactions, thus allowing scientists to explore large reaction spaces more thoroughly. The use of these AI techniques presents a pathway for developing robust technologies that not only facilitate increased yields but also reduce process costs significantly, which explicitly points to the indispensable role of AI in contemporary chemical R&D. In a real-world application of this technology, one experiment demonstrated machine learning-optimized design of a photoredox tertiary amine synthesis. The researchers navigated a high-dimensional space of continuous and discrete variables, such as solvent choice, to optimize reaction outcomes. In the research, a novel Bayesian optimization algorithm, nomadic exploratory multiobjective optimization (NEMO), was employed to establish key reaction parameters that affect yield and cost, leading to an improvement in productivity. As such, the flow chemistry process achieved a throughput of approximately 25 times greater than traditional batch reactions, demonstrating the ability of AI tools to facilitate rapid and effective chemical processes in the context of photocatalysis (N/A, 2025). Yet another instance of AI's potential, Materials Acceleration Platforms (MAPs) harness artificial intelligence and automation to speed up the discovery of next-generation materials in photocatalytic materials. By making the experimental protocols automatically efficient, MAPs can enhance the scalability of reactions dramatically while simultaneously addressing environmental problems such as climate change. The application of AI in processing catalysis data enables scientists to identify potential materials for solar chemical production, specifically in heterogeneous CO_2 photo(thermal) catalysis. The data-driven and systematic approach made possible by MAPs accelerates material discovery and optimization, resulting in green practice in chemistry (Aspuru-Guzik et al., 2023). This revolutionary synergy demonstrates the significant impact that AI technology has in optimizing operation in photocatalysis, ushering a new era of efficiency and innovation.

8.5.3. Predictions for the Future of Photocatalysis in Flow Chemistry

The evolution of photocatalysis within flow chemistry has a bright future, particularly with the introduction of innovative materials such as phosphorene. Forged developments envision a paradigm shift towards using two-dimensional materials for enhanced photocatalytic efficiency. Phosphorene, a derivative obtained from black phosphorus, has attractedinterest due to its unique surface reactivity and nanoscale behavior. This new substrate is set to facilitate new catalytic processes and improve reaction conditions, with significant implications for chemical synthesis. As the development of phosphorene functionalization and its use in diverse reactions continues, conjugating the photocatalyst with the two-dimensional material can reorient flow chemistry practice toward energy-efficient, environmentally friendly processes (Benito et al., 2019). Additionally, the integration of photocatalysis with online monitoring technology in flow chemistry will revolutionize the field. By applying state-of-the-art sensing technology, researchers can achieve precise control over reaction conditions to enhance photocatalytic processes for higher yields and selectivity. Artificial intelligence and machine learning will further strengthen this process, making predictive modeling of reaction pathways under various conditions more efficient. Such a paradigm shift has the potential to revolutionize the design of air purification systems to create cleaner indoor environments in urban areas. These advances would come a long way in mitigating indoor air quality issues and achieving sustainable living standards in the context of increasing urbanization (Chen et al., 2017). Last but not least, future development in photocatalysis using flow chemistry will likely address the sustainability factor by aiming at cleaner reaction routes. With the increasing social pressure to use environmentally friendly methods, scientists are expected to emphasize developing photocatalytic processes that rely on renewable energy sources, such as sunlight. It may result in a radical reduction in energy consumption and greenhouse gas emissions compared to conventional chemical processes. By advancing the science of flow reactor design and photocatalytic mechanisms, researchers can realize the maximum potential of photocatalysis to facilitate reactions that are not only efficient but also environmentally benign. The synergy of green standards and new technologies will undoubtedly dictate the future of flow photocatalysis.

SUMMARY

In summarizing the application of photocatalysis to flow chemistry, it is essential to note the paradigm-shifting ability of this technology towards addressing contemporary issues in chemical synthesis and environmental purification. Through the exploitation of the capabilities of photocatalysts, particularly titanium dioxide (TiO_2), chemists are poised to enhance reaction efficiency and selectivity to a great extent. The ability to apply continuous flow processes gives enhanced control over reaction conditions and improved mass transfer, both of which are important for photocatalytic reaction optimization. Research has shown that various preparation techniques of TiO2, such as sol-gel and hydrothermal synthesis, yield different morphologies and photocatalytic activities, highlighting the need for ongoing studies in this area (Králová et al., 2012). The integration of advanced methodologies offers more effective photocatalysts that can operate under mild conditions and also responds to the potential of green chemistry in photocatalysis. Furthermore, self-organized TiO_2 nanotubes have a promising way to enhance photocatalytic processes in flow chemistry. These nanotubes offer unique characteristics, such as increased surface area and improved electron mobility, which can potentially significantly enhance photocatalytic activity (Nguyen et al., 2016). By tuning the growth conditions and incorporating unique functionalities, researchers are creating tailor-made photocatalytic systems to better tackle various environmental and chemical challenges. Self-assembly of the nanotubes into individual morphologies, such as double-walled forms, has been shown to enhance the overall stability and activity of the photocatalysts. Hence, integration of flow chemistry platforms for the use of these advanced materials can lead to revolutionary solutions in energy conversion and pollutant degradation, enhancing the commercial viability of photocatalytic technologies. Short, the interface between flow chemistry and photocatalysis offers several routes for advancing sustainable green chemistry procedures. The development of methodologies for photocatalyst design and synthesis continues not only to refine reaction performance but also to contribute to longer-term agendas for sustainability and resource efficiency. Subsequent research should focus on the up-scaling of these complex systems and their integration into existing chemical processes, ensuring that photocatalysis becomes a cornerstone for green, innovative methods. By embracing multidisciplinary research involving materials science, chemical engineering, and photochemistry, the research community can further harness the full potential of photocatalytic technologies, ultimately paving the way for groundbreaking innovations in industrial applications and environmental protection.

The combination of photocatalysis with flow chemistry is a welcome advancement in the disinfection of wastewater, and particularly for drinking water disinfection. Traditional means of disinfection work well, but they are usually inefficient and too slow. The application of semiconductor photocatalysis, such as titanium dioxide, can enhance solar water disinfection by significantly reducing the exposure time. This technology is not only directed toward eliminating pathogenic microorganisms like E. coli. Both Cryptosporidium and Escherichia coli are also targeted, along with the growing problems presented by contaminated water sources worldwide. Emphasizing the effectiveness of sunlight as a power force for photocatalysis, the application of novel reactors can boost maximum photon flux and improve liquid mixing to improve the overall efficiency of disinfection processes further (Byrne et al., 2014). Thus, photocatalysis in flow chemistry represents a suitable solution to the pressing issue of the sustainability of potable drinking water worldwide. Additionally, recent scientific progress signals a close international collaboration in the study of photocatalysis and its uses. Major research activities are found in countries such as China, Spain, and the USA, signaling the international agreement towards addressing water shortage and pollution. Each of these nations has contributed significantly to the evolution of catalyst material and reactor architecture. Research studies in this field demonstrate a collaborative effort aimed at improving advanced oxidation processes, particularly in industrial and municipal wastewater treatment. Systematic research of research clusters indicates the emphasis on addressing some pollutants, such as pharmaceuticals and pesticides, thereby contributing to more sustainable water treatment (Agüera et al., 2019). Such a global scientific community supports innovation, which is fundamental in overcoming the adversity experienced in water safety and availability. Further, the choice of the catalyst and the reactor design also play an essential role in the success of photocatalytic processes. The application of immobilized catalysts in place of suspended particles enhances the operational efficiency by minimizing mass transfer and sedimentation issues in reactors. Proper optimization of parameters such as catalyst loading, flow rate, and illuminated volume is essential, and studies show their direct impact on the dynamics of the photocatalytic process. Innovations in reactor design not only improve disinfection rates but also aim to decrease operation costs, a requirement for commercial viability. Although numerous advances have been accomplished, mass production and commercialization remain problematic, reflecting the need for continued research and development in this multidisciplinary field. The resolution of these challenges will pave the way for greater application of photocatalysis as a clean technology for delivering safe drinking water in the future (Byrne et al., 2014).

REFERENCES

1. Abdullah, N., Mohd Padzil, S. A., Mohd Salleh, R., Pillay, T. S., et al. (2018). Model selection and model averaging on mortality of upper gastrointestinal bleed patients. https://core.ac.uk/download/18264202.pdf (accessed on 28th September 2025).
2. Garrido-Cardenas, J. A., Esteban-García, B., Agüera, A., Sánchez-Pérez, J. A., & Manzano-Agugliaro, F. (2019b). Wastewater treatment by advanced oxidation process and their worldwide research trends. *International Journal of Environmental Research and Public Health*, *17*(1), 170. https://doi.org/10.3390/ijerph17010170Aiga, N., Basu, M., Pattengale, B., Du, W. M., Gainov, R. R., Gu, J. B., Huang, Z. Q., et al. (2015). Ultrafast hole trapping and relaxation dynamics in p-type CuS nanodisks. https://core.ac.uk/download/213061727.pdf (accessed on 28th September 2025).
3. Aillet, T., Dechy-Cabaret, O., Loubière, K., Prat, L., et al. (2013). Photochemical synthesis of a "cage" compound in a microreactor: Rigorous comparison with a batch photoreactor. https://core.ac.uk/download/18540116.pdf (accessed on 28th September 2025).
4. Antonietti, M., Kurpil, B., Mazzanti, S., Pieber, B., et al. (2020). Dichloromethylation of enones by carbon nitride photocatalysis.
5. Armstrong, E., Collins, G., Geaney, H., McNulty, D., et al. (2018). 2D and 3D photonic crystal materials for photocatalysis and electrochemical energy storage and conversion. https://core.ac.uk/download/153478439.pdf (accessed on 28th September 2025).
6. Aspuru-Guzik, A., Bozal-Ginesta, C., Kumar, S. G. H., Ozin, G. A., et al. (2023). Designing materials acceleration platforms for heterogeneous CO_2 photo(thermal)catalysis. http://arxiv.org/abs/2308.03628 (accessed on 28th September 2025).
7. Bacchini, F., Greppi, G., Mura, S., Roggero, M., et al. (2013). Advances of nanotechnology in agro-environmental studies. https://core.ac.uk/download/16749559.pdf (accessed on 28th September 2025).
8. Bahnemann, D. W., Bloh, J. Z., Kasper, C., Wagner, J., et al. (2011). Toxicological issues of nanoparticles employed in photocatalysis. https://core.ac.uk/download/52919976.pdf (accessed on 28th September 2025).
9. Bals, S., Cott, D. J., Deng, S., Detavernier, C., et al. (2014). Atomic layer deposition-based synthesis of photoactive TiO_2 nanoparticle chains by using carbon nanotubes as sacrificial templates. http://arxiv.org/abs/1410.5942 (accessed on 28th September 2025).

10. Barata Vallejo, S., Cooke, M. V., & Postigo, A. (2018). Radical fluoroalkylation reactions. https://core.ac.uk/download/287880745.pdf (accessed on 28th September 2025).
11. Benito, I. I., Bini, R., Bolognesi, M., Caporali, S., et al. (2019). A perspective on recent advances in phosphorene functionalization and its application in devices. https://core.ac.uk/download/301577585.pdf (accessed on 28th September 2025).
12. Bonomo, M., Cavallo, C., Di Pascasio, F., Dini, D., et al. (2017). Nanostructured semiconductor materials for dye-sensitized solar cells. https://core.ac.uk/download/54538567.pdf (accessed on 28th September 2025).
13. Byrne, A. J., Dionysiou, D. D., Dunlop, P. S. M., Ibanez, P. F., et al. (2014). Solar photocatalysis for water disinfection: Materials and reactor design. https://core.ac.uk/download/pdf/60775675.pdf (accessed on 28th September 2025).
14. Capaldo, L., Ravelli, D., & Fagnoni, M. (2021). Direct photocatalyzed hydrogen atom transfer (HAT) for aliphatic C–H bonds elaboration. Chemical Reviews, 122, 1875–1924. https://doi.org/10.1021/acs.chemrev.1c00263 (accessed on 28th September 2025).
15. Chen, X., Clements-Croome, D., Gal, C., Liu, Y., et al. (2017). A review of air filtration technologies for sustainable and healthy building ventilation. https://core.ac.uk/download/84587706.pdf (accessed on 28th September 2025).
16. Cvejn, D., Foniok, K., Praus, P., Smýkalová, A., et al. (2020). Post-synthetic derivatization of graphitic carbon nitride with methanesulfonyl chloride: Synthesis, characterization and photocatalysis. https://core.ac.uk/download/323112788.pdf (accessed on 28th September 2025).
17. De Buysser, K., Van Driessche, I., Van Zele, M., Watté, J., et al. (2018). Recent advances in low-temperature deposition methods of transparent, photocatalytic TiO_2 coatings on polymers. https://core.ac.uk/download/154952507.pdf (accessed on 28th September 2025).
18. Durme, J. V., Ghodbane, H., Hamdaoui, O., Leys, C., et al. (2014). Degradation of AB25 dye in liquid medium by atmospheric pressure non-thermal plasma and plasma combination with photocatalyst TiO_2. https://core.ac.uk/download/55854100.pdf (accessed on 28th September 2025).
19. Falco, M., Loddo, V., Palmisano, G., et al. (2004). Photocatalytic degradation of dyes by using a membrane reactor. https://core.ac.uk/download/53229280.pdf (accessed on 28th September 2025).

20. Fedin, M. V., Gascon, J., Hendon, C. H., Houtepen, A. J., et al. (2016). Electronic origins of photocatalytic activity in d^0 metal organic frameworks.
21. Fernández Velasco, L., Jiménez Carmona, R., Laurenti, E., Maurino, V., et al. (2016). Carbon materials as additives to WO_3 for an enhanced conversion of simulated solar light. https://core.ac.uk/download/45446849.pdf (accessed on 28th September 2025).
22. Fiore, G. B., Gilbert, R., Rasponi, M., et al. (2009). Realization and efficiency evaluation of a micro-photocatalytic cell prototype for real-time blood oxygenation. https://core.ac.uk/download/9633402.pdf (accessed on 28th September 2025).
23. Freund, R., Zaremba, O., Arnauts, G., Ameloot, R., Skorupskii, G., Dincă, M., Bavykina, A., et al. (2021). The current status of MOF and COF applications. Angewandte Chemie International Edition, 60, 23975–24001. https://doi.org/10.1002/anie.202106259 (accessed on 28th September 2025).
24. Gregoire, J. M., Guevarra, D., Haber, J. A., Kan, D., et al. (2020). Combinatorial synthesis of oxysulfides in the lanthanum-bismuth-copper system. https://core.ac.uk/download/323084798.pdf (accessed on 28th September 2025).
25. Hakki, A., MacPhee, D. E., Wang, F., Yang, X., et al. (2017). The effect of interfacial chemical bonding in TiO_2–SiO_2 composites on their photocatalytic NO_x abatement performance. https://core.ac.uk/download/131024011.pdf (accessed on 28th September 2025).
26. Jang, S., Kim, D.-P., Parmar, J., Sanchez, C., et al. (2015). Nano-photocatalysts in microfluidics, energy conversion and environmental applications. https://core.ac.uk/download/81578008.pdf (accessed on 28th September 2025).
27. Kaplani, E., Panagiotaras, D., Papoulis, D., Rapsomanikis, S., et al. (2014). Nanocrystalline TiO_2 and halloysite clay mineral composite films prepared by sol-gel method: Synergistic effect and the case of silver modification to the photocatalytic degradation of basic blue-41 azo dye in water. https://core.ac.uk/download/41990680.pdf (accessed on 28th September 2025).
28. Králová, M. (2012). Self-Cleaning Properties of Thin Printed Layers of Titanium Dioxide. https://core.ac.uk/download/30310816.pdf (accessed on 28th September 2025).
29. Leung, D. Y. C., Leung, M. K. H., Luo, X., Luque, R., et al. (2014). *In situ* photogalvanic acceleration of optofluidic kinetics: A new paradigm for advanced photocatalytic technologies. https://core.ac.uk/download/152439003.pdf (accessed on 28th September 2025).

30. N/A. (2025). Machine Learning-Driven Optimization of Continuous-Flow Photoredox Amine Synthesis. https://core.ac.uk/download/659488056.pdf (accessed on 28th September 2025).

31. Nguyen, N. T., Riboni, F., Schmuki, P., So, S., et al. (2016). Aligned metal oxide nanotube arrays: Key aspects of anodic TiO_2 nanotube formation and properties. http://arxiv.org/abs/1610.04517 (accessed on 28th September 2025).

CHAPTER 9

GOLD IN PHOTOCATALYSIS

Contents

INTRODUCTION

The emergence of photocatalysis has revolutionized materials science and environmental chemistry, opening new avenues to utilize solar energy for sustainable processes. At the forefront of this revolution is gold, a noble metal generally known for its beauty and economic value, which has now proven to be an active participant in photocatalytic reactions. Gold nanoparticles possess unique optical and electronic properties, making them more effective in photocatalysis as catalysts capable of catalyzing chemical reactions under light irradiation. By using these properties, researchers are exploring the possibility of gold's application as a catalyst towards degrading pollutants, synthesizing valuable chemicals, and ultimately driving the innovation of renewable resources. Therefore, the use of gold in photocatalysis not only underscores its worth in theoretical research but also underscores its application in addressing fundamental issues in the environment. The addition of gold to photocatalytic systems is also supported by its compatibility with most semiconductor materials, which aid significantly in boosting photocatalytic efficiency. Semiconductors are the pillars of photocatalytic reactions, and when combined with gold, they create synergetic effects that enhance charge separation and light absorption. Recent advances in nanotechnology have enabled the synthesis of gold-supported semiconductor composites, utilizing the exceptional properties of gold to enhance photocatalytic activity. By examining these interactions, researchers can engineer advanced materials specific to specific photocatalytic applications, such as water splitting and pollutant removal. On that note, gold's catalytic versatility highlights the ongoing evolution of photocatalytic technologies aimed at opposing environmental pollution and enabling sustainable energy alternatives. Furthermore, research on gold as a photocatalyst holds potential for additional technological development, particularly in artificial photosynthesis. As the world grapples with the effects of global warming and depletion of natural resources, solutions to sustainable energy have become crucial. Gold's ability to catalyze artificial photosynthesis, mimicking nature, offers a very efficient solution for the conversion of solar energy to chemical fuel. The application of gold nanoparticles in photoelectrochemical devices not only enhances photoconversion efficiency but also facilitates the development of green agriculture, thereby continuing to narrow the gap between environmental management and technology. By focusing on the potential applications of gold in photocatalytic processes, researchers encourage a multidisciplinary approach to address the world's energy issues and assist in bringing about a more sustainable world, as contended in nanotechnology development debates and applications in agriculture and the environment (Bacchini et al., 2013; Deng et al., 2019).

Importance of Photocatalysis in Environmental Applications

Photocatalysis is now a central technology in solving various environmental problems, primarily the degradation of pollutants and the remediation of contaminated environments. An increase in the world's population and the synthetic demand it generates for agriculture has compelled the use of chemically derived fertilizers and pesticides. While these innovations enhance crop production, they are usually accompanied by the inadvertent release of poisonous by-products into the atmosphere, hence water and soil pollution. Photocatalytic processes offer an economical method for degrading contaminants by using light to activate photocatalysts, which catalyze chemical reactions to transform contaminants into less harmful products. Recent studies have demonstrated that the combination of photocatalysis and nanotechnology can increase the efficiency of such processes, with a focus on their use in enhancing environmental quality and promoting sustainability (Bacchini et al., 2013). In addition to the degradation of pollutants, photocatalysis plays a crucial role in addressing water-treatment problems. Contaminated water sources can be highly toxic to health, necessitating effective methods of purification. Photocatalytic systems can degrade organic pollutants and microorganisms in the presence of illumination by light and thus can be a good solution for wastewater treatment. Experiments have proven that titanium dioxide, a popular photocatalyst material, can significantly reduce the bacterial concentration that inflicts harm and organic contaminants in water (Benito et al., 2019). Photocatalysis is flexible in utilizing solar light, making it a highly plausible green technology. Through the photo-conversion of light into chemical energy, these activities not only clean water resources but also make energy more sustainable, thereby positioning photocatalysis as a leading technology in modern environmental remediation. The incorporation of nanomaterials, such as gold nanoparticles, within photocatalytic systems is also a promising future trajectory for the discipline. Gold nanoparticles enhance the photocatalytic activity of traditional semiconductors by dramatically increasing efficiency in applications ranging from air purification to hydrogen evolution. Their distinctive optical properties enhance electron transfer processes, which are central in photocatalytic reactions. This innovation highlights the convergence of materials science and photocatalysis, where the inclusion of nanomaterials can lead to heightened activity and selectivity in environmental processes. As research continues to investigate the use of various nanostructures in photocatalysis, the potential for developing cost-effective, sustainable solutions becomes increasingly real, bringing photocatalysis—and gold photocatalysis specifically—to the forefront of green environmental technology (Benito et al., 2019).

Overview of Gold's Role in Photocatalysis

Gold nanoparticles have been a topic of significant interest in photocatalysis due to their unique electronic and optical properties. As part of photocatalytic systems, gold possesses the potential to enhance charge separation and light absorption, which are of utmost significance for effective catalysis. The ability of gold to facilitate surface plasmon resonance enables the photocollecting of visible light for generating energetic carriers that can drive chemical reactions.

As noted by the findings by Demir et al. (2014), small gold nanocrystals are more efficient in the generation of energetic carriers due to their confined nanoscale. This diminishes momentum conservation challenges. This influence further demonstrates the capability of gold as a photocatalyst, particularly in visible light-dependent processes, thereby expanding its application beyond traditional ultraviolet activation.

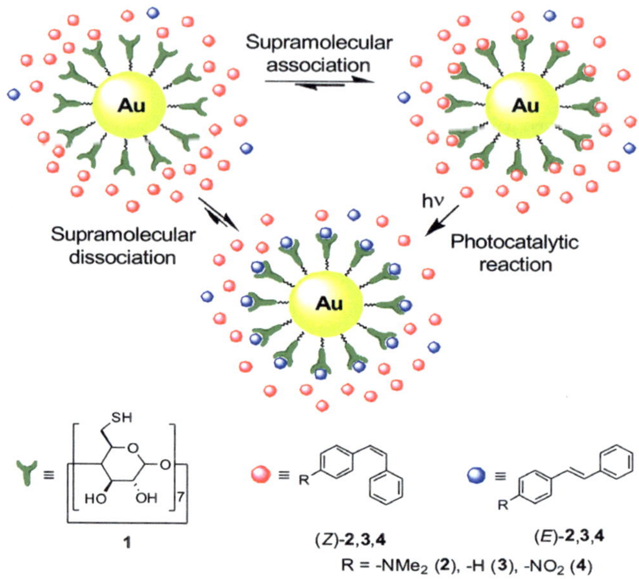

Figure 9.1. Strategy devised to supramolecularly enhance the photocatalytic activity of gold nanoparticles.

Source: https://pubs.rsc.org/en/content/articlehtml/2017/cc/c6cc09600j.

Aside from its optical characteristics, gold plays a significant role in enhancing the reaction kinetics of numerous photocatalytic reactions. Its catalytic performance is usually greater than that of most other metals, owing to its ability to promote the formation of intermediate species and stabilize reactive sites. For example, incorporating gold into metal oxide matrices greatly promotes the photocatalytic degradation of organic pollutants by preventing charge recombination, a universal limitation in the efficiency of photocatalysis. The performance of gold in activating molecular oxygen for oxidizing reactions also highlights the valuable contribution of gold in improving photocatalytic performance. These mechanisms highlight the importance of investigating the synergistic interactions between gold and various semiconductor materials to harness their full photocatalytic potential. This is supported by research on plasmon excitation dynamics and energetic carrier generation in metal nanocrystals, as discussed by Demir et al. (2014). In considering potential future uses, incorporating gold into hybrid photocatalytic systems is emerging as an effective method for addressing complex environmental issues, such as wastewater purification and energy conversion. The work at Tohoku University highlights innovative methods to improve gold-based photocatalysts, resulting in higher conversion ratios and reduced energy demands. The research of gold's activities with other materials, particularly in nanostructured forms, shows that its capability can be significantly enhanced. These enhancements can make the development of more efficient and sustainable photocatalytic systems possible. As photocatalytic technologies continue to emerge, the understanding and use of gold's function will be critical to enhancing energy sustainability solutions as well as environmental management, so it is now a focal point in research in this field (Figure 9.1).

9.1. PROPERTIES OF GOLD NANOPARTICLES (AUNPS)

The desirable characteristics of gold nanoparticles (AuNPs) render them essential materials for photocatalytic use. Due to their unique electronic framework and surface plasmon resonance, AuNPs possess a significant capacity for light absorption in the optical range, with high photocatalytic efficiency. This characteristic allows AuNPs to absorb light in specific wavelengths, exciting electrons that could participate in various redox processes. When coupled with photocatalytic systems, including titanium dioxide (TiO_2) composites, gold nanoparticles have been used to enhance the rate of reaction by enabling electron transfer processes. Studies show that titanium dioxide films with AuNPs significantly improve photocatalytic activity, particularly in the degradation of organic pollutants (Darr et al., 2011). This enhancement is attributable to the improved charge carrier dynamics, and the gold interrupts electron-hole recombination, a significant bottleneck in photocatalytic efficiency. Particle size and morphology are also of the utmost significance in determining the photocatalytic properties of gold nanoparticles. The particle size dictates the surface area-to-volume ratio of the nanoparticles, which has a direct effect on their catalytic activity. Smaller nanoparticles will typically have greater surface reactivity due to the greater number of surface atoms that can interact with reactants. AuNP geometry, such as spherical, rod-like, or triangular shapes, can influence their optical and electronic features, thereby enhancing their application in photocatalytic reactions. Experiments show that geometries can lead to various plasmonic responses, making it possible to use them selectively in photocatalytic applications (Atwater et al., 2020). These tunable properties demonstrate the versatility of gold nanoparticles, which can be adjusted for customized photocatalytic activities, ultimately leading to environmental remediation and energy conversion applications. Also, the combination of gold nanoparticles with other semiconductor materials enhances their efficiency in photocatalysis. For instance, by mixing AuNPs with p-type nickel oxide, they exhibit enhanced selectivity towards CO_2 reduction, showing their ability to provide for selective chemical conversion under illumination. This is due to hot electron generation via gold nanoparticles, which facilitates the transformation of CO_2 into valuable products, such as carbon monoxide and formate, and prevents unwanted hydrogen evolution reactions. By utilizing the characteristic properties of AuNPs, researchers can introduce significant modifications in the pathways and efficiencies of photocatalytic reactions, thereby improving the overall performance of photocatalytic systems. Such advances not only indicate the multi-functionality of the application of gold nanoparticles in photocatalysis but also indicate the central role of gold nanoparticles in the design of sustainable energy technology (Figure 9.2).

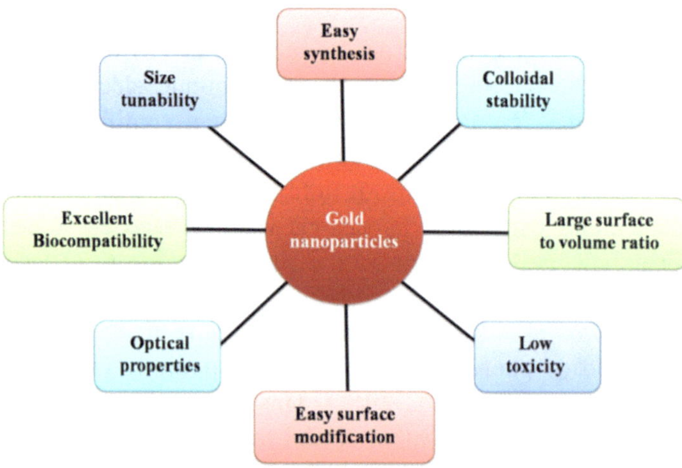

Figure 9.2. Desirable properties of gold nanoparticles.

Source: https://www.researchgate.net/figure/Desirable-properties-of-gold-nanoparticles_fig1_361093736.

9.1.1. Unique Optical Properties

The unique optical characteristics of gold nanoparticles are responsible for their functioning as photocatalysts. In particular, the characteristics are induced by the localized surface plasmon resonance (LSPR) phenomenon, in which conduction electrons on the nanoparticle surface oscillate when illuminated. Such an effect results in extensive absorption and scattering of light in specific wavelengths, allowing gold nanoparticles to harvest photons efficiently. Consequently, their ability to absorb broad-spectrum light provides scope for a range of photocatalytic applications, including energy conversion and chemical synthesis. Their shape and size have been predetermined by studies, which have indicated that optimizing these factors can further enhance optical properties and consequently improve catalytic efficiency. Thus, gold's new optical properties offer possibilities for designing next-generation photocatalytic systems that utilize light-driven processes for sustainable purposes (李 慧慧, 2012).

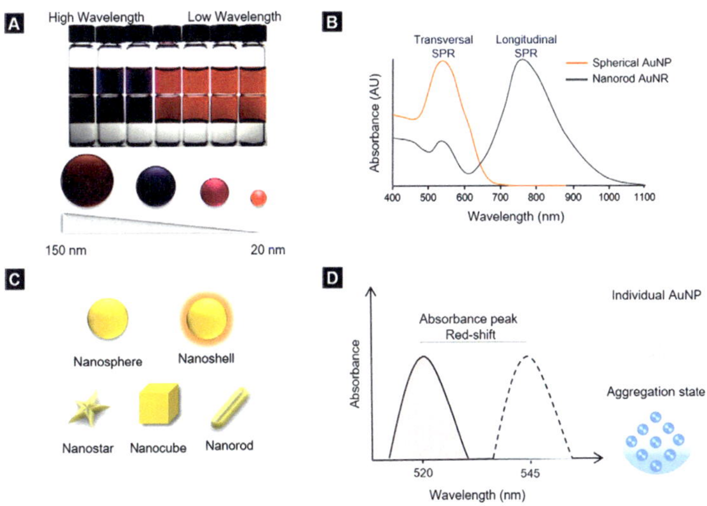

Figure 9.3. *Optical properties of gold nanoparticles (AuNPs).* (A) *Size-dependent optical properties of AuNPs. As the mean diameter of AuNPs decreases from 150 to 20 nm, the peak absorbance shifts to a lower wavelength, resulting in a brighter color of the AuNP solution.* (B) *Absorption profiles of two AuNPs (40 nm) of different shapes, corresponding to transverse (≈520 nm) and longitudinal (near the infrared region of the spectrum) surface plasmon resonances, showing the shape-dependent optical properties of AuNPs.* (C) *Different shapes of AuNPs.* (D) *Aggregation-dependent optical properties of AuNPs. Aggregation of AuNPs causes the peak absorbance to shift to a higher wavelength.*

Source: https://www.researchgate.net/figure/Optical-properties-of-gold-nanoparticles-AuNPs-A-Size-dependent-optical-properties-of_fig2_347168869.

In addition to size and shape, the surrounding environment also influences the optical properties of gold nanoparticles. The surrounding dielectric medium covering the gold nanoparticles modifies the LSPR features of the nanoparticles, and thus their photocatalytic activity can be different. For instance, encapsulating gold nanoparticles inside a metal-organic framework (MOF) creates a perfect microenvironment for supporting light-matter interactions. These positioning sites can lead to improved energy transfer processes that maximize the catalytic ability of gold under visible light, resulting in optimal catalytic activity (Fan et al., 2024). The research on the integration of gold nanoparticles with MOFs demonstrates the synergy achieved through coordination, which not only preserves but also enhances the inherent optical features of gold, thereby increasing photocatalyst activity (Fan et al., 2024). Besides, the combination of gold nanoparticles with other materials can be employed to engineer their optical behavior to be compatible with particular photocatalytic reactions. This is accomplished through hybrid systems, where gold is combined with a second semiconductor or catalytic matrix to create a composite that has higher photocatalytic performance. Gold electronic coupling with the assisting material is used to shift the absorption spectrum towards the red, allowing the composite to absorb light more effectively. This leads to increased charge separation and reduced recombination rates, leading to a better photocatalytic process. Such findings indicate how understanding and controlling the unique optical properties of gold lead to revolutionary advancements in photocatalysis, once again establishing its significance in both basic and applied sciences (李慧慧, 2012) (Figure 9.3).

9.1.2. Size and Shape Effects on Catalytic Activity

The size and shape of gold nanoparticles are significant factors in their photocatalytic reactivity and are crucial for optimizing photocatalysts. Nanoparticles with smaller sizes are likely to possess higher surface area-to-volume ratios, and this may result in higher reactivity. It is because there are more active sites available for catalysis. In addition, the electronic properties of gold nanoparticles are also size-dependent; smaller particles can exhibit higher electron delocalization; therefore, they will be more reactive towards electron transfer reactions. As with various studies, such as the photocatalytic degradation of organophosphorus compounds, the particle size of gold nanoparticles is of utmost significance in defining overall catalytic efficiency and selectivity in such applications, providing a route for enhanced remediation methods in environmental chemistry (Parker et al., 2016). Besides, the morphological characteristics of gold nanoparticles, such as shape, can also define their catalytic activity. Different geometrical structures of particles, like spherical, rod-like, or cubic ones, produce disparate distributions of surface atoms and thereby affect the catalysis mechanism. Specific shapes, for instance, will facilitate better light absorption or create specific electronic features that will enhance photocatalytic reactions. Research indicates that adding gold nanoparticles of engineered shapes to catalyst systems may lead to dramatic improvements in photocatalytic activity for various responses, particularly in CO_2 reduction reactions. Recently, there has been a rising interest in engineering these nanoparticles to achieve particular catalytic performances, taking advantage of their plasmonic behavior and unique surface characteristics (Atwater et al., 2020) (Figure 9.4).

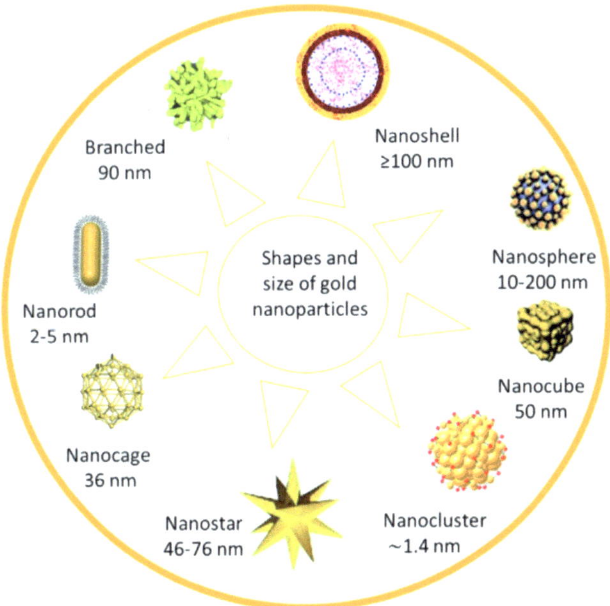

Figure 9.4. Shapes and size of gold nanoparticles.

Source: https://www.researchgate.net/figure/Shapes-and-size-of-gold-nanoparticles_fig1_355657305.

The interplay between size, shape, and catalytic activity calls for an advanced understanding of nanoparticle engineering in photocatalysis. By modulating these properties of gold nanoparticles in a controlled manner, researchers can optimize their efficiency in photocatalytic activities, particularly in environmental cleaning and energy conversion technologies. The development of photocatalysts that employ well-engineered gold nanoparticles could make significant contributions to achieving more efficient photocatalytic processes. As ongoing research uncovers the mechanisms of size and shape effects, the potential for tailoring gold nanoparticles for specific catalytic functionalities will broaden, ultimately providing more sustainable solutions to essential environmental problems. The incorporation of these advancements into this discipline promises to enhance our knowledge of photocatalytic systems and subsequent innovations within it (Parker et al., 2016).

9.1.3. Stability and Reactivity of Gold Nanoparticles

The stability and reactivity of gold nanoparticles (AuNPs) play a significant role in their effectiveness as photocatalysts. Their shape and size impart them with unique optical and electronic properties, making them highly valuable for application in photocatalysis. As a setback, the nanoparticles are susceptible to environmental factors such as temperature, pH, and ionic strength, which may impact their stability and, consequently, their catalytic activity. For instance, oxidation and agglomeration reactions can decrease their active surface area, reducing their photoactivity to a low extent. Dispersion of AuNPs on suitable supports, such as mesoporous silica, has been shown to stabilize them without compromising on their reactivity, thereby enhancing their performance in photocatalytic applications (Mart et al., 2018). Recognizing the reactivity of gold nanoparticles involves understanding how these materials interact with other materials of a photocatalytic system. The addition of different metal components, like copper in alloy materials, can influence the catalytic activity of AuNPs. The addition can enhance charge separation and overall reaction rate by providing added active sites and altering electron transport patterns. In a study where Au and Au-Cu alloy photocatalysts were synthesized, low-metal loadings, although small, contributed significantly to photocatalysis for reactions such as CO oxidation using solar light irradiation (Mart et al., 2018). Hence, meticulous designing of nanoparticle supports and compositions is of immense significance to fine-tune reactivity and stability for optimal photocatalytic performance. Additionally, there are some environmental implications of using gold nanoparticles for photocatalysis that must be examined. While nanomaterials offer the benefit of greater efficiency and selectivity, controversy has long revolved around their conceivable toxicity and persistence in the environment. Utilization of higher concentrations of nanoparticles is dangerous because it releases the materials into the environment, and hence degrades water and soil quality. Yet when utilized properly, nanotechnology can be used to clean up contaminated environments by assisting in the breakdown of the contaminating materials. This double-edged sword is used to emphasize the necessity for sustainable approaches to employing gold nanoparticles because their reactivity and stability need to be balanced against environmental considerations during photocatalytic reactions.

9.2. SYNTHESIS OF GOLD PHOTOCATALYSTS

The synthesis of gold photocatalysts represents a significant advancement in photocatalysis due to the exceptional properties of gold nanoparticles. Various approaches employed in the synthesis include chemical reduction, the sol-gel process, and deposition techniques. There are particular advantages of each method with respect to particle size control, dispersion, and surface functionality. For instance, the sol-gel method allows the intimate mixing of gold nanoparticles with supports such as titanium dioxide, leading to enhanced photocatalytic activity and stability. Furthermore, it has been found that strategically incorporating gold can significantly influence the electronic characteristics of composite materials, thereby improving their efficiency upon illumination. Since gold photocatalysts are presently under intense development, familiarity with these synthesis processes is essential for optimizing their application in various chemical reactions, such as CO oxidation and organic transformations.

The use of gold nanoparticles incorporated in mesoporous hosts has drawn widespread interest, leading to advancements in synthesis and functionalization processes. Experiments indicate that using supports like SBA-15, both pure and modified with titanium dioxide, achieves the highest dispersion of gold and copper nanoparticles in the pore structure. For instance, an influential study demonstrated that high-dispersion Au nanoparticles supported on the TiSBA-15 host exhibited outstanding catalytic performance in CO preferential oxidation reactions. These findings underscore the importance of support selection, as a suitable mesoporous structure can enhance the stability and usage of gold photocatalysts, particularly in environments that favor photocatalytic reactions. Consequently, continued exploration of these material interactions will unveil paths toward more efficient gold-based photocatalysts, which will facilitate their use in renewable energy and environmental remediation. Additionally, optimizing the thickness of films in gold photocatalyst syntheses has been another critical factor affecting their photocatalytic activity. Current studies have established that adjusting the thickness of titanium dioxide films with incorporated gold or silver nanoparticles might lead to significant fluctuations in photocatalytic performance.

A thickness of approximately 200 nm has been found to optimize photocatalytic degradation of organic substances and oxygen evolution in sacrificial media. This suggests that fine-tuning of synthesis conditions is needed for the development of desired photocatalytic behavior. By varying the film thickness, the scientists can explore the trade-off between light absorption and catalytic activity in gold-containing films. Finally, such observations of moviemaking provide important insights for constructing advanced gold photocatalysts to effectively address existing environmental issues, thereby representing a groundbreaking milestone in the field of photocatalytic science (Figure 9.5).

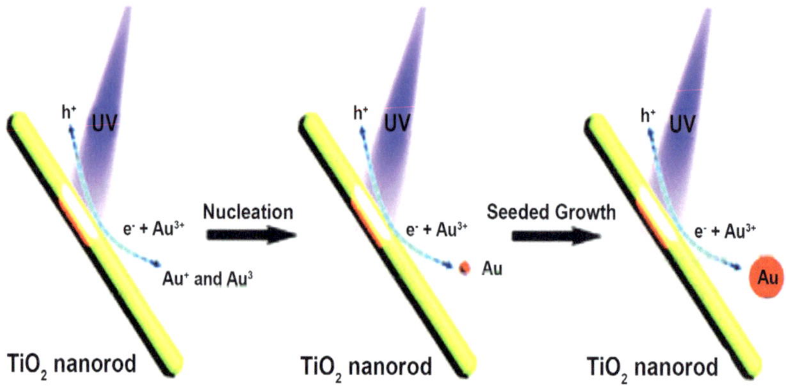

Figure 9.5. Photocatalytic synthesis of gold nanoparticles using TiO_2 nanorods.

Source: https://pubs.rsc.org/en/content/articlelanding/2019/cp/c9cp02493j.

9.2.1. Chemical Reduction Methods

The study of chemical reduction methods in the context of photocatalysis has significantly advanced the synthesis and performance of catalysts incorporating gold nanoparticles. The unique electronic properties of gold allow it to not only function as a catalyst but also as an essential component that enhances the photocatalytic activity of host materials like titanium dioxide. Upon integration in thin film systems using sol-gel dip-coating methods, gold nanoparticles facilitate the generation of reactive species that can perform chemical reductions effectively. This is heavily evident in studies comparing the photocatalytic degradation of organic dyes, where the interface between films of gold and titanium dioxide significantly enhances performance, subsequently reducing energy needs in such chemical processes (Darr et al., 2011). Understanding these interactions and balancing film thickness to maximum efficiency puts gold in a pivotal position in advanced photocatalytic systems. Moreover, the operation of plasmonic catalysts, particularly gold-based ones, is a crucial aspect in the selective reduction of substances such as carbon dioxide. Gold nanoparticles' ability to respond to light at specific wavelengths enables a form of photocatalysis that enhances the reduction pathway, favoring desired product formation over undesirable by-products. It was found that it is possible to tune the plasmonic capabilities of silver and gold catalysts to create enhanced selectivity in complex electrochemical reactions. For instance, plasmonic photocatalysts have demonstrated the ability to selectively generate carbon monoxide and methanol while simultaneously suppressing concomitant hydrogen evolution (Corson et al., 2019). This selective improvement under varied cathodic potentials reflects the potential of gold to optimize chemical reduction processes, overcoming efficiency and selectivity issues in photocatalytic reactions. Finally, integrating gold nanoparticles into chemical platforms for reduction is a groundbreaking approach to photocatalysis, emphasizing the design of materials as a central parameter in achieving optimal catalytic activity. Efficiency in catalysts can typically be hindered by side reactions, which need innovative ways to inhibit undesired processes such as hydrogen evolution. By exploiting the intrinsic characteristics of gold, researchers have managed to manipulate the electronic environment of photocatalysts to enhance common reaction pathways. For instance, by using sacrificial electron acceptors in collaboration with gold-loaded titanium dioxide films, oxygen evolution has been activated, and the product yield in water oxidation reactions has been enhanced (Darr et al., 2011). These subtle tactics, which blend chemical and physical principles, pave the way for more sustainable photocatalytic systems, emphasizing gold's key role in the foundation of chemical reduction methods in photocatalysis (Figure 9.6).

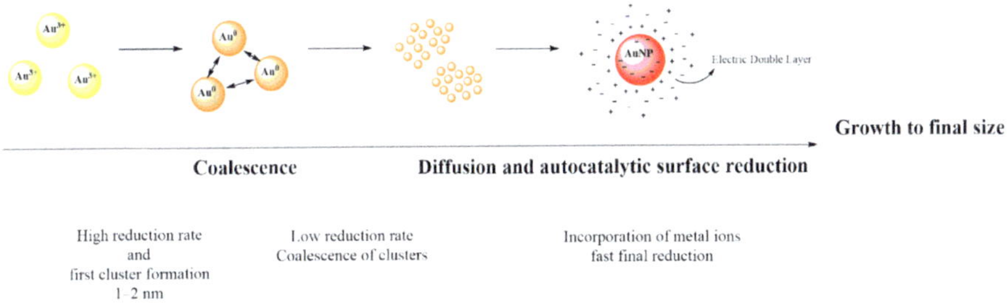

Figure 9.6. Colloidal gold nanoparticle (AuNP) formation process.

Source: https://www.mdpi.com/molecules/molecules-26-05823/article_deploy/html/images/molecules-26-05823-g002.png.

9.2.2. Physical Vapor Deposition Techniques

Physical vapor deposition (PVD) techniques are a foundation of material science, particularly in thin-film material development for photocatalyst use. Of these techniques, thermal evaporation, sputtering, and molecular beam epitaxy can facilitate the controlled layering of materials, which is important for increasing catalyst efficiency, such as that of gold in photocatalysis. The electronic and morphological characteristics of gold nanoparticles that are deposited using PVD have the potential to affect their catalytic activity significantly. This is especially relevant in situations where maximum interaction with reactants is meaningful and enhanced surface area and ideal electronic states are essential. These methods not only allow for better thickness and uniformity control of gold films but also enable the deposition of gold in various forms, enhancing its stability and effectiveness as a photocatalyst under varied irradiation conditions. Integration of PVD processes with other methods can even further improve the quality of gold photocatalysts. For instance, when combined with chemical vapor deposition (CVD) in the preparation of photocatalytically active TiO2 films modified by silver films, PVD facilitates controlled deposition, which is an important element in forming the desired nanostructure. Composite films not only are superior in their photocatalytic activity but also show durability, an important requirement for actual applications in environmental remediation processes. This seamless understanding of hybridization between PVD and CVD techniques can lead to new applications of gold, particularly in creating materials with both photocatalytic and antibacterial properties, as demonstrated in various studies on hybrid systems consisting of metal nanoparticles and semiconductor substrates (Brook et al., 2008). Furthermore, observing physical vapor deposition under varying conditions can provide beneficial nonlinear optical properties applicable for photocatalytic activity. For example, when the thermal response of metals under high irradiation is studied, models show that deviations from classical behavior suggest gold films grown by PVD can exhibit enhanced light absorption and thermal behavior. Such findings highlight the importance of comprehensive studies charting the interaction between the incident light and gold nanostructures, particularly at high temperatures, as this information applies to maximizing their photocatalytic performance. Understanding the thermodynamic and optical properties of PVD films provides deeper insight into utilizing gold for photocatalyst applications, as it offers a better understanding of improved photoreactivity mechanisms, particularly in potential environmental or energy-related applications (Chu et al., 2016).

9.2.3. Green Synthesis Approaches

The emergence of green synthesis techniques is a significant departure in the manufacture of nanomaterials, particularly in photocatalysis. This is driven by the imperative of lowering environmental footprints as enshrined in sustainable development strategies. Green synthesis differs from conventional approaches, which often rely on toxic chemicals and high-energy input, as it uses natural precursors like plant extracts and microorganisms to synthesize nanoscale materials. These bio-based agents not only facilitate the reduction of metal ions to nanoparticles but also enhance the functional properties of the obtained materials. Hence, green synthesis not only represents an environmentally friendly protocol but also results in enhanced photocatalytic activity, which is necessary to obtain efficient energy transformation and waste removal in systems like gold-based photocatalysis. The use of plant-based compounds in green synthesis of gold nanoparticles has been found to have significant advantages, particularly in photocatalytic technology. Some phytochemicals exhibit reducing and stabilizing properties that are accountable for the favorable morphology and nanoparticle size, both of which play a key role in elucidating photocatalytic performance. For instance, experiments demonstrated that controlled biological template synthesis of gold nanoparticles would produce surface properties optimized, thus improving charge separation and transfer dynamics. This has a direct impact on the photocatalytic activity as efficient charge dynamics are required for optimal efficiency in reactions such as hydrogen evolution and pollutant degradation. Thus, the application of green synthesis in gold nanoparticle synthesis is not only a methodological advancement but also a strategic advancement in the area of photocatalytic technology. Investigating the semiconductor-metal hybrid structure properties further establishes that its composition and synthesis via green methods can play a critical role in photocatalytic outcomes. Specifically, the positioning of gold between semiconductor matrices allows for regulated electronic interaction, most significantly at the nanoscale. Experiments have indicated that variations in the size and shape of gold nanoparticles produce dramatic modifications in charge transfer kinetics, as demonstrated by studies on cadmium sulfide-gold nanorods. This tunability enables the optimization of photocatalytic activity by correlating the derived size-dependent relationships with the efficacy of the photocatalytic system. Thus, the confluence of green synthesis techniques and photocatalytic efficiency is a significant area of future research. Understanding these processes can lead to more effective and environmentally sound energy conversion technologies, reducing environmental impacts compared to traditional synthesis techniques.

9.3. APPLICATIONS OF GOLD IN PHOTOCATALYSIS

The use of gold in photocatalysis provides a modern solution to environmental issues, especially in organic pollutant degradation. The utilization of gold nanoparticles, which possess superior electronic and optical properties, as co-catalysts for widely used photocatalysts like titanium dioxide (TiO_2) renders them competent co-catalysts. Incorporation of gold enhances the light-harvesting capability of TiO_2, particularly in the visible region, which is vital given that TiO_2 will have limited activity under such circumstances. This blend not only improves photocatalytic activity but also aids in charge separation, reducing recombination losses, which is crucial for sustained catalysis. With continued research, these advancements unlock gold's potential for application in various environmental remediation processes and render it a valuable commodity in photocatalytic technology. Furthermore, the catalytic ability of gold is not limited to serving as a co-catalyst; it is employed in stabilizing other catalytic materials.

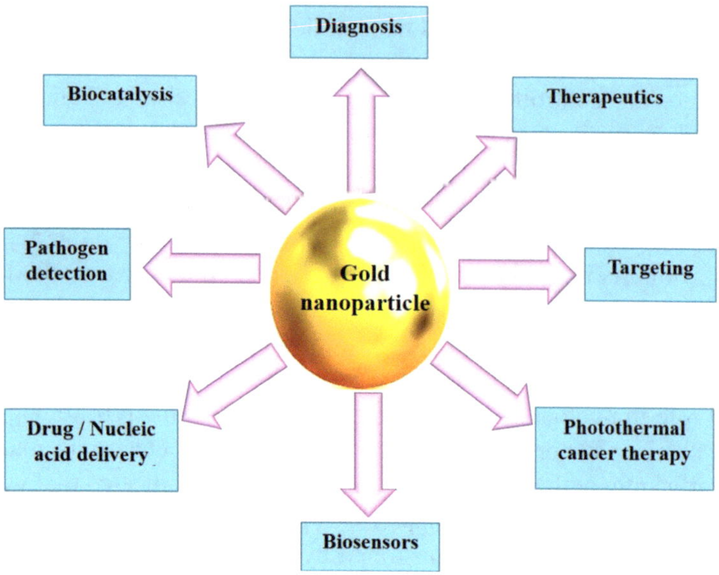

Figure 9.7. Multifunctional applications of gold nanoparticles.

Source: https://www.mdpi.com/2624-8549/4/2/26.

When used with semiconductors such as TiO_2, gold enhances the creation and stabilization of reactive species involved in pollutant degradation. It has been shown that gold nanoparticles can induce increased production of hydroxyl radicals, which are responsible for the photocatalytic degradation of organic contaminants. The ability of gold to catalyze the reaction under visible light conditions is particularly desirable for environmental remediation applications. Recent studies show that gold-based photocatalytic systems are highly efficient, indicating the importance of properly designed photocatalytic materials for effective pollutant degradation. In addition, the incorporation of gold into photocatalytic materials is found to have significant potential in energy functions, such as hydrogen generation through water-splitting reactions. These systems utilize the ability of gold to harvest sunlight efficiently and convert it into chemical energy. Methods such as doping and the creation of heterostructures have been employed to optimize gold-supported photocatalysts for increasing hydrogen generation rates. The stability and non-toxic nature of gold also make it a safer choice compared to other heavy metals posing potential environmental threats. These aspects highlight the importance of gold in accelerating photocatalysis, which drives innovations in environmental remediation and clean energy production, leading to a greener future (Cho et al., 2008; Bacchini et al., 2013) (Figure 9.7).

9.3.1. Water Splitting for Hydrogen Production

Hydrogen production through the water-splitting process is a key focus in renewable energy studies, particularly for its potential to address the world's energy needs. It is the dissociation of a water molecule into hydrogen and oxygen, a process that can be effectively catalyzed by photocatalysts. Some of these include applying titanium dioxide (TiO_2), which is commonly used due to its high photocatalytic properties and stability. The introduction of metal nanoparticles, such as gold, enhances the photocatalytic activity of TiO_2 by improving charge separation, which is necessary for successful chemical reactions. The nuance between the surface properties and electronic features of these composite materials underlies the ultimate achievement of maximizing hydrogen production rates, thereby paving a path to not only sustainable energy application but also the mitigation of environmental implications associated with fossil fuels. New advances in gold nanoparticle combinations with TiO_2 provide more efficient water oxidation photocatalysis. Studies validate that the thickness of TiO_2 film strongly determines photocatalytic activity, with certain thicknesses being optimal for oxygen evolution, which is essential for optimizing hydrogen yields. Specifically, a 200 nm film thickness has performed better with persulfate as an electron acceptor for water oxidation, emphasizing the material's capability to facilitate reactions under visible light illumination. This optimization emphasizes the need to enhance material properties to confer specified photocatalytic effects, a critical element in current efforts to engineer greater practicality and scalability in hydrogen production approaches. Awareness of these parameters can provide the key to more advanced uses in renewable energy technology. In addition to physical properties of photocatalysts, chemical environments and feedstock used for photocatalytic reactions also contribute significantly toward optimizing hydrogen production. Recent research on complex organic compounds such as polyols and cyclic alcohols has unveiled interesting dynamics regarding hydrogen evolution rates. The presence of hydroxyl groups and accessible α-hydrogens in these molecules significantly influences catalytic activity, with every element in the structure contributing to the efficiency of hydrogen generation. Even though such organic substrates are promising for sustainable hydrogen generation, the inherent steric bulkiness of certain sugars limits their use. Exploring the synergy between diverse photocatalyst structures and new feedstock applications holds promising avenues to optimize hydrogen evolution and the advancement of photocatalysis in a sustainable manner, hence consolidating gold's role in this revolutionary technology (Darr et al., 2011; Bahruji et al., 2018) (Figure 9.8).

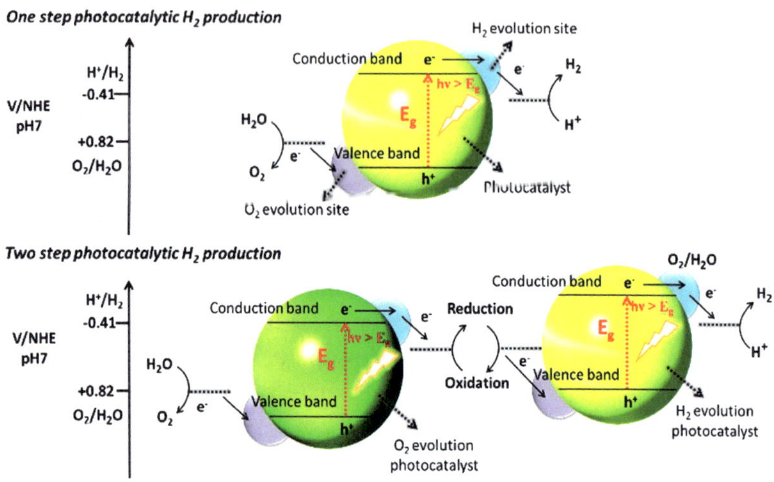

Figure 9.8. Schematic illustration of photocatalytic water splitting via one-step and two-step photoexcitation systems.

Source: https://www.researchgate.net/figure/Schematic-illustration-of-photocatalytic-water-splitting-by-one-step-and-two-step_fig1_343620581.

9.3.2. Degradation of Organic Pollutants

Degradation of organic pollutants has emerged as an important area in environmental science in recent years, thanks to the development of new photocatalytic processes. Gold nanoparticles have entered this domain due to their unique optical and electronic properties, making them excellent catalysts for photocatalysis. By incorporating gold into typical semiconductor materials, researchers have observed an enhancement in the production of reactive species, such as hydroxyl radicals, which are critical for breaking down difficult organic pollutants. The enhancement is important and significant in addressing pollutants and contaminants from agricultural runoff, where excess use of fertilizers and pesticides can lead to enormous contamination. Thus, the utilization of gold in photocatalytic reactions is an encouraging method for organic pollutant remediation, essentially eradicating their detrimental influence on the environment as discussed in recent research studies (Bacchini et al., 2013). In addition, the study of the photodegradation of various organic pollutants under photocatalyst conditions indicates how gold facilitates the process. For instance, PAH experiments demonstrate that the adsorption of polycyclic aromatic hydrocarbons on gold-modified titanium dioxide has a profound effect on photocatalytic activity.

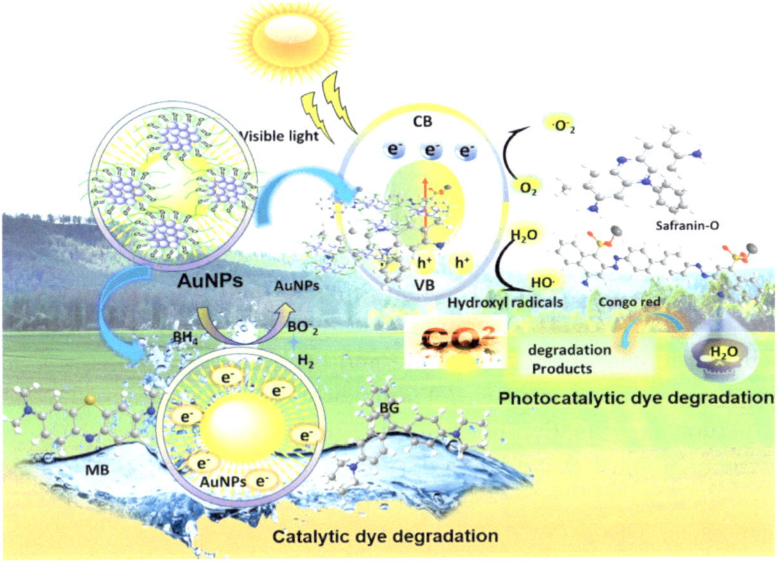

Figure 9.9. Enhanced photocatalytic degradation of organic pollutants by green-synthesized gold nanoparticles using polysaccharides for environmental remediation.

Source: https://ars.els-cdn.com/content/image/1-s2.0-S0141813024026710-ga1_lrg.jpg.

9.3.3. CO_2 Reduction to Valuable Chemicals

Converting carbon dioxide (CO2) into valuable chemicals is a potential way of achieving energy sustainability and environmental cleansing. Photocatalysis, with materials such as gold nanoparticles, has been gaining prominence as a possible method of enhancing the reduction. The unique electronic properties of gold render it a very suitable co-catalyst able to enhance light adsorption and charge separation in semiconductor materials significantly. Such synergy not only enhances the efficiency of conversion processes but also aids in the synthesis of high-value chemicals, ensuring economic viability in renewable energy ventures. Through efficient imitation of natural photosynthetic processes, photocatalytic devices can synthesize organic compounds such as methanol and formic acid, both of which have immense industrial potential and utility in addressing the world's energy crisis. Recent research highlights the role of semiconductor nanowires in enhancing photocatalytic activity for CO_2 reduction. The one-dimensional nanostructures are particularly advantageous because they provide a high surface area and new optical and electronic properties, which facilitate enhanced light harvesting and charge carrier transfer. Their ability to interface with other materials also makes it possible to develop integrated nanosystems that effectively simulate the cycles of natural photosynthesis. As described in recent reviews, exploring the application of semiconductor nanowires for photoelectrochemical half-reactions paves the way for constructing more efficient systems with higher selectivity towards valuable products, such as 2,5-furandicarbaldehyde (FDC), a bio-based material platform chemical (Deng et al., 2019). Such a shift in focus towards nanostructured photocatalysts represents a promising avenue for carbon utilization and green chemical synthesis (Alagöz et al., 2013). The ongoing research in photocatalysis emphasizes the need to optimize catalyst structure and operating conditions to attain the highest CO_2 reduction efficiencies. Particularly, the choice of metal oxides and their structures is critically important in regulating the photocatalytic activity. For instance, laboratory-synthesized TiO_2 nanoparticles show high selectivity towards valuable chemicals compared to their commercial counterparts. This enhancement is because of their primarily amorphous nature, which provides discrete active sites for the catalytic reaction. Moreover, the sol-gel method of preparation applied to these nanoparticles allows for precise control over their physical and chemical properties, enhancing their performance in photon adsorption and subsequent chemical conversion. Such innovations not only guarantee a better conversion rate but also create the development of a circular economy, consistent with a chemical production environment (Alagöz et al., 2013; Deng et al., 2019).

9.4. COMPARISON WITH OTHER PHOTOCATALYSTS

The science of photocatalysis has seen significant development, particularly in the comparison of various photocatalysts used in environmental purification. Of these, titanium dioxide (TiO_2) has been highly popular due to its chemical inertness, non-toxicity, and low cost. However, the only significant disadvantage of TiO_2 is its poor activity under visible light irradiation, which makes it less applicable in diverse environments. Recent advancements have attempted to overcome the drawback by modifications such as noble metal deposition and cationic doping. Both processes attempt to enhance the photocatalytic activity of

TiO_2 under visible light, making it a potential candidate in the field of pollutant degradation applications. Nonetheless, while TiO_2 remains the dominator in photocatalytic studies, researchers are now shifting towards other materials that might offer improved performance. For instance, gold-based photocatalysts use localized surface plasmon resonance to realize enhanced light absorption. Compared to common photocatalysts like TiO2, gold nanoparticles (AuNPs) offer several advantages in photocatalytic reactions, particularly due to their unique optical characteristics. Surface plasmon resonance of AuNPs at the microscopic level facilitates a more effective absorption of light across the visible range, significantly enhancing their photocatalytic activity. Experiments have demonstrated the possibility of combining gold with other semiconductors to synergistically improve overall catalytic activity, thus acting as a better substitute in the degradation of organic pollutants. The tunability of gold in composite formation with materials such as TiO_2 or other metal oxides can significantly enhance the yield of photocatalytic reactions for real-world applications, addressing challenges realized in the review of single-treatment technologies for pollution removal. This tunability in coupling with other materials realizes gold's ability to propel photocatalyst technology beyond its traditional limits. Besides, new integrated photocatalytic processes illustrate the potential of gold photocatalysts over conventional methods. Combining photocatalysis with other related techniques, such as electrohydraulic discharge or sonophotocatalysis, has shown great potential for better removal of persistent pollutants from wastewater. Experiments show that the effectiveness of photochemical degradation is greatly enhanced by integrated systems, indicating that ordinary solo processes may not be sufficient for effective environmental remediation. By introducing gold-based photocatalysts into such systems, researchers can capitalize on the increased light absorption and catalytic properties of gold, thereby optimizing operational efficiencies, reducing costs, and improving the ruggedness of the treatments. With the continued search for effective and sustainable environmental solutions, the use of gold in photocatalysis will only increase further, becoming a key player in future applications that emphasize efficiency and practicality in addressing global pollution problems (Cho et al., 2008; Abou-Ghazala et al., 2014).

9.4.1. Advantages of Gold Over Traditional Photocatalysts

The unique optical properties of gold nanoparticles offer a greater array of advantages over traditional photocatalysts in enhancing photocatalytic reactions. Traditional photocatalysts such as TiO_2 and ZnO are usually limited by aspects relating to light absorption and electron-hole pair recombination. Gold nanoparticles can collect a broader range of light, including visible wavelengths, thus being more active under natural sunlight conditions. Compared to conventional photocatalysts, which typically operate under ultraviolet light, the localized surface plasmon resonance (LSPR) of gold allows the excitation of electrons at lower energies, subsequently improving photocatalytic efficiencies. This property not only enhances the functional utility of gold in various environmental remediation processes but also aligns with the growing calls for sustainable practices in photocatalytic technology. The utility of gold nanoparticles consequently announces a significant possibility for optimizing pollutant degradation in different environments, hence making an effective alternative to traditional photocatalysts (Cui et al., 2024). In addition to that, the stability and long life of gold offer some advantages over conventional photocatalysts. Unlike most traditional catalysts that degrade and lose their potency over time, gold nanoparticles offer significant stability under various environmental conditions. This stability reduces replacement frequency and translates into lower operating costs in real-world scenarios. Furthermore, the incorporation of gold in hybrid photocatalysts is known to facilitate synergic interactions with other compounds to enhance general photocatalytic activity. Such composite systems can achieve greater pollutant degradation rates compared to single-component traditional catalysts. For instance, if gold is blended with host materials like TiO_2, solar charge separation and enhanced electron mobility induce higher photocatalytic activity. Synergistic enhancement combined with increased solar stability thus strengthens gold's leading role in the shifting paradigms of photocatalysis (D'Amato et al., 2018) (Figure 9.10).

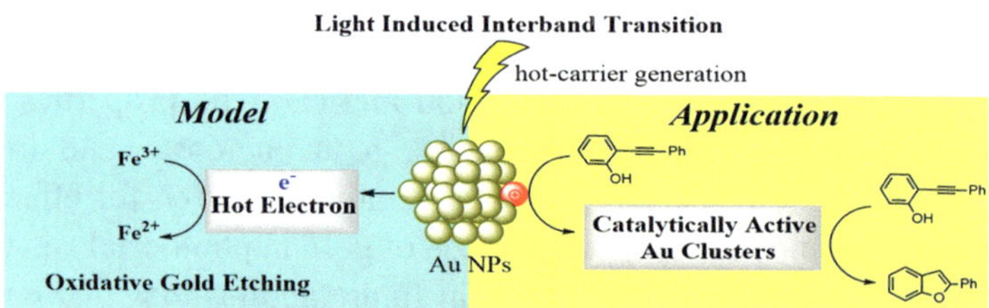

Figure 9.10. Light-driven excitation of gold nanoparticles (GNPs) has emerged as a potential strategy to generate hot carriers for photocatalysis through the excitation of localized surface plasmon resonance (LSPR).

Source: https://pubs.acs.org/cms/10.1021/acscentsci.7b00122/asset/images/medium/oc-2017-001226_0006.gif.

Moreover, the role of gold in photocatalysis extends beyond degradation processes, offering new applications in energy conversion and environmental responsibility. The catalytic property of gold makes it a valuable component in various photocatalytic processes, such as water splitting and CO2 reduction, which are important for sustainable energy production. By mixing gold with traditional photocatalysts or employing it as a single material, researchers have begun to explore novel paths for effective energy generation. Its ability to enable effective chemical reactions without the limitations that plague traditional catalysts attests to the role gold plays in revolutionizing photocatalysis. With the intellectual and industrial focus shifting towards cleaner and more effective technologies, gold's multi-functional advantages lead the way in advancing photocatalytic processes to achieve global sustainability goals (Cui et al., 2024).

9.4.2. Limitations of Gold in Photocatalytic Applications

The application of gold in photocatalysts is of advantages but with inherent drawbacks that must be addressed. The main drawback is the intrinsic low absorption by gold nanoparticles within the visible light range, restricting their activity under solar irradiation conditions. Although gold has excellent catalytic characteristics, its solar energy utilization ability is inherently disadvantaged compared to other materials such as titanium dioxide. This limitation can prevent gold-based photocatalysts from competing with more effective alternatives in actual applications. Furthermore, the stability of gold nanoparticles is also a problem, as agglomeration may occur over time, significantly affecting catalytic efficiency. This is one of the causative factors of poor overall performance in applications in photocatalysts, and additional research is needed to improve gold's optical properties and stability (Braslavsky et al., 2011). Another weakness of gold in photocatalysis is its susceptibility to certain environmental conditions, which influence its catalytic activity. For instance, the presence of impurities or side reactions can be detrimental to gold's capability to catalyze photocatalytic reactions efficiently. Furthermore, the operating conditions, such as pH and temperature, can influence the interaction of gold nanoparticles with reactants as well. Such sensitivity can complicate the design of stable and reproducible photocatalytic systems, possibly resulting in erratic results. As photocatalytic technology is increasingly applied in areas ranging from environmental remediation to renewable energy, it is imperative to recognize and overcome limitations for efficient gold-based catalyst utilization. This points towards the necessity of ongoing research to further explore gold's behavior under various operational scenarios and to explore hybrid systems that may render it more stable and efficient (Alessandri, 2024). Moreover, its economic viability as a photocatalyst is another issue for its widespread application. As a precious metal, the cost factor involved in gold seriously limits its applicability at a large scale in photocatalytic systems. The economic barrier of gold usage is complemented by the relatively low gold concentration, which makes it unsuitable for most applications that require cost-efficient alternatives. Transition metal oxides, being lower in price and higher in concentration, offer comparable photocatalytic activity to that of gold but without the associated economic drawbacks. Such limitations emphasize the need to explore a diversified pool of catalysts that offer efficiency, stability, and affordability to meet the requirements of photocatalytic applications. Therefore, while gold undoubtedly has superior merits, its limitations create challenges that require caution and innovation in photocatalysis (Braslavsky et al., 2011) (Figure 9.11).

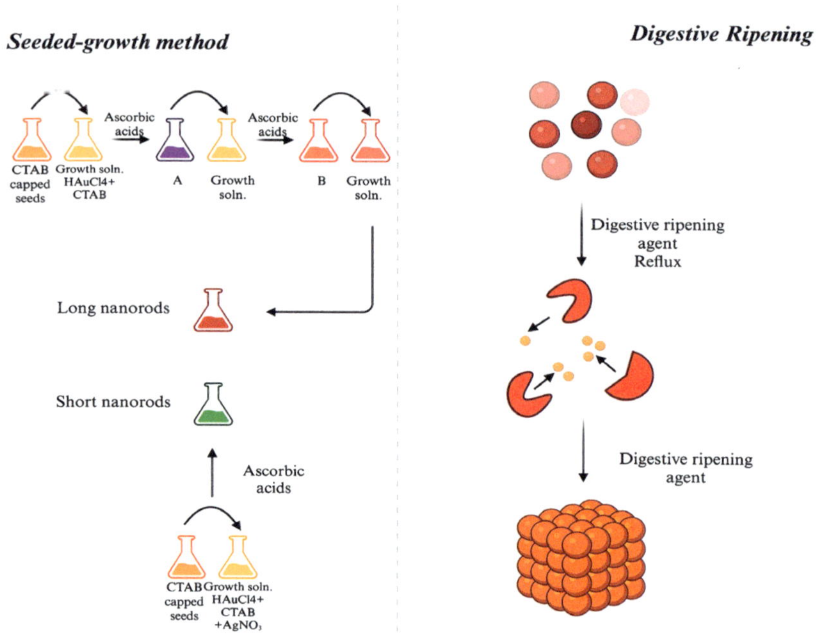

Figure 9.11. Steps in the manufacture of gold nanoparticles (NPs): (A) The seeded-growth method; and (B) the digestive ripening technique.

Source: https://www.mdpi.com/2079-4991/14/22/1805.

9.4.3. Synergistic Effects with Other Materials

The photocatalytic ability of gold is greater when gold is blended with other compounds, leading to a synergistic effect that enhances catalytic activity. Gold nanoparticles (AuNPs), when supported over a substrate like titanium dioxide (TiO_2), exhibit enhanced photocatalytic activity due to greater surface plasmon resonance. This increases light absorption, resulting in more charge carriers available for photocatalytic reactions. Further studies have shown that the physical features of such composites, such as the size and distribution of the metal cluster, play a crucial role in their performance. Specifically, one pre-treatment process involving sulfate has been found to enhance dispersion and metal cluster size over TiO_2, resulting in higher photoactivity in degradation processes such as phenol degradation, which validates the role of synergistic interaction in enhancing photocatalytic systems (Colón et al., 2011). Merging gold with other photocatalytic materials not only augments its inherent advantages but also provides new functionalities. One of the most optimistic ones is the utilization of plasmonic nanoantennas that are hybridized with photocatalysts like ruthenium complexes. With these types of systems, the presence of gold or silver nanoparticles can concentrate light close to the photocatalyst, thereby significantly increasing its activity under visible light. This interaction maximizes the generation of reactive species, which play a crucial role in catalyzing oxidative reactions that degrade pollutants. The principle behind lies in the synergistic influence by the plasmonic nature of the gold nanoparticles and the photocatalytic nature of the complex, leading to significantly enhanced capabilities for bacterial inactivation, which may even be several orders of magnitude better than that of individual components. These intricate interactions reinforce the necessity of research on hybrid systems in photocatalysis (An et al., 2020). These synergy effects in gold-based photocatalysts reinforce the pivotal importance of materials engineering in defining photocatalytic performance. As different materials introduce unique properties in their interactions, scientists are constantly discovering the best combinations that harness these synergies for practical applications. For instance, the combination of AuNPs with metal oxides presents new opportunities in solar energy harvesting, making photocatalytic reactions under sunlight more efficient. The enhancement in activity and stability resulting from the introduction of gold with diverse supports and co-catalysts provides a promising avenue for investigation aimed at solving environmental issues. Such attempts not only provide insight into the fundamental photocatalytic processes but also enable the introduction of novel advanced materials that can transform waste materials into valuable resources through sustainable pathways. Exploration of these synergistic effects is thus an essential frontier in gold photocatalysis research.

9.5. RECENT ADVANCES IN GOLD PHOTOCATALYSIS

The area of photocatalysis has witnessed remarkable growth, particularly with the addition of gold nanoparticles (AuNPs). The improvements have caused an enormous impact on energy conversion and environmental purification processes. Latest studies indicate that gold's unique electronic properties and vast surface area act as a platform to enhance photocatalytic reactions. Apart from this, gold's ability to absorb visible light and facilitate localized surface plasmon resonance has made it a good catalyst in sunlight. This has led people to employ it in many processes, such as the generation of hydrogen and breaking down pollutants. All these findings align with the already established significance of gold in developing sustainable technologies that maximize the use of renewable energy sources, highlighting its pivotal role in today's photocatalytic scene. Current studies on the optimization of gold photocatalysis have explored its combination with other materials to optimize catalytic activity. For instance, studies have focused on optimizing hybrid systems consisting of gold nanoparticles and semiconductor materials with improved charge separation and reduced electron-hole pair recombination rates. This composition design facilitates improved photocatalytic activity, and solar energy is utilized effectively for chemical conversions. The potential application of such hybrid systems in artificial photosynthesis represents a significant step toward simulating natural processes, as gold enables the transformation of light into chemical fuel. With the creation of an interface between gold nanostructures and semiconductor nanowires, researchers are in a position to study the intricate behavior of photoelectrochemical reactions, paving the way for next-generation energy conversion devices that will be able to offer gains to society at large (Deng et al., 2019). Secondly, the use of nanotechnology in agro-environmental applications is also indicative of gold photocatalysis's potential for revolutionizing farming practices. The dual functionality of gold nanomaterials to enhance agricultural production while simultaneously remediating contaminated sites is a significant advancement. As the world faces the growing demand for sustainable agriculture, the use of gold nanoparticles in modern farming systems could reverse environmental degradation caused by excessive chemical use. Research has proven that nanomaterials can improve the effectiveness and safety of pesticides and fertilizers, reducing their harmful effects on soil and water systems. The crux of this emphasis is on the changing potential of gold photocatalysis, not only in energy fields but also in agriculture, with special reference to its ability to improve environmental quality and food security (Bacchini et al., 2013).

9.5.1. Innovations in Gold Nanoparticle Design

The field of photocatalysis has also witnessed remarkable advances in gold nanoparticle engineering, primarily underpinned by their exceptional optical and electronic properties. The synthesis of gold nanoparticles has resulted in particles that exhibit size-dependent behavior, where variations in diameter can significantly affect catalytic efficiency. Scientists have explored strategies such as seed-mediated growth and polyol methods, which allow size and shape to be regulated with high precision. These nanoparticles are especially tailored and have a higher surface area and reactivity, which makes their capacity to catalyze photoreactions at optimal levels.

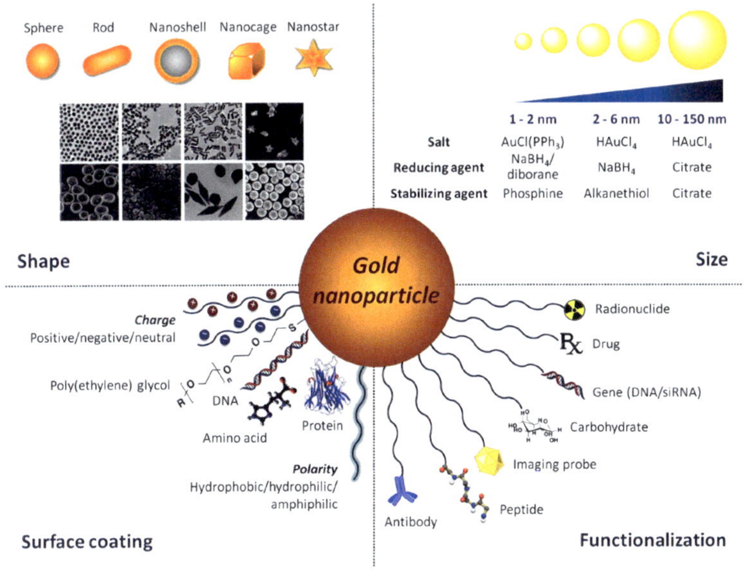

Figure 9.12. The versatility of gold nanoparticles. Gold nanoparticles exhibit tunable sizes and shapes, along with straightforward surface modification, enabling fine-tuning of their properties and functions.

Source: https://www.mdpi.com/1422-0067/24/4/4109.

Functionalization of such nanoparticles with various ligands can enhance their interaction with reactants and light, thereby improving their photocatalytic performance. The synthesis techniques continue to be optimized, and this opens up approaches that minimize environmental footprint while ensuring maximum effectiveness, thus changing the way nanotechnology is being implemented for photocatalytic applications. Other than traditional synthesis methods, the use of hybrid systems has been a breakthrough in gold nanoparticle development. By combining gold nanoparticles with other nanomaterials, researchers have created composite photocatalysts that leverage the synergistic nature of both materials. Synergy tends to produce more efficient charge separation and improved light absorption capabilities, addressing some of the most significant challenges in conventional photocatalysis. For instance, titanium dioxide and gold nanoparticles together have shown an immense potential in accelerating the degradation rate of organic pollutants under UV light. Moreover, hybrid systems can have many metal oxides or carbonaceous compounds as constituents, allowing for extensive applications. This multifaceted application highlights again the flexibility and versatility of gold nanoparticles in assuming the complex task of photocatalysis. Recent developments in the synthesis of gold nanoparticles also underscore the importance of green routes, given growing concerns for the sustainability of chemical reactions. The application of green chemistry principles has motivated studies on biologically inspired and green routes to the synthesis of gold nanoparticles. The utilization of plant extracts or microorganisms, in addition to facilitating the synthesis process, also enhances nanoparticle stability and biocompatibility. These methods have demonstrated potential in synthesizing nanoparticles of defined size and morphology and minimizing toxic byproducts. As research continues to advance, these new green synthesis processes will undoubtedly become predominant in using gold nanoparticles on an expansive scale in photocatalysis, ensuring they positively impact the environment and adhere to current sustainability measures. The continuous exploration of these advancements solidifies gold nanoparticles as an essential component in the future of efficient and sustainable photocatalytic systems (Figure 9.12).

9.5.2. Enhanced Photocatalytic Systems

The improvement of photocatalytic systems has been drastically enhanced by the integration of noble metals like gold into semiconductor matrices. The modern systems benefit from the particular electronic properties of gold nanoparticles, which are capable of charge separation and increasing the photosensitivity of semiconductor materials. Titanium dioxide (TiO_2) remains the most popular photocatalyst due to its stability and large band gap. When mixed with gold, TiO_2 has increased photocatalytic activity, especially in reactions for the degradation of environmental pollutants. The incorporation of gold nanoparticles alters the band structure, enabling the absorption of visible light, which is especially essential for sustainable uses of photocatalysis. As shown by Darr et al. (2011), the nanoparticle composite films not only demonstrate excellent performance in oxygen evolution but also improve the efficiency of organic dye degradation, thus highlighting the significance of gold in photocatalytic innovation. Besides gold, the metal modification of TiO_2 further opens the possibilities for improved photocatalytic systems. A comparative study focused on the photodeposition of noble metals such as platinum and palladium has delineated the significance of metal characteristics on photocatalytic activity. The varied extents of metal dispersion and particle size—based on surface treatment—may significantly influence catalysis performance. Specifically, Colón et al. (2011) indicate that the sulfation process leading to greater dispersion is related to a synergistic effect on photocatalytic activity for phenol degradation. The study highlights that morphology and the interaction of the metal deposits and TiO_2 significantly control the efficiency of the photocatalytic reaction. These multi-metallic designs hold promise for tailoring photocatalytic systems to specific applications, an avenue of research with the potential to influence environmental remediation. Furthermore, scalability and practical applications of enhanced photocatalytic systems are contingent upon effective integration of these noble metals into commercial nanostructured systems. Stability and recovery problems of noble metal-doped photocatalysts remain challenges. Improvements in synthesis methods, such as uniform metal incorporation using sol-gel techniques, play essential roles in maintaining the effectiveness of these photocatalytic systems over extended periods of time. Practical utility and theoretical comprehension need to be balanced in the development of these new materials. Better photocatalytic systems are not only one step towards green chemical processes but also have the potential to solve major environmental issues. By developing and refining these systems through the synergistic combination of metals like gold and their composites, scientists can significantly contribute to the future of photocatalysis and green technologies.

9.5.3. Future Trends in Research and Development

With the development of science and technology in photocatalysis, some major trends are being observed that point towards incorporating nanotechnology into overall applications. Research on nanomaterials has identified their ability to enhance photocatalytic processes significantly, primarily due to their influence on reaction efficiency and selectivity. Among these developments, gold nanoparticles are gaining fame due to their unique electric and optical properties, which are major drivers of light absorption and electron transfer. Control of these nanoparticles at the nanoscale allows for unprecedented property manipulation, and therefore, the significance of knowing how they behave and interact in photocatalytic systems. With increased agricultural and environmental problems, the incorporation of nanotechnology is bound to reverse pollution as well as boost agricultural production, and thus place emphasis on the dual benefit of photonic technological advancements in photocatalytic applications alongside environmental sustainability (Bacchini et al., 2013). Development of semiconductor-metal hybrid nanostructures is a frontrunner area in photocatalytic study, particularly in charge separation and enhancing catalytic activity. Synthesis techniques have been advanced over the past few decades, such that it is now possible to have exact control over their shape, morphology, and size. This not only facilitates tailoring their optical and electrical properties but also provides invaluable information on how these affect the photocatalytic process. For instance, research using cadmium sulfide-gold nanorods has explained the relationship between tip size and photocatalytic function in gold tips, highlighting how these factors are critical to optimizing hydrogen production. Understanding the size-dependent charge transfer dynamics highlights the need for current studies to de-mystify these complex behaviors to inform more efficient photocatalytic applications (Banin et al., 2016). In the future, interdisciplinary research will undoubtedly be a crucial aspect of photocatalytic technology development. By integrating techniques from chemistry, material science, and engineering, researchers are able to devise new solutions that combine theoretical modeling and practical implementation. The ongoing research into hybrid nanostructures, particularly those involving noble metals like gold with semiconductors, presents fertile ground for breakthroughs. Practitioners and researchers must both focus on optimizing interaction between material properties and photocatalytic activity to meet the world community's growing energy and environmental requirements. Thus, an innovative method will transfer lab advances to scalable technology, ensuring that future research aligns well with technological innovation and photocatalytic sustainability objectives (Figure 9.13).

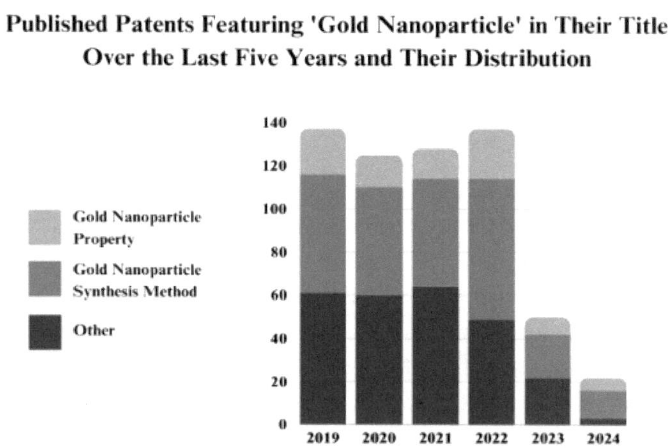

Figure 9.13. Graph representing the number of patents containing keywords related to properties and synthesis methods, in comparison to the total number of patents published from 2019 to 2024.

Source: https://www.mdpi.com/2079-4991/14/22/1805.

SUMMARY

The application of gold in photocatalysis presents a multifaceted approach to enhancing the efficacy of photocatalytic reactions. The unique properties of gold nanoparticles, including their capability to absorb light at specific wavelengths and to provide for the separation of charges, make them invaluable for taking photocatalytic processes to new levels. With continued research, the understanding of how gold can be successfully incorporated into various systems, including titanium dioxide, is constantly evolving. This is demonstrated by recent research showing the ability to immobilize small quantities of materials like RuO2 on titanium dioxide nanorods to improve the evolution of hydrogen from water using visible light. These developments emphasize the importance of hybrid materials in achieving proper functionality in environmental cleanup and energy conversion, realizing the effective fusion of form and function in photocatalytic processes. The functional potential of gold photocatalysts is also enhanced by applying the materials to self-cleaning and pollution clean-up technologies. As discussed in recent reviews, photocatalytic materials have the potential to be central to environmental sustainability while simultaneously serving aesthetic functions. The surge in the popularity of self-cleaning surfaces, inspired by nature, signifies a trend towards novel, multifunctional materials in building design and urban planning. These innovations not only address logistical issues related to maintenance and lifespan but are also integrated into sustainability goals. In addition, the integration of photocatalytic materials into building systems enhances energy demand reduction, which represents a systemic approach to the mitigation of modern issues in energy consumption as well as pollution control. Smooth integration of these technologies has the potential to transform environmental engineering and urban planning paradigms in the coming years. Lastly, the most promising fields of research in future photocatalysis have significant potential for gold and its alloys. The possibility of designing material properties and maximizing photocatalytic performance under different environmental conditions promises to open the gates to more effective solutions. Research on the behavior of hybrid material properties and their interactions at the nanoscale is highly promising for achieving ultimate cutting-edge breakthroughs. For example, the photoresponse between rutile TiO_2 and RuO_2 highlights the roles of surface defects and heterointerfaces in enabling photocatalytic activity at visible light. This reinforces the hypothesis of a strategic combination of nanoparticles and emphasizes the need for continuous research into material developments to meet energy demand. Overall, these trends indicate that continued research into gold in photocatalysis will play an essential role in future environmental solutions and energy technologies, ultimately leading to a cleaner and sustainable planet.

The use of gold in photocatalysis has been well-received due to its unique electronic properties and its ability to enhance photocatalytic reactions. When gold nanoparticles are used as catalysts, they can initiate the process of molecular change by light irradiation, thereby making photocatalysis more efficient. In environmental uses, this is particularly crucial, as it can be employed to break down harmful pollutants and provide improved water quality. Specifically, the inclusion of gold in various semiconductor matrices has been shown to broaden the light absorption window, allowing a wider range of photon energy to be utilized for chemical reactions. This information indicates that gold not only acts as a reaction enhancer but also broadens the window of operation for photocatalytic processes, further accentuating its significance in effectuating sustainable means for environmental detoxification. In addition to improving catalytic performance, operational stability of gold nanoparticles is an indispensable determining factor in their applicability in photocatalysis. Evidence for gold's superior stability arises from its non-leachability, which is otherwise a common detrimental trait related to other metal-based catalysts that can lower efficacy upon prolonged interaction with reactive systems. Research has shown that gold nanoparticles are stable for extended periods, highlighting their potential for long-lived applications such as wastewater treatment and air purification in the future.

Their stability plays a significant role in making the entire photocatalytic process sustainable, minimizing replacement frequency, and highlighting the contribution of gold to enhancing the efficiency and economic viability of photocatalysts for addressing environmental problems. Besides, the technological advances in the synthesis of gold nanoparticles have made their application in photocatalysis, where their shape and size can be controlled, more pronounced. These are crucial parameters because they control the electronic and surface area features of the nanoparticles, directly influencing the catalytic activity. Recent developments in synthesis technologies, including controlled hydrolysis of precursors and the use of stabilizing agents, have yielded high-quality nanoparticles with extremely uniform sizes and high photocatalytic activity. Notably, it has been revealed that highly optimized gold nanoparticles efficiently facilitate charge separation and reduce recombination rates in semiconductor photocatalysts, leading to high efficiency. Their use in photocatalytic systems is one of the developing fields in nanotechnology. Observations on the impact of nanomaterials on agriculture and farming processes demonstrate how they can address challenges of contemporary environmental interest, along with their environmental implications (Alphonse et al., 2010; Bacchini et al., 2013).

REFERENCES

1. Tijani, J. O., Fatoba, O. O., Madzivire, G., & Petrik, L. F. (2014c). A Review of Combined Advanced Oxidation Technologies for the Removal of Organic Pollutants from Water. *Water Air & Soil Pollution*, *225*(9). https://doi.org/10.1007/s11270-014-2102-yAlagöz, O., Augugliaro, V., Loddo, V., Palmisano, L., et al. (2013). Photocatalytic selective oxidation of 5-(hydroxymethyl)-2-furaldehyde to 2,5-furandicarbaldehyde in water by using anatase, rutile and brookite TiO_2 nanoparticles. https://core.ac.uk/download/53281107.pdf (accessed on 28th September 2025).
2. Alessandri, I. (2024). Impact of surface enhanced Raman spectroscopy in catalysis. https://core.ac.uk/download/621420010.pdf (accessed on 28th September 2025).
3. Alphonse, P., Tendero, C., & Varghese, A. M. (2010). Stable hydrosols for TiO_2 coatings. https://core.ac.uk/download/12042515.pdf (accessed on 28th September 2025).
4. An, X., Liu, P., Naowarojna, N., Reinhard, B. M., et al. (2020). Hybrid plasmonic photoreactors as visible light-mediated bactericides. https://open.bu.edu/bitstream/2144/40614/3/manuscript_1021_XA.pdf (accessed on 28th September 2025).
5. Anderson, A. L., Binions, R., Chen, S., Romero, M. J., et al. (2016). Thin films for advanced glazing applications. https://core.ac.uk/download/77042660.pdf (accessed on 28th September 2025).
6. Atwater, H. A., Cheng, W.-H., DuChene, J. S., Li, Y., et al. (2020). Optical excitation of a nanoparticle Cu/p-NiO photocathode improves reaction selectivity for CO_2 reduction in aqueous electrolytes. https://core.ac.uk/download/287633875.pdf (accessed on 28th September 2025).
7. Bacchini, F., Greppi, G., Mura, S., Roggero, P., et al. (2013). Advances of nanotechnology in agro-environmental studies. https://core.ac.uk/download/16749559.pdf (accessed on 28th September 2025).
8. Bahruji, H., Bouleghlimat, E., Bowker, M., Davies, P. R., et al. (2018). Hydrogen generation by photocatalytic reforming of potential biofuels: Polyols, cyclic alcohols and saccharides. https://core.ac.uk/download/151176424.pdf (accessed on 28th September 2025).
9. Bai, J., Fernandez Sanz, J., Fujita, E., Graciani, J., et al. (2016). Visible light-driven H_2 production over highly dispersed Ruthenia on rutile TiO_2 nanorods. https://core.ac.uk/download/81582504.pdf (accessed on 28th September 2025).

10. Barroso Martín, I., Infantes-Molina, A., Moretti, E., Rodriguez-Castellon, E., et al. (2018). Au-Cu/SBA(Ti) based catalysts for photocatalytic applications. https://core.ac.uk/download/214838638.pdf (accessed on 28th September 2025).
11. Benito, I. I., Bini, R., Bolognesi, M., Caporali, S., et al. (2019). A perspective on recent advances in phosphorene functionalization and its application in devices. https://core.ac.uk/download/301577585.pdf (accessed on 28th September 2025).
12. Braslavsky, S., Braun, A., Cassano, A., Emeline, A. V., et al. (2011). Glossary of terms used in photocatalysis and radiation catalysis (IUPAC recommendations 2011). https://core.ac.uk/download/53303225.pdf (accessed on 28th September 2025).
13. Brook, L. A., Ditta, I. B., Evans, P., Foster, H. A., et al. (2008). Biocidal silver and silver/titania composite films grown by chemical vapor deposition. https://core.ac.uk/download/1667180.pdf (accessed on 28th September 2025).
14. Cho, D. L., Kim, J. B., Kim, J. H., Kim, H., et al. (2008). Visible light responsive titanium dioxide (TiO_2). https://opus.lib.uts.edu.au/bitstream/10453/9199/1/2008005324.pdf (accessed on 28th September 2025).
15. Chu, S. W., & Sivan, Y. (2016). Nonlinear plasmonic at high temperatures. http://arxiv.org/abs/1607.03481 (accessed on 28th September 2025).
16. Colón Ibáñez, G., Hidalgo López, M. del C., Maicu, M., Navío Santos, J. A., et al. (2011). Comparative study of the photo deposition of Pt, Au and Pd on pre-sulphated TiO_2 for the photocatalytic decomposition of phenol. https://core.ac.uk/download/187420252.pdf (accessed on 28th September 2025).
17. Cordeiro, D. S., & Corio, P. (2009). Electrochemical and photocatalytic reactions of polycyclic aromatic hydrocarbons investigated by Raman spectroscopy. https://core.ac.uk/download/pdf/37460184.pdf (accessed on 28th September 2025).
18. Corson, E. R., Creel, E. B., Eichhorn, J., Kostecki, R., et al. (2019). Directing selectivity of electrochemical carbon dioxide reduction using plasmonic. https://core.ac.uk/download/323069705.pdf (accessed on 28th September 2025).
19. Cui, J., Luo, S., Meng, G., Wang, Y., et al. (2024). The research progress on photocatalytic materials for pollutant degradation: A review. https://core.ac.uk/download/636476025.pdf (accessed on 28th September 2025).

20. D'Amato, C. A., Di Cicco, A., Giovannetti, R., Gunnella, R., et al. (2018). Band gap implications on nano-TiO_2 surface modification with ascorbic acid for visible light-active polypropylene coated photocatalyst. https://core.ac.uk/download/pdf/162429261.pdf (accessed on 28th September 2025).
21. Darr, J. A., Hyett, G., Kafizas, A., Kundu, J., et al. (2011). An investigation into the effect of thickness of titanium dioxide and gold-silver nanoparticle titanium dioxide composite thin films on photocatalytic activity and photo-induced oxygen production in a sacrificial system. https://core.ac.uk/download/9040748.pdf (accessed on 28th September 2025).
22. Demir, H. V., Govorov, A. O., Gunko, Y. K., Zhang, H., et al. (2014). Photogeneration of hot plasmonic electrons with metal nanocrystals: Quantum description and potential applications. https://core.ac.uk/download/52923658.pdf (accessed on 28th September 2025).
23. Deng, J., Liu, B., Liu, C., Liu, Y., et al. (2019). Nanowire photo electrochemistry. [DOI/link missing]
24. Fan, Y. (2024). Rational design of multifunctional framework materials for synergistic catalysis. https://core.ac.uk/download/614775415.pdf (accessed on 28th September 2025).
25. Khoperia, T. N., & Zedginidze, T. I. (2012). Competitive nanotechnologies for fabrication of thin films and powderlike nanomaterials. https://core.ac.uk/download/20360145.pdf (accessed on 28th September 2025).
26. Parker, D. M. (2016). Photocatalytic degradation of an organophosphorus compound by porous Au- and WO_3-modified TiO_2. https://core.ac.uk/download/215152772.pdf (accessed on 28th September 2025).

CHAPTER 10

APPLICATIONS IN SUSTAINABLE AND GREEN CHEMISTRY

Contents

INTRODUCTION

The field of green and sustainable chemistry has now become a key aspect of contemporary scientific work, aimed at minimizing environmental impact using new chemical processes and materials. As environmental challenges like global warming and depletion of natural resources continue to reach ever-critical levels, the application of green chemistry principles is required for the production of eco-friendly solutions. This area of study not only aims to reduce waste and energy consumption from traditional chemical methods but also demands the creation of non-toxic and degradable compounds. With lifecycle assessments (LCA) and process design for sustainability, researchers face the challenge of rethinking how chemicals are produced and utilized in various industries. The value of integrating sustainability into chemical practice is at the forefront of developments that will facilitate a circular economy mindset and eventually lead to a healthier world for future generations. Key developments indicate the potential of green chemistry, emergent materials, and new green chemistry methodologies. One of them is the modification of pervious concrete by means of applying graphene oxide, thereby improving its functionality in stormwater management while removing potential pollutants. Experiments have revealed that GO-modified fly ash pervious concrete significantly enhances the durability and stormwater infiltration ability of the materials, suggesting that innovative materials are able to address environmental concerns in cities adequately. The use of sophisticated analytical instruments, such as nuclear magnetic resonance and X-ray diffraction, facilitates deeper exploration of interactions between modified materials and the environment. As these developments illustrate, the convergence of sustainable chemistry and material science not only provides practical application but also highlights the importance of technological innovation in combating pressing environmental issues (Shi et al., 2018). Furthermore, the application of life cycle assessment (LCA) for process systems engineering is a significant step towards a more sustainable future for chemical engineering. Current literature indicates that LCA methodology can be effectively utilized to analyze and design chemical processes to optimize their environmental impacts throughout their lifecycle. By analyzing unit processes as whole systems, rather than individual operations, practitioners can identify areas of increased efficiency and lowered waste. As more sectors place greater emphasis on environmental considerations in their planning, LCA tools combined with traditional engineering procedures appear to be a sure bet. Not only does it enhance the potential for sustainable design, but it also integrates environmental performance into process optimization, bridging operations and ecological health (Jacquemin et al., 2011).

Definition of Sustainable and Green Chemistry

Definitions in green and sustainable chemistry serve as guiding principles for designing and applying chemical processes with a lower environmental footprint. Sustainable chemistry emphasizes the need to develop chemical products and processes that are benign to the environment, economically viable, and socially justifiable. It encourages practices that consider not only the technical needs of chemistry but also ecological and ethical factors. This combined approach aims to harmonize the development of technology with the stewardship of the environment, ensuring that chemical practices support the general goals of sustainability. Green chemistry seeks to design chemical processes that reduce or eliminate harmful materials, building on the concept of a circular economy where materials are continuously cycled through systems to maximize utility and sustainability.

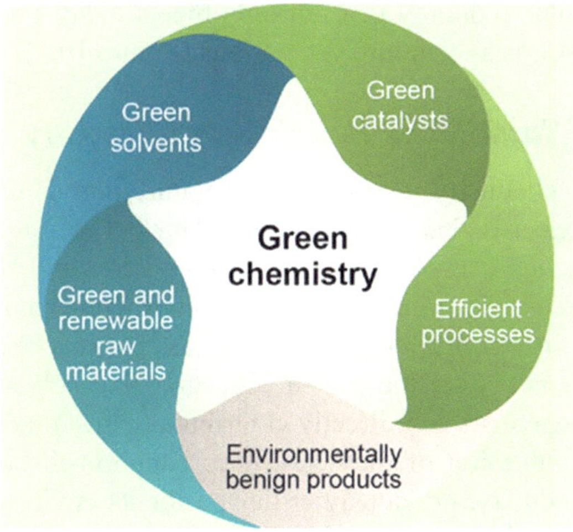

Figure 10.1. Green chemistry.

Source: https://webfiles.amrita.edu/2022/11/green-chemistry-1.jpg.

When examining the intersection of green and sustainable chemistry, one can see how these terms are made operational in practice. For example, industries are embracing the principles of green chemistry to develop processes that minimize waste and energy use, demonstrating a commitment to responsible chemical production. As noted, while the Circular Economy sets forth a core framework for the alignment of economic activities with environmental aims, its implementation is subject to inherent conflicts and limitations. These are potential subordinations of social considerations critical to ethical sustainability. A fresh understanding, therefore, allows for a broadened definition of the Circular Economy that not only acknowledges material cycling but also scales in the quest for optimal human wellbeing and ecosystem integrity. This integral understanding is essential in guiding upcoming trends of chemistry use in green environments. These definitions and paradigms are critical to shaping policies and practices in various sectors, from pharmacy to agriculture. Businesses can evolve to accommodate sustainable and green chemistry concepts to ensure their operations align with global sustainability goals, thereby enhancing their image and bottom line. The relevance of such principles can be demonstrated in the development of biodegradable products, sustainable solvents, and energy-efficient production processes that reduce their environmental footprint. For N/A (2017), innovations in such directions prove that chemistry is capable of supporting the enhancement of global sustainability. By reinventing and redefining traditional practices in green and sustainable chemistry, stakeholders can ensure a circular economy that not only meets today's societal needs but also conserves resources for future generations (Figure 10.1).

Importance of Sustainable Practices in Chemistry

In contemporary chemistry, greening is the unifying theme to addressing environmental concerns that endanger our planet. The integration of green chemistry approaches centers on waste minimization and the utilization of renewable resources, ensuring progress without compromising ecological harmony. By shifting the emphasis towards approaches with minimal environmental impact—like the use of biodegradable solvents or renewable energy sources—chemists can directly contribute to sustainable development. Furthermore, the adoption of these practices supports the public reputation of the chemical industry, previously criticized for its environmental miscues. Reports of greenwashing suggest demands for transparency and accountability within corporations. Brands can no longer afford insincere green washing, as consumers increasingly demand real commitments to sustainable practices that feed into broader environmental principles (Wagner et al., 2014). Combining further momentum with the importance of sustainability in chemistry is the emerging paradigm of the Cradle to Cradle (C2C) design philosophy, which seeks to create a circular economy. The model replaces traditional linear production and waste management methods by ensuring the constant reuse of materials through design that prevents waste from the outset. In practice, this involves designing products that do not just function but can also be fully reclaimed or recycled when their working life ends. Such principles open up new possibilities for pollution and resource use reduction without harming economic growth. The recent vision research identifies current challenges and opportunities to influence the move towards a Cradle to Cradle practice, and it suggests that industry stakeholders must collaborate effectively to trigger change (Dewulf et al., 2010). Ultimately, the role of green practices in chemistry cannot be overstated, as they serve as a springboard for further development in the field. The potential for new solutions to emerge from sustainable uses is vast, from the pharmaceutical sector through to materials science. By emphasizing ecologically friendly alternatives and optimized methods, chemists can contribute to worldwide sustainability, creating a climate that enhances both economic and ecological well-being. The challenge falls to both academia and industrial leaders to develop a next generation of chemists well-equipped with the knowledge and resources necessary to embrace these sustainable ideals. To become greener is, in the end, a promise not just to the art of chemistry but to the health and wellness of the world and its inhabitants.

10.1. PRINCIPLES OF GREEN CHEMISTRY

Green chemistry principles are a guiding philosophy for the development of more sustainable chemical processes and products. Based on 12 guiding principles, the principles encourage the design of chemical substances and their synthesis processes to consider environmental impact as well as efficiency of resources. Of special mention is an emphasis on reducing waste; by addressing waste at the source, chemists can inherently create processes that produce less waste. Not only does this preventive step help reduce environmental pollution, but it also generates significant economic benefits, such as saving and preserving resources. Implementation of such principles in chemical manufacturing is an essential path towards ultimate sustainability, further on the path towards the circular economy and reduction of overall environmental footprint around chemical manufacturing (Singh et al., 2023). Moreover, the use of green chemistry principles extends beyond waste minimization and impacts the lifecycle of chemical products in general. The safer chemicals design principle aims to replace hazardous chemicals with safer alternatives to reduce health risks to employees and consumers. This element forms the basis for more sustainable product design, promoting materials that not only perform their intended function optimally but also harm the environment as little as possible. By emphasizing product longevity and degradability, green chemistry enables the grand objectives of sustainable development and permits materials to be adequately recycled or reused after their useful life (Adeli et al., 2024). Mutual compliance with these principles allows for a shift towards sustainable production practices that are good for human and environmental health. Application of the principles of green chemistry in research and industrial operations must be tackled through a multi-faceted strategy that transcends several challenges and opportunities. The convergence of academia, industry, and governments is required to overcome the barriers to implementation. The integration of Responsible Research and Innovation (RRI) with green chemistry provides a comprehensive framework for evaluating the social, ethical, economic, and political dimensions of sustainability. By adopting responsible road mapping methodologies, scientists can build interdisciplinary agendas that not only focus on green chemistry's technical advancements but also consider their social ramifications. Such a holistic approach optimizes green chemistry's impact not only in developing new solutions but also in significantly contributing to global sustainability goals, while prioritizing ethical issues as a primary concern for the chemistry community (Adeli et al., 2024).

10.1.1. 12 Principles of Green Chemistry

In pursuing sustainable chemistry within the chemical industry, the 12 Principles of Green Chemistry form a fundamental blueprint for emerging methods that aim to reduce environmental impacts. Developed by Anastas and Warner in 1988, the principles aim to reduce the use and creation of hazardous substances and enhance efficiency in chemical processes. They highlight key fields, such as waste reduction, atom economy maximization, and the use of safer solvents, which all address the urgent environmental requirements posed by traditional chemical syntheses. The precepts not only encourage chemists to revisit their practices but also enable the implementation of new approaches consistent with sustainability goals. With the chemical industry increasingly being called to decrease its environmental impact, these guidelines represent a sorely needed shift toward a green orientation, encouraging innovations that concurrently balance economic viability with environmental care. Catalysis, as highlighted in the 12 Principles of Green Chemistry, exemplifies a practical approach in shifting organic synthesis towards sustainability. By employing catalytic instead of stoichiometric reagent equivalents, the chemists can minimize waste and maximize yields. Having an efficient catalyst is the secret; it conserves resources for chemical reactions while maximizing product capacity. Besides, catalysis adheres to the atom economy principle, which guarantees that most reactants remain in the product. This is especially true in drug synthesis, where minimizing waste is critical. Thus, the integration of catalysis into synthetic processes not only illustrates the green chemistry ideology but also maximizes the commercial viability of green processes to a broad spectrum of industries, from pharmaceutical to materials science (Yang et al., 2021). Besides, the combination of pot, atom, and step economies (PASE) has been referred to as a revolutionary approach in optimizing synthetic routes in organic chemistry. This combined approach emphasizes the need for reaction design that minimizes steps and optimizes material efficiency, thus reducing total resource consumption and waste product. The preparation of complex molecules, such as highly substituted tetrahydropyran-4-ones, is an example of how PASE can lead to significant environmental benefit. By quantifying efficiency in the form of widely accepted measurements, chemists can record clear-cut benefits over conventional methods. This not only validates the usefulness of the 12 Principles of Green Chemistry but also suggests a pathway for future research and design. As a guiding principle, embracing PASE underscores the strength of green chemistry in forging new avenues in sustainable chemical practice (Clarke et al., 2007) (Figure 10.2).

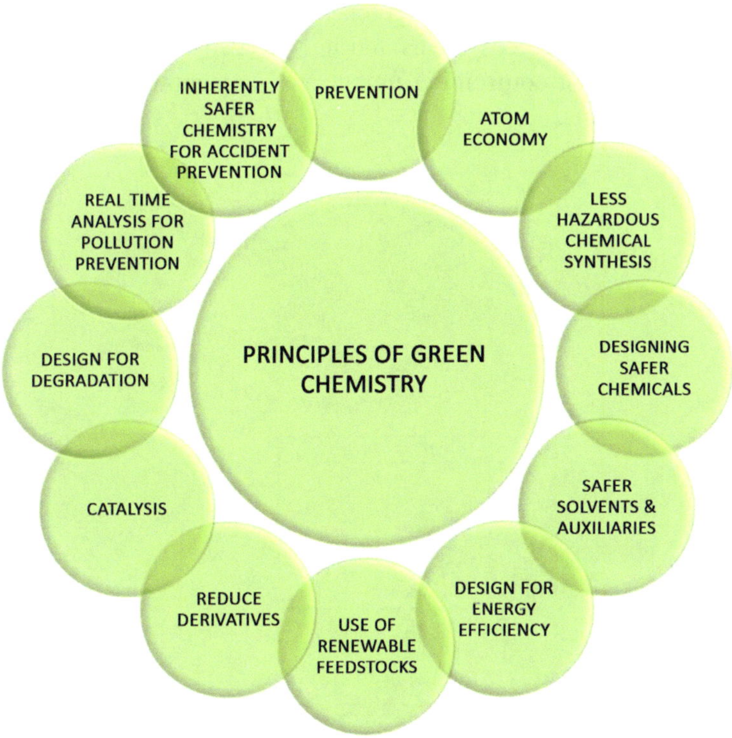

Figure 10.2. Principles of green chemistry.

Source: https://www.researchgate.net/profile/Srishti-Dwivedi-3/publication/358727889/figure/fig2/AS:1125237693399042@1645288980591/Principles-of-green-chemistry-1.png.

10.1.2. Role of Green Chemistry in Reducing Environmental Impact

The role of green chemistry in reducing environmental effects is emphasized by its focus on minimizing waste and maximizing resource utilization in chemical production. By focusing on the design of chemical products and processes that limit the creation of toxic substances, green chemistry lays the ground for sustainable industrial processes. This futuristic approach not only addresses present-day environmental concerns but also shapes the innovation culture for chemical research and development. Concepts like the Circular Economy, which redefine waste as value and promote sustainability at every level of production, have gained popularity under the banner of green chemistry. Haynes et al. (2015) identify the manner in which sustainable practices can revolutionize industries by establishing the necessity of reprocessing and recycling, subsequently contributing to healthier ecosystems and human well-being. Green chemistry plays a critical role in confronting urgent environmental issues, like coastal conservation problems.

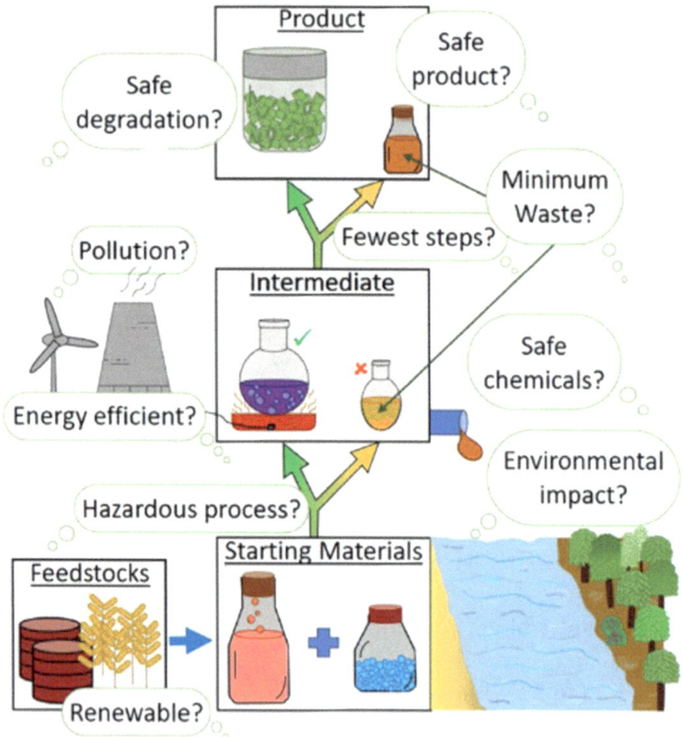

Figure 10.3. When creating a product from raw materials, scientists must ask multiple questions throughout the process to ensure they are following the 12 Principles of Green Chemistry.

Source: https://kids.frontiersin.org/articles/10.3389/frym.2023.1045345.

Recent studies have established 10 Grand Challenges in this area, highlighting the imperative need for sustainable practices in chemical applications. For instance, controlling sea waste and open seafood sourcing can go a long way in reducing environmental footprints from conventional fishing and aquaculture activities. N/A (2015) acknowledges that by embracing the principles of green chemistry, such as developing biodegradable products and eco-friendly fishing gear, the harmful impact on marine ecosystems can be reduced. This reflects how green chemistry not only addresses problems at hand but also equips industries with the strategies they must undergo to adapt to shifting environmental demands. Aside from real-world applications, integrating the values of green chemistry into school curricula and policy recommendations is essential to support a sustainable future. As governments and organizations start to enforce comprehensive environmental policies, the value of green chemistry is further emphasized. Aligning educational programs with the above guiding principles can create a generation of scientists and engineers attuned to green practices. By co-designations of ethical obligation and ecological stewardship, learning institutions can develop a workforce capable of innovating within the paradigms of green chemistry. These programs are consistent with broader goals of the Circular Economy, promoting processes and materials that optimize the functions of ecosystems while meeting the needs of society. This philosophy of inclusiveness ultimately illustrates the potential for change in green chemistry to facilitate sustainable industrial development (Figure 10.3).

10.1.3. Innovations Driven by Green Chemistry Principles

The implementation of green chemistry principles in numerous industries has been the force behind monumental innovations, most notably in the automobile sector. Manufacturers and researchers are using more sustainable methods to develop materials and processes that not only minimize waste but also save energy. For instance, the use of bio-based polymers from renewable feedstocks has become a more viable alternative to the prevailing petroleum-based plastics. These innovations not only reduce reliance on fossil fuels but also enable the circular economy, where materials are reused and recycled. As discussed in research with the motor sector, such innovations highlight the potential of green chemistry to revolutionize product lifecycles and promote sustainability by adopting innovative materials and processes (Maïder et al., 2025). Additionally, green chemistry laws have birthed groundbreaking developments in synthesis and manufacturing processes. Shifting towards using fewer hazardous materials and improving reaction conditions simultaneously not only enhances safety but also the efficiency of chemical reactions. For example, solventless reactions and the use of alternative, non-harmful solvents can successfully reduce the environmental footprint of chemical production. These methodological changes align with the growing concerns about the detrimental effects of traditional chemical approaches on health and the environment. Emphasizing the urgency for process intensification and conservation of resources, these innovations symbolize the broader push toward sustainable approaches in numerous diverse fields (Maïder et al., 2025). Additionally, the role of green chemistry in the development of sustainable energy cannot be overlooked. Technologies such as efficient catalytic converters and biofuels are the best examples of green chemistry practice in limiting greenhouse gas emissions. Transition to new sources of energy requires novel ways of energy capture, storage, and conversion compatible with green stewardship. By imitating nature's processes and materials, green chemistry not only enables energy sustainability but also optimizes energy efficiencies across a range of applications. Thus, the relentless evolution of technology aimed at environmental integrity is a crucial segment of the broader narrative on sustainable development in the frame of green chemistry innovations (Maïder et al., 2025).

10.2. RENEWABLE RESOURCES IN CHEMICAL PROCESSES

The application of renewable sources in chemical conversions is a milestone towards pursuing sustainable and green chemistry. The utilization of renewable feedstocks from biomass, e.g., agricultural residues and by-products, allows the chemical industry to reduce its dependency on fossil fuels. This change not only diminishes the environmental footprint of chemical production but also enhances resource circulation. The transition towards biobased products places more importance on the development of new methodologies, such as solid-state fermentation (SSF), which effectively utilizes waste as raw material for high-added value commodities, including biofuels and biocatalysts. The development towards a bioeconomy is thus not just an environmental requirement; it is also an economic driver for industries that want sustainable development amidst stricter environmental regulations (Cammarota et al., 2017). Merging renewable materials with chemistry requires transforming existing industry processes to update manufacturing models for increased sustainability. In the use of biocatalysis, for example, biological routes facilitate enzymatic conversion of biomass to value-added chemicals and fuels, in preference to cleaner production routes. The bioprocessing not only optimizes the efficiency of the product but also minimizes toxic by-products traditionally associated with traditional chemical production. The apparent environmental benefit of bioproduction demonstrates a strong commitment to the tenets of green chemistry. However, widespread adoption of such methods is far from simple in light of the industry's established conventions. Bridging such a gap requires an infusion of significant research and development inputs, along with an organizational culture that demands renewable resource-based use as the norm rather than an exception (Cammarota et al., 2017). The promotion of renewable resources in chemical reactions enhances not only environmental outcomes but also the potential for innovation in new products across diverse industries.

Utilization of agro-industry by-products has the potential to create a host of sustainable products, from biofuels, pharmaceuticals, and specialty chemicals, hence offering a myriad of environmentally friendly alternatives. Academics and industries have to collaborate in maximizing the potential of renewable resources, such that technological advances contribute towards enabling the transition to a circular economy. Through this collective effort, greater scope for automating sustainable processes opens up, enhancing their efficiency and scalability in chemistry. With this paradigm further expanding, it is essential that stakeholders recognize the synergies among bioprocessing advancements and their significant contribution to transforming the chemical world towards future sustainability (Cammarota et al., 2017) (Figure 10.4).

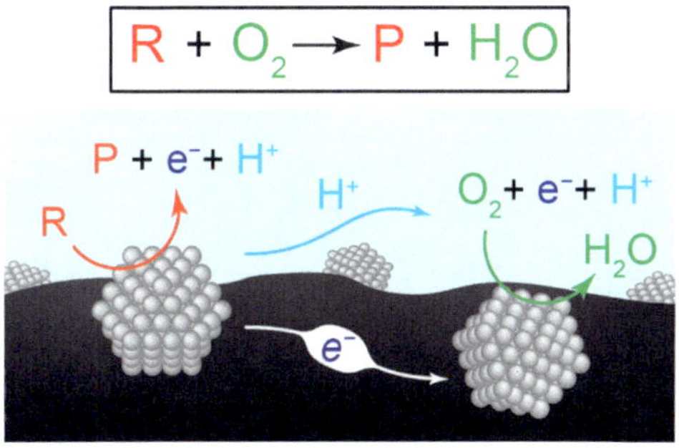

Figure 10.4. Two perspectives on the chemical reactions used to produce renewable fuels and chemicals.

Source: https://energy.mit.edu/news/chemical-reactions-for-the-energy-transition/.

10.2.1. Use of Biomass as a Feedstock

The use of biomass as a feedstock presents a groundbreaking opportunity in the discipline of green and sustainable chemistry. Biomass, consisting of organic compounds of plant and animal origins, is a renewable energy source and a source of chemical intermediates. By utilizing biomass, industries can reduce the use of fossil fuels, thereby improving ecological sustainability. Conversion of biomass into bioproducts and biofuels can occur using various biochemical and thermochemical processes. These not only yield renewable energy sources, but also valuable byproducts that can replace petrochemical-derived materials in numerous applications. Innovation is key in this regard because biomass feedstocks used in industrial chemical processes have the potential to enhance resource use while reducing carbon emissions, directly addressing significant climate challenges. R&D works lie at the heart of the disintegration of problems related to the use of biomass. The roadmap for the bioraffinage strategy is short-term and long-term in approach to incorporate biomass into a bio-based economy. It combines early commercial rollouts with pilot schemes and demonstrations to drive development of a sustainable industry. Also, strategic research aims to develop future-proof methods of biomass conversion and to maximize the economics of bioraffinage processes. By developing scalable technology and maximizing methods for processing biomass, stakeholders can optimize efficiency and broaden applications for biomass-derived products. This is vital for actualizing theoretical principles in practical applications, culminating in an improved chemical landscape (Annevelink et al., 2009) (Figure 10.5).

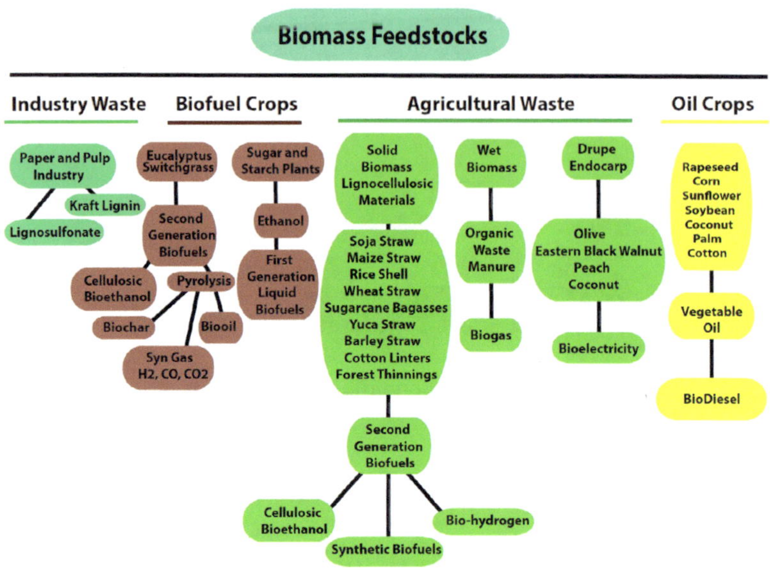

Figure 10.5. Biomass feedstocks and their utilization in the production of biofuels, bioenergy, and bioproducts.

Source: https://www.researchgate.net/figure/Biomass-feedstocks-and-their-utilization-in-the-production-of-biofuels-bioenergy-and_fig3_280494105.

Global energy research also offers significant evidence towards the employment of biomass as a feedstock. Research by institutions such as the International Energy Agency (IEA) outlines the important contribution that bioenergy can make towards achieving energy security and mitigating climate change impacts. Through analysis of the prospects of biomass and biomass conversion technologies, the IEA highlights the necessity of policy measures in stimulating investment in bioenergy opportunities. Such policies can stimulate market growth, encourage research funding, and facilitate cross-sectoral cooperation to promote the use of biomass as a viable feedstock. Scholars and professionals should understand these dynamics because the integration of biomass into sustainable chemistry not only translates into economic benefit but also contributes towards global efforts in managing the transition process towards renewable energy systems (Annevelink et al., 2010).

10.2.2. Advantages of Renewable Resources Over Fossil Fuels

The switch to alternative resources offers numerous advantages over traditional fossil fuels, primarily in terms of environmental sustainability. Alternative resources like wind, solar power, and hydroelectricity emit little to no greenhouse gases when utilized, making their role in climate change a negligible one (Cenex et al., 2010). Alternatively, burning fossil fuels releases a significant amount of carbon dioxide and other toxic materials, which have detrimental effects not only on the atmosphere but also on health and the environment. By utilizing renewable materials, societies can reduce harmful effects and enact the principles of green and sustainable chemistry, aiming to minimize environmental impact while maximizing energy generation efficiency. In addition to their ecological benefits, renewable energy sources enhance energy security and economic stability. As nations invest more in local renewable energy technologies, reliance on foreign fossil fuels decreases, supporting enhanced energy independence (Tani et al., 2018). This shift can generate jobs in new green sectors and stimulate local economies through the establishment of sustainable infrastructure. Moreover, renewable resources are likely to incur lower long-term operating expenses since sun and wind are free commodities that do not incur periodic extraction expenses like fossil fuels. In turn, the move to renewables can not only curb the strain on the environment but also enhance socio-economic resilience. Furthermore, the scalability and diversity of renewable resources present real-world applications across various sectors, from transport to industrial processes. For instance, the integration of hydrogen fuel cells, made possible by the advancement in pure hydrogen technology, illustrates the potential of renewable power to drive the future of clean transport (Cenex et al., 2010). Such a system can be designed to ensure maximum utilization of energy and minimize waste, the principles of green chemistry. Moreover, the bio-based economy, as specified in the current literature, showcases the role of renewable resources in advancing towards a unified approach for sustainability (Tani et al., 2018). By establishing a bio-based economy, we can promote sustainable behavior that realigns consumption and production patterns, thereby generating a sustainable and equitable future (Figure 10.6).

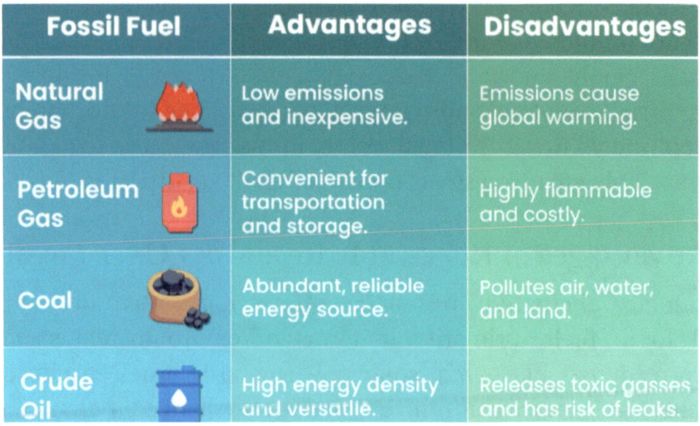

Fossil Fuel	Advantages	Disadvantages
Natural Gas	Low emissions and inexpensive.	Emissions cause global warming.
Petroleum Gas	Convenient for transportation and storage.	Highly flammable and costly.
Coal	Abundant, reliable energy source.	Pollutes air, water, and land.
Crude Oil	High energy density and versatlie.	Releases toxic gasses and has risk of leaks.

Figure 10.6. Advantages and disadvantages of fossil fuels.

Source: https://cdn.educba.com/academy/wp-content/uploads/2024/04/Advantages-and-Disadvantages-of-Fossil-Fuel.png.

10.2.3. Case Studies of Successful Renewable Resource Applications

The transition to sustainable resources is demonstrated in several successful case studies that provide accounts of pioneering applications of sustainable methods within the framework of green chemistry. One such example is the European Community's involvement in developing products based on renewable sources through collaborative projects such as Biostruct, Forbioplast, and Woody. These projects contribute towards valorizing forest products and resources, focusing on the production of bio-based products like polymers and additives as alternatives to conventional petrochemical derivatives. Synergy between these projects resulted in conferences, such as the International Conference on Bio-based Polymers and Composites (BiPoCo, 2012), which facilitated the exchange of state-of-the-art work, enabling researchers in the field to collaborate on resolving challenges and advancing technology. Lastly, such case studies suggest the promise of research-based interventions to create sustainable alternatives and enable a circular economy. Another significant field of renewable resource use in green chemistry is seen through the shift towards biodegradable materials with low environmental impact. Sustainable polymer materials, as described at BiPoCo (2012), have helped pave the way to innovations in packaging, agriculture, and biomedical applications. Using natural fibers and biopolymers, scientists have created composite materials with enhanced properties as environmentally friendly alternatives to traditional plastics. The application of bio-based parts not only contributes to reducing fossil fuel use but also addresses the world's common issue of plastic waste. Conference lectures and debates at forums like BiPoCo present an overview of current advances in this area, identifying feasible applications that solve economic and environmental goals with the potential to revolutionize material science. While case studies involving renewable resources are plentiful, ongoing research from cooperative undertakings reflects the ingenuity and perseverance present in the field of green chemistry. The diverse range of applications showcased at BiPoCo (2012) spanned from innovations in green polymer blends that merge green processes with commercial sustainability. With and interdisciplinary way, these projects not only offset the ill effects of conventional materials but also propel eco-innovation in various industrial sectors. The impacts of these developments extend beyond environmental sustainability, as they also drive economic growth by creating markets and technologies for bio-based products. In essence, these case studies represent a basic analysis of the future, an effort to balance industrial need with ecological wholeness that ultimately results in a sustainable society.

10.3. WASTE MINIMIZATION TECHNIQUES

The implementation of waste minimization techniques lies at the heart of the precepts of green and sustainable chemistry. These techniques need to be applied to reduce the environmental impact of chemical processes and facilitate effective resource utilization. Source reduction is one of the fundamental waste minimization approaches that emphasize altering product and process design to reduce waste generation at its source. By focusing on developing fewer toxic materials and streamlining production processes, industries can significantly reduce the amount of waste produced. Not only does this reduce the environmental load, but it also contributes to economic efficiency since companies can lower disposal costs and better utilize resources. The overall goal is still to create an environmentally friendly model that prioritizes ecological health while remaining stimulating for industrial profit, as elaborated by the holistic green chemistry principles defined by Singh et al. (2023). Recycling and reuse initiatives also further support the waste minimization model of sustainable operations. These initiatives strive to transform wasteful products into valuable resources, thereby inducing a circular economy with minimal reliance on virgin materials. For instance, recycling of solvents used in chemical manufacturing not only conserves waste but also reduces the use of raw materials from other sources, thereby making the entire production cycle more sustainable. Emphasizing the use of such measures can have significant economic and environmental benefits, as companies using recycling measures indicate a level of commitment not only towards ecological sustainability but also towards minimizing operational costs. In addition, successful implementation of such programs in most cases requires collaboration by different stakeholders, i.e., the industry, government, and education sector, as realizing shared interests can double the effectiveness of recycling strategies discussed by Bjelkemyr et al. (2013). In addition to source reduction and recycling, creative product design is also a central part of waste minimization practice. Sustainable product design involves considering end-of-life factors, aimed at maximizing biodegradability or designing for disassembly and recycling. This preventive approach not only addresses waste arising during the manufacturing process but also considers the whole life of the product, ensuring that when it is discarded, it produces minimal waste. Encouraging the use of sustainable materials brings companies closer to eradicating the harmful impacts of product waste. The practical implications of these strategies have been observed in various business segments, where companies implementing these actions experience increased customer satisfaction and loyalty, as well as improved environmental performance. These observations provide evidence of the crucial role of conflating corporate social responsibility with sustainable strategies in minimizing waste, as presented within the broader framework by Singh et al. (2023).

10.3.1. Strategies for Reducing Chemical Waste

Minimizing chemical waste is crucial in developing sustainable and green chemistry, where a concerted approach must be adopted to reduce the environmental impact of industrial activity. A strategy that has been extremely popular of late is the adoption of circular economy principles, which aim to redesign production processes to maximize material reuse and recycling. Through promoting resource-saving designs and creating closed-loop systems, companies can reduce chemical waste while strengthening economic resilience. Their application supports the assertion that environmental well-being is inseparable from economic activity, as demonstrated in the campaign for integrating sustainable development methods into agricultural, industrial, and policy platforms (Haynes et al., 2015). This synergy reinforces the value of chemical waste minimization as an inherent part of a new economic paradigm. Another effective way to minimize chemical waste is by manufacturing and using bio-based alternatives in industrial processes. Bio-refinery technologies that convert renewable biomass to added-value products provide affordable avenues to decrease reliance on petroleum-based chemicals. This strategy not only focuses on minimizing waste but also presents greater economic opportunities, paving the way for a shift to a bio-based economy. For instance, the Netherlands has outlined strategic directions by its roadmap that explore both short- and long-term applications of bio-refinery programs, thereby facilitating a sustainable process of chemical production (Annevelink et al., 2009) (Figure 10.7).

Figure 10.7. Strategies for reducing chemical waste.

Source: https://thirdview.info/wp-content/uploads/2023/03/4._Chemical-waste-can-cause-serious-health-problems-for-humans-and-animals-as-well-as-harm-eco-systems.jpg.

By drawing on advances in this field, industries can not only dramatically minimize their chemical footprints but also contribute to the creation of a sustainable economic framework that respects ecological boundaries. Implementing effective waste management systems within industry environments also poses challenges to traditional linear production models. Informing stakeholders about the benefits of waste reduction and implementing technologies that track and manage chemical inputs and outputs can produce remarkable reductions in waste creation. Incorporating processes such as green solvents and innovative alternative pathways assists in creating a culture of accountability and sustainability within the producers. Having quantifiable parameters to determine chemical waste, combined with incentives for meeting waste-reduction goals, encourages continuous process improvement. Lastly, these steps not only reduce the quantity of hazardous waste but also establish a more responsible approach to managing the environment in chemistry. They demonstrate the capability for change provided by adopting sustainability in many areas.

10.3.2. Importance of Waste Reduction in Industrial Processes

The significance of waste minimization in industrial processes cannot be overstated, as it is central to sustainable manufacturing methods. By minimizing the generation of waste products, industries not only adhere to regulatory requirements but also become more efficient in their operations. Waste minimization measures, such as process improvement and recycling of materials, lead to reduced raw material use as well as energy usage. The implementation of these strategies conforms to green chemistry principles, which encourage the development of inherently safer processes and products with a lower environmental impact. Moreover, companies focused on waste management typically gain through cost reduction, improved product quality, and enhanced reputation, leading to greater competitiveness in the general marketplace.

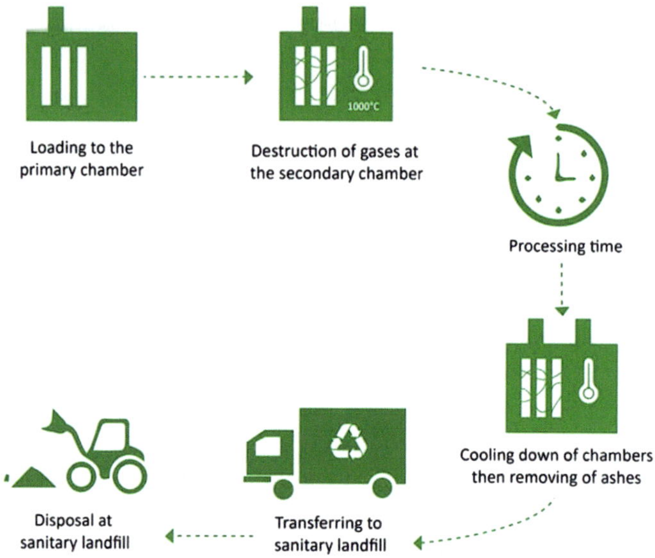

Figure 10.8. Process of industrial waste treatment.

Source: https://www.wastechindia.com/wp-content/uploads/2022/03/WhatsApp-Image-2022-03-31-at-6.24.16-PM-1.jpeg.

Thus, in the case of green and sustainable chemistry, waste minimization is the paramount driver that promotes research and encourages environmental responsibility in the case of industrial processes. Technical advancements in chemical processing have significantly contributed to efforts to minimize waste in industrial settings. For instance, the application of advanced reactors and separation techniques has helped produce high-value products from waste feedstocks. The FAME production methods from waste cooking oil are indicative of this trend, where pressure in the form of reactors, such as cavitational and microwave reactors, is utilized to enhance efficiency and reduce waste production. The procedures not only ease the chemical conversions required—ranging from transesterification to esterification—but also ensure that the choice of catalysts and separations of the product are optimized to reduce waste formation. By using such tailor-made technologies, companies can create a circular economy in their production setups, harmonizing production with eco-friendly processes while simultaneously fulfilling the market's need for greener products (Aubin et al., 2013) (Figure 10.8).

Further, the growing interest in producing bioenergy continues to highlight the benefits of waste reduction in propelling sustainability. Reports by the International Energy Agency (IEA) highlight the use of biomass and biorefinery applications as feedstocks for converting energy from waste, thereby supporting waste-to-energy initiatives and climate change protection measures. By employing agricultural and industrial waste materials, these technologies catalyze the formation of new energy sources, reducing fossil fuel usage and greenhouse gas emissions. The focus on bioraffinage highlights the potential of novel chemistry applications to facilitate waste reduction and create value from materials that would otherwise be wasted (Annevelink et al., 2010). This multi-faceted approach not only increases environmental sustainability but also enhances the robustness of industrial practices in an increasingly resource-constrained world.

10.3.3. Examples of Companies Implementing Waste Minimization

Waste reduction practices are not a passing trend but part of environmentally friendly business processes. Companies across different industries are employing innovative methods to minimize waste, thereby adhering to the principles of green chemistry. For instance, large companies like Unilever have committed to sustainability by optimizing their supply chains to reduce the generation of waste. They are further involved in industrial symbiosis, wherein they exchange by-products with neighboring industries, converting waste into valuable products. This not only saves the environment but also encourages financial savings through reduced disposal costs and raw material usage, showing how waste minimization is environmentally and financially beneficial (Marca et al., 2015). Similarly, global automobile brands like BMW have integrated waste minimization into their organizational system. By adopting circular economy principles, BMW is eager to cover the entire product's life cycle, ranging from design to end-of-life. Some of its practices involve salvaging materials from wrecked cars and using them in production. The systemic practice reduces the environment's imprint and optimizes resource efficiency, further highlighting the company's sustainability drive. By promoting waste reduction through modern recycling processes, BMW not only lowers its carbon footprint but also sets a benchmark for the automobile industry, demonstrating that waste reduction is essential in the quest for sustainable development (Bjelkemyr et al., 2013). Technology giant Apple is a good example too, as it has implemented several measures towards minimizing electronic waste. By producing products that are built to last and be repaired, Apple has reduced the levels of electronic waste. The company has also developed a recycling process using robots called Daisy that dismantle discarded devices to recover valuable components. This effort embodies effective waste reduction and promotes a sustainable product life cycle. Apple's efforts confirm that companies can apply technological innovation to support waste reduction, ultimately leading to economic and environmental rewards. Initiatives for waste minimization not only verify corporate accountability but also promote the creation of sustainable and green chemistry techniques in industries (Marca et al., 2015).

10.4. GREEN CATALYSIS

Green catalysis is one of the pillars of green and sustainable chemistry, promoting chemical reactions with environmentally friendly processes. Not only does this process enhance the reaction efficiency, but it also reduces the environmental impact of traditional catalytic processes. By utilizing renewable feedstocks and producing minimal harmful waste, green catalysts advocate for the values of sustainability. In particular, the employment of biocatalysts, which are natural enzymes themselves, represents a revolutionary shift to more environmentally acceptable processes. Biocatalysis is often carried out under mild conditions, reducing energy needs and enabling one to synthesize complex molecules that are not readily accessible by conventional means. This union of efficiency and ecological consciousness reflects the revolutionary power of green catalysis to reduce the carbon footprint of the chemical industry. Beyond biocatalysis, another essential aspect of green catalysis is the use of metal catalysts that are not only efficient but also environmentally friendly. The employment of less toxic and more available transition metals, such as iron and copper, is an area with great promise for the future of sustainable chemical synthesis. These metals have the potential to catalyze various organic conversions with superior selectivity and yield while providing minimal toxic emissions. The development of methodologies using solvent-free reactions is another example of the quest for sustainability. Such new approaches are called for to reduce dangerous waste in connection with traditional catalytic processes. Green catalysts aid in propelling root adjustments needed for an even more sustainable chemical industry, aligning with global efforts toward sustainability and reduced environmental impact, as reported by Annevelink et al. (2010). Moreover, the advancement in green catalysis is not only of theoretical importance but also has enormous industrial use in sectors such as pharmaceuticals and biofuels. The integration of green catalytic strategies into designing production processes has proven to enhance the efficacy and cost-effectiveness of producing essential chemical intermediates. For instance, the application of green catalysis in biofuel production enhances the sustainability of energy generation by converting biomass into fuels using efficient catalytic pathways. Such processes enable us to reduce carbon footprints through the use of fossil fuels, all in a bid to achieve energy security and sustainability. As industries adopt these greener approaches on a larger scale, they enhance overall environmental goals while remaining economically feasible. The blending of industrial growth with sustainability reflects the growing realization of green catalysis as essential for achieving sustainable processes, supported by evidence introduced by Cukalovic et al. (2009).

10.4.1. Definition and Importance of Catalysis in Chemistry

Catalysis, a central concept in chemistry, is the acceleration of a chemical reaction through a material known as a catalyst. The catalyst does not get consumed in the reaction, thus making it reusable. Catalysis' significance stems from the fact that it lowers the activation energy of reactions and hence accelerates them without altering the overall thermodynamics. This feature makes catalysts essential for industrial and lab use, where efficiency and cost are of extreme importance. Catalysis is particularly crucial in green and sustainable chemistry since it enables reactions to be performed under mild conditions, reduces energy consumption, and minimizes waste. Research has emphasized the growing requirement for new catalytic processes in terms of environmental sustainability goals, highlighting the significance of catalysis going beyond its conventional applications in chemical production N/A (2017) and Aznar et al. (2018). Moreover, the importance of catalysis is also reinforced through its extensive application across a wide range of fields, from environmental decontamination to renewable energy harvesting. Catalysts enable the transformation of poisonous pollutants into harmless substances, hence serving a vital purpose in the fight against environmental problems. For instance, catalytic converters used in motor vehicles play a significant role in reducing poisonous emissions, depicting the use of catalysis in air purification. In addition, in renewable energy, catalysts serve a vital role in reactions such as water splitting and biomass conversion, which provide prospects for renewable energy sources (Figure 10.9).

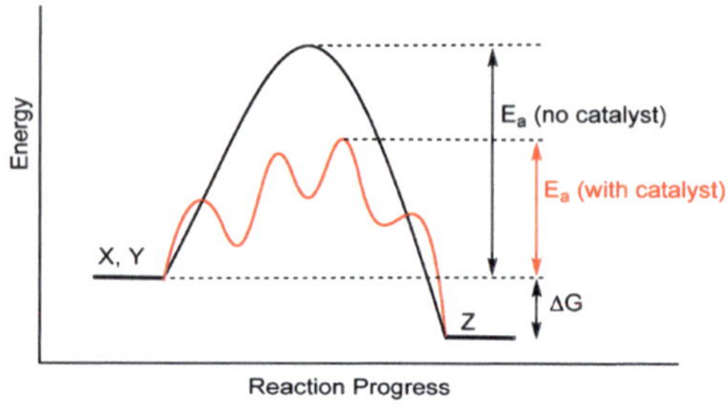

Figure 10.9. The effect of a catalyst.

Source: https://library.fiveable.me/inorganic-chemistry-ii/unit-10/principles-catalysis/study-guide/DUerBtCT9YhqpvA8.

Recent bibliometry describes a rising tendency toward sustainable catalysis, by a new paradigm of more efficient and environmentally friendly chemical processes (Aznar et al., 2018). On this basis, catalysis is thus a cornerstone of modern chemistry, confirming its role in the transition towards sustainable processes. The future prospects for catalysis in green and sustainable chemistry are very bright, ranging from traditional methods to sophisticated nanotechnologies and biomimetic approaches. Scientists are becoming increasingly interested in developing catalysts that not only work well but are also green, such as seeking new materials and reaction paths that generate little harmful byproduct. Interdisciplinary research methods must be combined to develop knowledge within this branch of science, with the work of chemists, engineers, and environmental scientists encouraging in-depth examinations leading to innovative breakthroughs. The exponential growth of research articles, as embodied in the present study, is a concurrent recognition of the key role of catalytic innovation in achieving sustainability, as well as environmental, economic, and social benefits (Aznar et al., 2018). Catalysis is thus at the forefront of the avenues for transforming chemistry to meet the demands of sustainable development.

10.4.2. Types of Green Catalysts

In the context of sustainable chemistry, green catalysts are key to minimizing environmental impact while maximizing efficiency in chemical reactions. Among the various green catalysts, heterogeneous catalysts have attracted considerable attention because they can catalyze reactions without being soluble within the reaction medium. This allows for a much easier separation and recovery upon completion of the reaction, as well as a reduction in the inevitable secondary risk of pollution. For instance, in biodiesel production, heterogeneous catalysts significantly enhance the transesterification process, changing waste cooking oil to fatty acid methyl esters (FAME) by optimizing the reaction conditions. As recognized in recent studies, novel equipment integrating these catalysts, such as microstructured reactors and reactive distillation columns, can improve and streamline the biodiesel production process, indicating the potential for heterogeneous catalysts to contribute to both sustainability and efficiency in green chemistry (Aubin et al., 2013). A further promising category of green catalysts includes the layered double hydroxides (LDHs), a class of materials that have proven to be highly diverse in catalysis owing to their structure tunability and high surface areas. Of particular note, cerium-based LDHs have demonstrated vast potential in numerous chemical conversions, including lignin-derived compound hydrogenolysis.

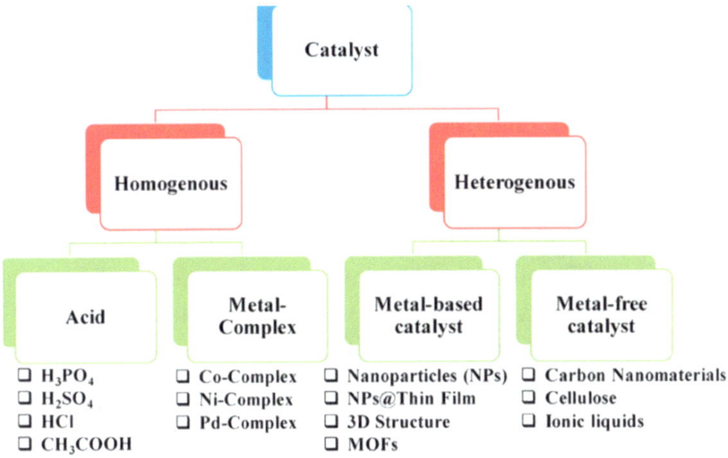

Figure 10.10. Types of catalysts.

Source: https://external-content.duckduckgo.com/iu/?u=https%3A%2F%2Fwww.researchgate.net%2Fprofile%2FHani-Abdelhamid%2Fpublication%2F346884387%2Ffigure%2Ffig3%2FAS%3A972957757231105%401608982614983%2Fe-Types-of-catalysts-used-for-the-hydrogen-generation-using-the-hydrolysis-of-NaBH-4.jpg&f=1&nofb=1&ipt=8d4dfd65e04fff3bfeabe4587ff15db0e00dde3faf27872f0a601406ffc99d85.

The research has revealed that loading cerium layered double hydroxides with transition metals like palladium and nickel provides more catalytic activity compared to traditional catalysts like ceria (CeO_2). This is attributed to the unique compositional and morphological characteristics of the LDH structure, which provides optimal positions for the reaction while reducing unwanted side products. These advancements highlight the importance of LDH-based catalysts in facilitating green chemical reactions, particularly biomass valorization (Bruneel et al., 2020). Another crucial category of green catalysts includes biocatalysts, which show the biochemical efficacies of nature's processes for industrial applications. Enzyme and whole-cell catalysts are used in various reactions, ranging from biosynthesis to degradation, by capitalizing on their selectivity and mild operating conditions. The application of biocatalysts not only reduces energy demand and waste generation but also eliminates the toxic by-products that are traditionally typical of conventional chemical catalysts. By way of example, enzymatic reactions tend to be operated at ambient temperature and pressure, significantly reducing the carbon footprint of chemical manufacture. In conjunction with emerging biotechnology, the synthesis of genetically modified enzymes is imminent to speed and maximize biocatalysis efficiency and selectivity, and further establish these green catalysts in the perspective of sustainable chemistry procedures. With continued innovation and discovery, biocatalysts are poised to be central to future sustainable chemical processes, consistent with the overall global push toward greener industrial processes (Figure 10.10).

10.4.3. Impact of Green Catalysis on Reaction Efficiency

Green catalysis significantly enhances the reaction efficiency while minimizing the environmental impacts in chemical reactions. Traditional catalysts typically incorporate heavy metals and synthetic chemicals that are toxic to human health and the environment. Green catalysis instead aims at exploiting natural resources as well as gentle reaction conditions to catalyze chemical transformations. For instance, the application of nanocellulose as a catalyst support defines this shift towards sustainable methods. Nanocellulose is a renewable, inexpensive, and biodegradable material, endowed with desirable characteristics that enable intensified reaction rates and selectivity, thus enhancing the overall process. As the principles of green chemistry are being adopted in companies, the use of sustainable catalysts is gaining prominence not only to achieve higher yields but also to comply with stringent regulatory requirements that ensure the destruction of chemical waste. Another key aspect of green catalysis is its efficiency in exploiting alternative sources of energy, thus improving reaction conditions and energy conservation. Traditional catalytic reactions commonly involve high energy input and therefore add to the carbon footprint of chemical production as a whole. By employing the likes of microwave-assisted catalysis or photocatalysis, reaction conditions can be set at lower temperatures and shorter reaction times, ultimately leading to more efficient routes. These unconventional energy inputs not only enhance reaction efficacy but also align with the overall aims of green chemistry, which encourages reducing energy dependence on fossil fuels. The development and deployment of green catalysts, particularly those exploiting nanomaterials, are also indicative of the rapid development of catalytic science towards enhanced sustainability, enabled by the need to address environmental challenges in an effective way. The commercialization of green catalysts highlights their role in enhancing the efficiency of reactions in sustainable chemistry. Nanocellulose-derived material fabrications described in current research illustrate how such materials can be designed for specific catalytic functions, thus improving their applications in industry. These catalysts not only satisfy the criteria of renewability and effectiveness but also facilitate the synthesis of functional material with applications from drug delivery to energy storage and environmental remediation. Phosphorylated nanocellulose, for example, is a multiregulatory candidate that enhances the effectiveness of many catalytic reactions while encouraging ecological wholeness. This shift toward the utilization of such cutting-edge catalysts is a step forward in green chemistry, which balances efficiency with environmental footprint minimization, as required by the times for sustainable production processes (Thomas et al., 2018; Trache et al., 2020).

10.5. SUSTAINABLE SOLVENTS

Increasing realization of the environmental impact of traditional solvents has spurred a massive trend toward eco-friendly alternatives in green chemistry. Sustainable solvents based on renewable resources are an essential factor in reducing the environmental impact of chemical processes. Their application not only minimizes reliance on petrochemicals but also supports the green chemistry philosophy in waste prevention, optimal synthesis, and the use of safe, non-toxic materials. Industries can enhance their sustainability without compromising performance by switching to solvents such as bio-based ethanol, propylene glycol, and ionic liquids. The implementation of sustainable solvents across various industries, from pharmaceuticals to agrochemicals, mirrors their uses and ability to reform conventional practices, addressing both environmental challenges and economic limitations. The transition to sustainable solvents is not straightforward; it requires joint effort in multiple disciplines and areas of use. Adopting new solvents means not only changing the material but also altering processes and procedures in an industrial context. Compatibility of the green solvent with the current systems, regulations, and economic cost of such a transition must be carefully weighed. Additionally, the risk of "greenwashing"—when companies may falsely market the use of sustainable methods—makes all such genuine efforts at risk of being defaced. The stakeholders, including corporations and regulators, must conduct rigorous tests to ensure that green solvent uptake is not a marketing ploy but leads to real environmental and public health gains, as highlighted by ongoing criticisms in the industry (Wagner et al., 2014). In the future, the challenge is to create a general strategy that will allow the large-scale adoption of sustainable solvents through collaborative research, development, and policy initiatives. Innovative approaches, like those initiated in the Cradle to Cradle strategy, can help reframe product life cycles to promote systems that are regenerative and restorative. This comprehensive strategy aligns with sustainable solvent uptake, focusing particularly on a multi-dimensional solution centered on governance, industrial practice, and public participation. By establishing an enabling atmosphere for the effective utilization of sustainable solvents, both the economic viability and the environmental benefits of green chemistry can be realized. Only through comprehension of the interrelation of these factors can the barriers encountered be overcome, and the path to a more sustainable chemistry future be pursued (Dewulf et al., 2010) (Figure 10.11).

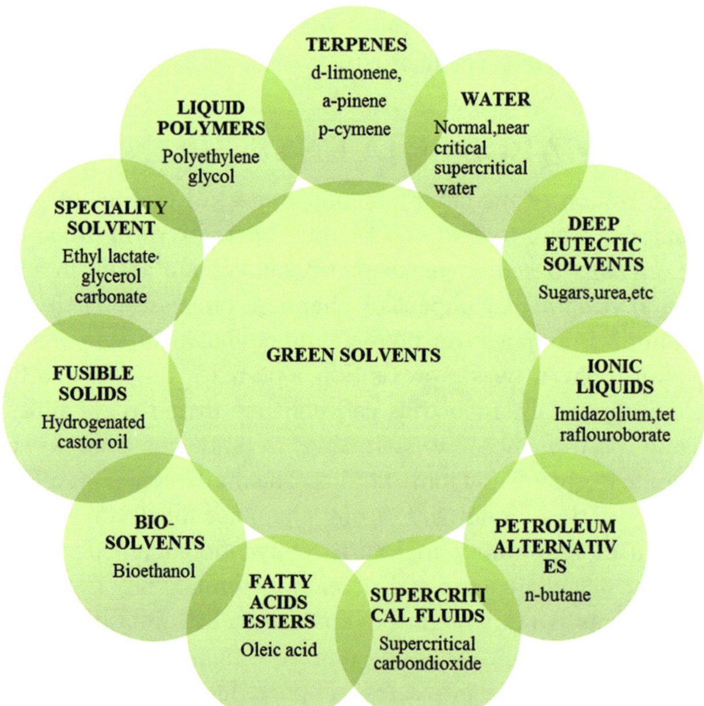

Figure 10.11. Different types of green solvents.

Source: https://www.researchgate.net/profile/Srishti-Dwivedi-3/publication/ 358727889/figure/fig3/AS:1125237693390850@1645288980624/Different-types-of-green-solvents.png.

10.5.1. Characteristics of Green Solvents

The characteristics of green solvents are central to their role in promoting eco-friendly policies in the field of chemistry. To begin with, a green solvent must not cause much damage to the environment and should have low toxicity and biodegradability. The solvent must not only create an effective medium for the reaction but also work to reduce the production of toxic waste. For example, the utilization of solvents derived from renewable biomass can significantly reduce petrochemical dependence and thereby alleviate fossil fuel depletion. Prioritizing solvents that show chemical stability and low vapor pressure further enhances the process safety and overall operating efficiency. Compliance with these demands makes green solvents capable of being integrated in any form of chemical applications with ease, eventually contributing to the transition towards an environmentally friendly approach in chemical manufacturing and industrial processes (Maïder et al., 2025). In addition to their reduced environmental footprint, green solvents are recognized for their flexibility in various applications. The types of operations undertaken by these solvents include activities in reaction media, extraction processes, and formulation blends, making them an essential constituent in various industries such as pharmaceutical and agrochemical. Due to their innocuous nature, the majority of green solvents can be employed in sensitive situations such as food and cosmetic manufacture. This versatility facilitates the creation of more efficient, safer chemical processes with a commitment to green chemistry principles. Additionally, research and use of solvent selection guides simplify the assessment and use of these solvents. By clearly delineating solvent use potential risk and environmental impact, these guides facilitate solvent selection decision-making with intelligence, elevating the spirit of green chemistry (Gissi et al., 2016). One of the most significant aspects of green solvent development is the regulatory settings and guidelines that drive it. Driven by legislation and shifting environmental attitudes, the development of green solvents for extractions, separations, formulations, and reaction chemistry has become increasingly vital as a science. The characterization of the green solvents relies on factors that gauge their toxicity, sustainability, and effectiveness, which are paramount in meeting stricter environmental regulations. As industries transition to meet these regulations, the demand for green solvents continues to grow, prompting innovative research into new formulations. The developments serve not only to satisfy regulatory demands but also towards global sustainability goals in chemical manufacturing. In general, the use of green solvents reflects a continued commitment to environmental chemistry, which can result in more sustainable industry practices (Gissi et al., 2016) (Figure 10.12).

Figure 10.12. Key properties of the green solvents.

Source: https://encyclopedia.pub/entry/48075.

10.5.2. Comparison of Traditional versus Sustainable Solvents

The application of traditional solvents in chemical operations has led to significant environmental and health problems. Such traditional solvents, typically petrochemical-based, are unsafe due to their hazardous nature, including toxicity, volatility, and potential as environmental pollutants. Industries have thus been increasingly pressured to re-engineer their processes for minimizing reliance on these usual compounds. The synthesis of 2D transition metal dichalcogenides (2D TMDs) exemplifies this trend, where traditional solvents like N-methyl-2-pyrrolidone and dimethylformamide have dominated due to their effective exfoliation power. The implications of employing such solvents have become increasingly unsustainable, necessitating green alternatives that do not compromise on performance but set new standards for sustainability. Greener methods are essential for materials science innovations that have no adverse ecological impacts, aligning with the general objectives of sustainable chemistry as outlined in research work toward greener approaches (Abduvakhidov et al., 2023). Sustainable solvent development is a turning point in chemical practice. Natural Deep Eutectic Solvents (NADES), for example, have been extensively seen as promising alternatives to traditional solvents. These solvents encompass combinations of naturally occurring compounds that impart efficient extraction characteristics, environmental compatibility, and lower toxicity. In crop farming environments, NADES have been successful extractants of bioactive plant compounds with minimal harmful use of pesticides. This is particularly observed in the study of *L. cuneifolia* extracts mixed with NADES, such as lactic acid and glucose, which exhibited high antifungal activity against *Botrytis cinerea*, the pathogen causing gray mold. The application of plant-based solvents not only enhances food safety but also contributes to the principles of green chemistry by linking extraction methodologies with eco-friendly approaches (Boiteux et al., 2019). The use of such solvents makes new applications possible while not sacrificing ecological wholesomeness and human health. Finally, the comparative assessment of traditional and sustainable solvents presents the need for a paradigm shift in the practice of chemistry. Traditional solvents, while effective, are becoming ever more incompatible with the principles of sustainability guiding current research and industry efforts. Sustainable alternatives such as NADES and biodegradable solvents demonstrate the ability of chemical reactions to be redirected towards greener methods without compromising effectiveness. Their application not only addresses concerns related to harmful substances but also promotes greater implementation of environmentally friendly practices across various sectors. With scientists continually opening the door to sustainable solvents, the compatibility of chemical innovation and ecological conservation becomes the basis of green chemistry, which consequently results in sustainable developments in different sectors (Boiteux et al., 2019; Abduvakhidov et al., 2023).

10.5.3. Applications of Sustainable Solvents in Industry

The industrial world is increasingly recognizing the potential of sustainable solvents to minimize environmental impact. The traditional solvents, typically petrochemical-derived, are highly prone to volatile organic compounds (VOCs) that contribute to air pollution and health risks. Sustainable solvents such as bio-based solvents derived from renewable resources, such as plant biomass, are a cleaner and healthier choice. Pharmaceutical and coating industries have begun using these solvents in their operations, decreasing toxicity and carbon footprint. Adopting green solvents, companies are not only complying with environmental regulations but also becoming more attractive as socially responsible businesses. This shift is significant, particularly given the skepticism directed at corporate sustainability initiatives in literature, which suggests that corporations may lean towards greenwashing as a response to external pressure (Wagner et al., 2014).

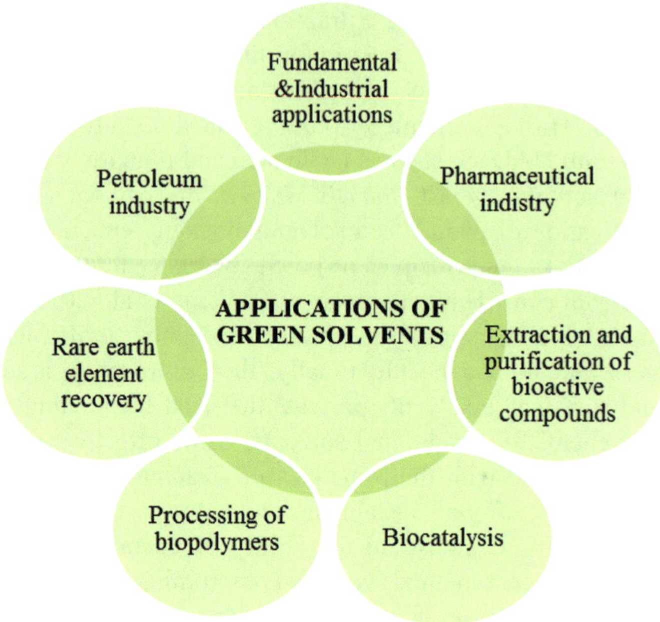

Figure 10.13. Applications of green solvents.

Source: https://www.researchgate.net/profile/Srishti-Dwivedi-3/publication/358727889/figure/fig4/AS:1125237693390851@1645288980654/Applications-of-green-solvents.png.

The use of sustainable solvents transcends environmental control; it is a game-changing step throughout the lifecycle of chemical manufacture. Current research indicates that developments in solvent technology can significantly enhance the eco-effectiveness and eco-efficiency of industrial processes (Huber, 2025). Supercritical carbon dioxide solvent extraction technologies have been identified as effective processes with reduced waste and energy consumption. Such emerging technologies not only increase the yields of products but also reduce the undesirable ecological effects associated with traditional solvents. Therefore, the transition to green solvents creates a ripple in the industrial landscape, spurring innovations that optimize greater ecological consistency along production lines. It makes industries focus on optimizing the upstream stages of product development, where the most significant scope for waste reduction and efficiency improvement resides. Besides, the push towards the use of sustainable solvents in traditional industrial processes is spurred by advances in chemical technology and research. As industries begin to embrace the advantages of green solvents, public-private partnerships and academia collaborations are becoming ever more significant. The collaborations not only foster the development of new materials but also the exchange of knowledge and best practices across industries. Conclusion based on a careful analysis of environmental innovations indicates that technological changes based on sustainability are more effective when initiated early in the product lifecycle (Huber, 2025). Accordingly, policies that encourage innovation and research and development of sustainable options are crucial for long-term environmental improvement. By prioritizing these developments, businesses can ensure that their processes align with initiatives to reverse ecological degradation and move towards sustainable economic growth (Figure 10.13).

SUMMARY

In the rapidly changing field of green and sustainable chemistry, the importance of integrating sustainability into chemical processes cannot be overstated. The potential to reduce environmental impact while supporting economic growth has convinced the scientific community to take the lead in visionary action. Ultimately, sustainable chemistry applications play a key role in creating materials and products that not only serve human needs but also preserve ecological balance. This holistic approach is best illustrated by a series of attempts to transition to greener methods, exemplifying how innovation can align with environmental ethics to lead to a more sustainable future. Furthermore, the business and activities of sustainable and green chemistry extend beyond ecological benefits; they also encompass economic viability and social stewardship. The development of biorefining strategies, as provided in contemporary bio-based economy roadmaps, is a classic demonstration of how sustainable practices can propel economic growth. They not only enhance the effectiveness of resources but also support circular economies by maximizing biomass utilization and minimizing waste. An effective implementation of these strategies can propel the transition towards a more sustainable chemical industry, ultimately benefiting society as a whole by creating jobs while minimizing the footprint on the environment (Annevelink et al., 2009). Overall, addressing the pressing challenges of environmental deterioration requires unprecedented cooperation in the realms of science, policy, and industry. Green sustainable chemistry is central to this endeavor, a principle guiding appropriate solutions that preserve equilibrium with nature. In the march toward sustainable approaches, continued interaction among researchers, policymakers, and industry participants will prove crucial in dismantling innovation barriers. With consistent hard work and targeted investments in green chemistry, we can envision a world where sustainability is not only a goal but a second nature of the chemical world (Buckley et al., 2014).

The emergence of green and sustainable chemistry highlights key points that emphasize its relevance in contemporary applications. One of these is bioenergy, specifically through the generation and consumption of biomass. As elaborated in the report of the International Energy Agency (IEA), the diversified attempts to harness bioenergy provide energy systems and ecosystem sustainability with short- and long-term benefits. By integrating renewable biomaterials into existing paradigms, we can produce a cleaner, more sustainable energy landscape. This not only alleviates present environmental issues but also opens up opportunities for future energy efficiency and resource innovation, towards a more sustainable economy. Another compelling argument placed within the debate revolves around bioraffinage, a cornerstone to the bio-based economy. According to the stipulated Roadmap Bioraffinage, the various proposed avenues highlight how short- and medium-term approaches can complement long-term research-driven strategies. The roadmap underscores the importance of demonstration plants and pilot projects as test sites for new chemical processes, which can assist researchers and industry stakeholders in minimizing the adverse effects on the environment. All of these endeavors can collectively lead to significant advances in sustainable practice, resulting in improved resource utilization efficiency and chemical process waste minimization. Finally, the methodology of Moonshots in bioraffinage strategies encapsulates the visionary high aims of sustainable chemistry advancement as defined in the roadmap outlined. These Moonshots are innovative directions with high potential for economic impact, representing a revolutionary approach to addressing global challenges such as climate change and resource depletion. By focusing on revolutionary solutions, stakeholders in the green chemistry industry can concentrate on high-risk, high-reward projects that push the boundaries of current technological capabilities. The combination of such strategies aligns with the primary objective of creating a resilient, sustainable economy that balances economic and environmental needs. This approach encapsulates the key issues to be addressed in applications of sustainable and green chemistry.

REFERENCES

1. Adam, J., Singh, M., Abduvakhidov, A., Del Sorbo, M. R., Feoli, C., Hussain, F., Kaur, J., Mirabella, A., Rossi, M., Sasso, A., Valadan, M., Varra, M., Rusciano, G., & Altucci, C. (2023). The effectiveness of cyrene as a solvent in exfoliating 2D TMDs nanosheets. *International Journal of Molecular Sciences*, *24*(13), 10450. https://doi.org/10.3390/ijms241310450Adeli, M., Amoneit, M., Haag, R., Mai, T., et al. (2024). Green chemistry and responsible research and innovation: Moving beyond the 12 principles. https://core.ac.uk/download/639037446.pdf (accessed on 28th September 2025).
2. Alessi, M., Egenhofer, C., Núñez Ferrer, J., Teusch, C., et al. (2012). Which economic model for a water-efficient Europe? Report of a CEPS Task Force. CEPS Task Force Report, 27 November 2012. https://core.ac.uk/download/10592231.pdf (accessed on 28th September 2025).
3. Annevelink, E., & van Ree, R. (2010). Country Report The Netherlands 2010: IEA Bioenergy Task 42. https://core.ac.uk/download/pdf/29240740.pdf (accessed on 28th September 2025).
4. Annevelink, E., Broeze, J., van Ree, R., et al. (2009). Opportunities for Dutch biorefineries. https://core.ac.uk/download/pdf/29248632.pdf (accessed on 28th September 2025).
5. Aubin, J., Mazubert, A., & Poux, M. (2013). Intensified processes for FAME production from waste cooking oil: A technological review. https://core.ac.uk/download/18540047.pdf (accessed on 28th September 2025).
6. Aznar Sánchez, J. Á., García Gómez, J. J., López Serrano, M. J., Velasco Muñoz, J. F., et al. (2018). The sustainable management of metals: An analysis of global research. https://core.ac.uk/download/286590215.pdf (accessed on 28th September 2025).
7. Barker, M., Cowie, B., Jones, A., Yueh, L., et al. (2010). What influences the emergence of a new subject in schools? The case of environmental education. https://core.ac.uk/download/29198038.pdf (accessed on 28th September 2025).
8. Bjelkemyr, M., Jönsson, C., Kurdve, M., Shahbazi, S., et al. (2013). Industrial waste management within manufacturing: A comparative study of tools, policies, visions and concepts. https://core.ac.uk/download/42143949.pdf (accessed on 28th September 2025).
9. Boiteux, J. J., Espino, M. B., de Fernández, M. L. Á., Pizzuolo, M., et al. (2019). Control sustentable poscosecha: Extractos de Larrea cuneifolia mediados por NADES frente a Botrytis cinerea. https://core.ac.uk/download/323211905.pdf (accessed on 28th September 2025).

10. Boone, J. W., Carson, A., Hecklau, S. F., Hughes, M., et al. (2011). Gettysburg College sustainability proposal. https://cupola.gettysburg.edu/cgi/viewcontent.cgi?article=1022&context=student_scholarship (accessed on 28th September 2025).
11. Bruneel, E., De Buysser, K., De Saegher, T., Hanssen, D., et al. (2020). Monometallic cerium layered double hydroxide supported Pd-Ni nanoparticles as high performance catalysts for lignin hydrogenolysis. https://core.ac.uk/download/287941993.pdf (accessed on 28th September 2025).
12. Buckley, C., Cloete, E., Godfrey, L., Hildebrandt, M., et al. (2014). The state of green technologies in South Africa. https://core.ac.uk/download/30610831.pdf (accessed on 28th September 2025).
13. Cammarota, M. C., Fernández-Lafuente, R., Ferreira-Leitão, V. S., Freire, D. M. G., et al. (2017). The protagonism of biocatalysis in green chemistry and its environmental benefits. https://core.ac.uk/download/80862874.pdf (accessed on 28th September 2025).
14. Cenex, & Speers, P. (2010). A hydrogen framework for the East Midlands. https://core.ac.uk/download/30624505.pdf (accessed on 28th September 2025).
15. Clarke, P. A., Martin, W. H. C., & Santos, S. (2007). Combining pot, atom and step economy (PASE) in organic synthesis. Synthesis of tetrahydropyran-4-ones. https://core.ac.uk/download/52398.pdf (accessed on 28th September 2025).
16. Cukalovic, A. (2009). Fifth International Conference on Renewable Resources and Biorefineries, 10–12 June 2009, Ghent, Belgium. https://core.ac.uk/download/55849830.pdf (accessed on 28th September 2025).
17. Gil Pérez, D., & Vilches Peña, A. (2013). Creating a sustainable future: Some philosophical and educational considerations for chemistry teaching. https://core.ac.uk/download/93038114.pdf (accessed on 28th September 2025).
18. Gissi, A., Xia, A. B., Curzons, A. D., Corrêa, A. G., Sicaire, A. G., Matlack, A. S., Zhou, B., et al. (2016). Tools and techniques for solvent selection: Green solvent selection guides. https://core.ac.uk/download/42626903.pdf (accessed on 28th September 2025).
19. Gupta, A., Kowalski, S., & Mehra, I. (2014). Overcoming India's food security challenges: The role of intellectual property management and technology transfer capacity building. https://core.ac.uk/download/72053513.pdf (accessed on 28th September 2025).

20. Haynes, K., Murray, A., & Skene, K. (2015). The circular economy: An interdisciplinary exploration of the concept and application in a global context. https://core.ac.uk/download/151157363.pdf (accessed on 28th September 2025).
21. Huber, J. (2025). Key Environmental Innovations. https://core.ac.uk/download/pdf/6264166.pdf (accessed on 28th September 2025).
22. Imre, B., & Pukánszky, B. (2013). Recent advances in bio-based polymers and composites: Preface to the BiPoCo 2012 Special Section. https://core.ac.uk/download/18405899.pdf (accessed on 28th September 2025).
23. Jacquemin, L., Pontalier, P. Y., & Sablayrolles, C. (2011). Life cycle assessment (LCA) applied to the process industry: A review. https://core.ac.uk/download/12043304.pdf (accessed on 28th September 2025).
24. La Marca, F. (2015). Treatment and valorization plants in materials recovery supply chain. https://core.ac.uk/download/54525197.pdf (accessed on 28th September 2025).
25. N/A. (2011). Green jobs in a sustainable food system. https://core.ac.uk/download/71345683.pdf (accessed on 28th September 2025).
26. N/A. (2015). Launching the grand challenges for ocean conservation. https://core.ac.uk/download/84911929.pdf (accessed on 28th September 2025).
27. Revill, J. (2013). Education and outreach activities within the Biological Weapons Convention. https://core.ac.uk/download/19568774.pdf (accessed on 28th September 2025).
28. Saint Jean, M. (GREThA), & Oltra, V. (GREThA) (2025). Incrementalism of environmental innovations versus paradigmatic change: A comparative study of the automotive and chemical industries. https://core.ac.uk/download/pdf/6297038.pdf (accessed on 28th September 2025).
29. Shi, X., & Xu, G. (2018). Environmentally friendly pervious concrete for treating deicer-laden stormwater: Phase II. https://core.ac.uk/download/185623460.pdf (accessed on 28th September 2025).
30. Singh, G., & Khullar, T. (2023). Green chemistry strategies for waste minimization: From waste to wealth. https://core.ac.uk/download/660318997.pdf (accessed on 28th September 2025).
31. Tani, A. (2018). A strategic niche management approach for shaping bio-based economy in Europe. https://core.ac.uk/download/188831414.pdf (accessed on 28th September 2025).
32. Thomas, B., Raj, M. C., Athira, K. B., Rubiyah, M. H., Joy, J., Moores, A., Drisko, G. L., et al. (2018). Nanocellulose, a versatile green platform: From

BioSource's to materials and their applications. Chemical Reviews, 118, 11575–11625. https://doi.org/10.1021/acs.chemrev.7b00627 (accessed on 28th September 2025).

33. Trache, D., Tarchoun, A. F., Derradji, M., Hamidon, T. S., Masruchin, N., Brosse, N., Hussin, M. H. (2020). Nanocellulose: From fundamentals to advanced applications. Frontiers in Chemistry, 8. https://doi.org/10.3389/fchem.2020.00392 (accessed on 28th September 2025).

34. Wagner, W. E. (2014). Imagining corporate sustainability as a public good rather than a corporate bad. https://core.ac.uk/download/211333149.pdf (accessed on 28th September 2025).

35. Yang, L. (2021). Selective C–H activation: Ruthenaelectro-catalysis and carborane cage activation. https://corc.ac.uk/download/182129142.pdf (accessed on 28th September 2025).

INDEX

D

E

F

G

N

O

P

Q

R

U

V

W

Z